L-1

RKA

Advanced Ecology

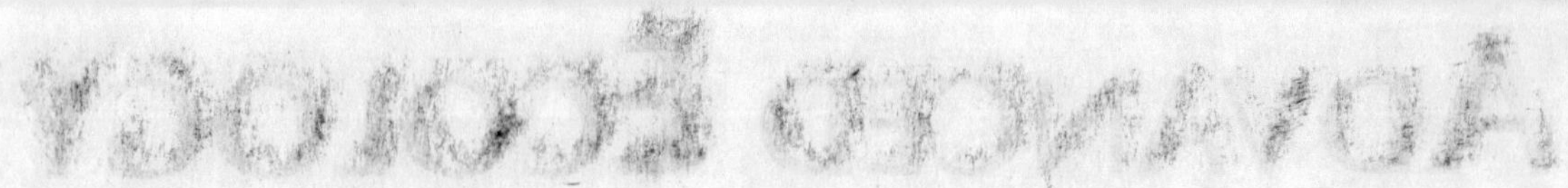

ADVANCED ECOLOGY

Editors
Professor (Dr.) Arvind Kumar
Pro-vice Chancellor
S.K.M. University, Dumka – 814 101 (Jharkhand)
&
Dr. Lalan Kumar Singh
Lecturer
Department of Chemistry, Mahila College, Jamtara

2006
DAYA PUBLISHING HOUSE
Delhi - 110 035

ISBN 81-7035-428-5

Published by : **Daya Publishing House**
1123/74, Deva Ram Park
Tri Nagar, Delhi - 110 035
Phone: 27383999
Fax: (011) 23244987
e-mail : dayabooks@vsnl.com
website : www.dayabooks.com

Showroom : 4760-61/23, Ansari Road, Darya Ganj,
New Delhi - 110 002
Phone: 23245578, 23244987

Laser Typesetting : **Classic Computer Services**
Delhi - 110 035

Printed at : **Chawla Offset Printers**
Delhi - 110 052

PRINTED IN INDIA

Preface

Ecology is a vast and encyclopaedic biological subject. To a beginner, at the outset, ecology may appear to be quite confusing and discouraging because it seems so diffuse and incoherent. However, the ecology is the study of ecosystems or the totality of the reciprocal interactions between living organisms and their environment. It is a dynamic biological unit which is greatly influenced by an enveloping and fluctuating environment.

Ecology is a distinct science because it is a body of knowledge not similarly organized in any other division of biology, because it uses a special set of techniques and procedures; and because it has unique point of view. The essence of ecology lies in the comprehensive understanding of the meaning of various phenomena such as: the local and geographical distribution and abundance of organisms; temporal changes in the occurrence, abundance and activities of organisms, the inter-relations between organisms in populations and communities; the structural adaptations and functional adjustments of organisms to their physical environment; the behaviour of organisms under natural conditions; the evolutionary development of all these inter-relations; the biological productivity of nature and how this may best serve mankind and the conservations and management of natural resources and pollution.

The scope of ecology is quite vast. The study of ecological principles provides a background for understanding the fundamental relationships of the natural community and also the sciences dealing with particular environment such as forest, soil, ocean and inland waters. Many practical applications of this subject are found in agriculture, horticulture, forestry, limnology, oceanography, fishery biology, pest control, public health, toxicology, pollution control, conservation etc.

Therefore, the aim of this book is to open an applied vista in various fields of environmental science so that people will avail all parameters for management of natural disaster.

My special thanks and appreciation go to the scientists whose contributions have enriched this volume. I wish to express my sincere gratitude to Dr. P.C. Hembram Hon'ble Vice Chancellor, S.K.M. University, Dumka who has been a source of constant inspiration. I owe my special thanks to Professor M. C. Dash, Hon'ble Vice Chancellor of Sambalpur University, Professor H.R. Singh, Hon'ble Vice Chancellor of Allahabad University, Professor K.C. Pandey, Former Vice Chancellor of Meerut University,

Professor N.C. Datta of Calcutta University, Dr. S.N. Kaul of NEERI, Professor U.S. Bagde of Mumbai University, Professor S.K. Konar of Kalyani University, Professor D.K. Belsare of Bhopal University, Professor U.C. Goswami of Gauhati University, Professor P.C. Mishra of Sambalpur University, Professor P. Natarajan of Kerala University, Professor P.S. Murthy of Bangalore University, Professor Ajit Varma of J.N.U., New Delhi, Professor P.P. Sood of Saurashtra University, Professor A.L. Bhatia of Jaipur University, Professor A.K. Mittal of B.H.U., Professor S.P. Hosmani of Mysore University, Professor K. Kapoor of Udaipur University, Professor K.B. Reddy of Nagarjuna University, Professor M. Vikram Reddy of Pondicherry University, Professor B.K. Tiwari of NEHU, Professor G. Tripathi of Jodhpur University, Professor R. Ramalingam of Annamalai University, Professor Sharif U. Ahmad of Nagaland University, Professor G.K. Kulkarni of Aurangabad University, Professor S.U. Mehram of Nagpur University, Professor S. K. Battish of PAU, Professor B. Manihar Sharma of Manipur University, Professor B.D. Joshi of Hardwar University, Professor G.C. Pandey of Faizabad University, Professor K.C. Sharma of Ajmer University, Professor A.K. Dobryal of Garhwal University, Professor Kaushal Kumar of Gorakhpur University, Professor M. Raziuddin of Hazaribag University, Professor Gurdeep Singh of I.S.M., Dhanbad, Dr. P.K. Goel of Karad, Dr. Ram Kumar of Delhi University, Professor S.P. Roy and Shri Tribhuwan Poddar of Bhagalpur University, Bhagalpur as well as Dr. Chandan Bohra of B.S.K. College, Barharwa (Jharkhand).

I also express my deep sense of gratitude to my parents whose blessings have always prompted me to pursue academic activities deeply. I am also thankful to my sweet wife, *Kumari Bimla* and my two lovely sons, *Kumar Pallav Shivshankaran* and *Kumar Prasun Ramakrishnan* whose natural smiles extended to me relief all through this tiresome endeavour.

Last but not the least, I am also thankful to Mr. Anil Mittal, Proprietor, Daya Publishing House, Delhi for taking keen interest in bringing out of this book. Finally, I will always remain a debtor to all my well-wishers for their blessings, without which this book would not have come into existence.

Dr. Arvind Kumar

Dr. Lalan Kumar

Contents

Chapter 1

Community Ecology: A Critical Review

Arvind Kumar

Environmental Science Research Unit, Post Graduate Department of Zoology
S.K.M. University, Dumka – 814 101, Jharkhand, India

Introduction

Organisms of a particular species form a population, which has characteristics like natality, mortality, age structure, growth, dynamics etc. But when several populations share a common habitat and its resources and interact among themselves, they develop into a biotic community or simply a community. Thus, a community is a longer unit than a population, since it includes more than one species population. In other words, a community is a local association of several population of different species or a naturally occurring assemblage of plants and animals belonging to different species which live in the same environment, are mutually sustaining and interdependent, constitute a biotic community or biocenose. Microorganisms, plants and animal populations sharing a common habitat and interacting among themselves develop into biotic communities. A biotic community has its own characteristics. Animal populations sharing a common habitat and interacting among themselves form an animal community and plant populations of an area form a plant community, but the concept of biotic community includes all populations of living organisms of a common habitat, ranging from a tract of forest to the whole of the forest, from a small pond to a large lake, and so on.

The biological potential of each species population determines a tolerance range for environmental conditions. A taxon can tolerate a certain range of environmental condition which is called as its ecological amplitude. The composition of a biotic community in any habitat is dependent upon the prevalence of environmental conditions in that habitat and the ecological amplitude of species populations. Therefore, the climate and other abiotic and biotic conditions of a habitat determine the type of community which survives and develops. Thus, large number of biotic communities are found in nature due to: (*i*) existence of diverse habitats with characteristic environmental conditions and (*ii*) co-occurrence of different species whose tolerance ranges overlap with the environmental conditions obtained in that habitat. When similar habitat conditions are repeated at another location, the same biotic community gets established there (Dash, 1993).

When territorial ranges of several species overlap, populations of the different species may co-exist within a given limited area. The organisms in such an area represent a community. A community always contains plants as well as animals, because both are very necessary for the survival of the community. Thus, there exists in nature an interaction between organisms and environmental, between plants and plants, between animals and animals, and between animals and plants and the biotic community is the net result of such interactions. The organisms of a community usually exhibit trophic relationships among themselves. They also interact in sharing the space and there may be interactions at a reproductive and behavioural level. Animal populations differ in their trophic requirements and therefore a particular plant community may be associated with a particular animal population. Many biological activities of plants, like periodicity, phenology, etc. are strongly associated with animal activities and vice versa.

Community Characteristics

Each biotic community possesses ecological characteristics which differentiate it from other communities. Major community characteristics are described below:

Species Diversity

Some communities, such as tropical rain forest and coral reef community, show high species diversity with many different kinds of species living at each trophic level. In other communities, like a desert, there may be relatively few species in the entire community. Species diversity includes the total number of species present in a community and the relative abundance of these species. Diversity is recognized as an important functional attribute of biotic community. Species diversity of a community can be classified into three categories:

Alpha Diversity

It refers to the diversity of organisms sharing the same habitat. A combination of species richness and equitability/evenness is used to represent diversity within a community or habitat.

Beta Diversity

Species frequently change when habitat or community changes. The rate of replacement of species along a gradient of habitats or communities is called beta diversity. Higher the heterogeneity in the habitats in a region or greater the dissimilarity between communities, higher is the beta diversity.

Gamma Diversity

Diversity of the habitats or community over the total landscape or geographical area is called gamma diversity.

Dominance

Only one or few species in a community are in sufficient abundance (having high density) to dominate and influence other species in terms of numbers and biomass production. They also determine animal distribution. For example, in terrestrial communities, generally the species of tall trees are dominants. Communities are generally named after their dominant species. For example, a forest community in which pine trees are dominant is called a pine forest. Grassland represents a community which has grass species dominating over the other herbaceous species. Such dominant species are called ecological dominants. In the ecological cycle of a community, the removal of a dominant species causes disturbances. Thus, some species populations exert more influence than others in the community because they control a higher percentage of energy flow. Senapati and Dash (1981) have found that grasses are the dominant producers in a grassland community of Orissa, India, and that cattle and earthworms are respectively the dominant populations among the above ground and below ground animal consumers. This study was based on the amount of primary and secondary productivity and energy flow, in a grassland community.

Ecotone and Edge Effect

Each community has its spatial limits. Some times the boundary between two communities may be very sharp (as between a forest and a lake) or gradual (as between a forest and a grassland). This transitional zone or junction zone between two or more diverse communities is called ecotone. The ecotone harbours a community including organisms of each of the overlapping communities and, in addition, organisms peculiar to it. Generally, the number of species and their population densities are greater in the ecotone than in the communities flanking it. The occurrence of increased diversity and density of organisms at the ecotone is due to the edge effect of two distinct communities. Man, by his activities, creates an ecotone. Human settlement creates ecotonal communities. If man settles in a forest, he reduces the forest to scattered small areas interspersed with grasslands, agricultural lands and other open habitats.

Community Structure, Composition and Stratification

Communities may be small, consisting of few species populations in a small space, or large area Elton (1927) observed that the communities of certain British rivers includes 131 species of invertebrates, in addition to fish, amphibians, populations of algae, protozoa, bacteria and rooted aquatic plants while on a sandy beach only five species of invertebrates. Dash *et al.* (1980, 1981 and 1984) studied the composition of a biotic community in some tropical pastures from Sambalpur, India. Dash and Guru (1980), and Das and Pradhan (1984) also studied the composition of a biotic community in a tropical hill ecosystem of Sambalpur, Orissa. Kar *et al.* (1987) and Dash *et al.* (1988) studied the community structure of river Ib. These studies indicate that a biotic community may be composed of very few to very many species. Besides, there are many genera in a large community consisting of one species only (80-90 per cent) while extremely few genera consist of more than one. Since the species of a genus may have more or less similar requirements of environmental conditions, in a particular habitat they will be competing for resources. During the course of evolution, this competition has eliminated all but very few species from each genus represented in a biotic community in an area.

A community is first noticed by its physiognomy. Physiognomy refers to the external appearance or "look" of the community. The external appearance is the total effect created by the combination of vertical structure and architecture of dominant species of vegetation. For instance, the high physiognomy of a forest differs markedly from a low physiognomy of a grassland. However, several communities may have similar physiognomy, yet they differ sharply on the basis of species composition and dominants (*e.g.* different forest types).

Community structure, composition, shape, and so on one called qualitative features. The measurement of density, frequency of occurrence, coverage, height, and growth represent the quantitative features. Since seasonal changes occur in the appearance of plant structure and growth, periodicities and phenology are important community parameters. Another important feature is the sociability or the nature of grouping or the associations of individuals of different species.

Stratification is a characteristic feature of a biotic community. Stratification of a community depicts vertical layering of the vegetation. Different layers are occurring by different species. The vertical stratification provides physical structure to the plant community, in which many forms of plant and animal life are adapted to live in. The simplest stratification includes two strata:

(*i*) an upper euphotic zone consisting mainly of autotrophs, and (*ii*) a lower consumer and decomposer zone consisting mainly of heterotrophic organisms. These two strata can be divided into many substrata, which increase the habitats and microhabitats for different organisms. Stratification also helps in niche separation. Increased habitats and niche separation reduce competition and thus enable a large number of species to occupy a community. A well developed forest ecosystem exhibits a highly stratified structure, consisting of several layers of vegetation. These layers include the canopy, the under story tree layers, the shrub layers and the herb layers. The forest floor also consists of leaf litter (L layer), a partially decomposed leaf litter layer (F layer), a humus layer and so on. These layers are associated with different organisms, as each stratum provides a different type of habitat. The stratification helps in the segregation of niches.

In a grassland community, the euphotic zone consists of the herbaceous stratum consisting of grasses and other herbs as producers, and rodents, cattle and insects as consumers. The subterranean stratum consists of different layers of soil containing litter, humus, mineral fraction of soil, plant roots and organisms, such as bacteria, fungi, protozoa, microarthropods, insect larval, pupae, oligochaets and mollusks. Most small organisms are found in the top 10 cm of soil. Oligochaetes can burrow deeper as can comphibians, snakes and rodents. Thus, a stratification in the distribution of organisms can easily be seen in a grassland community. Similarly, a pond community has surface dwellers and bottom dwellers. In aquatic ecosystems, stratification from surface to bottom is determined by light penetration, temperature profile and oxygen profile.

Qualitative Features of the Community

Raunkiaer (1934) recognized that plant growth is limited by unfavourable environmental condition and depends upon the ecological amplitude of the species. Keeping these facts in mind, he considered the position and degree of protection to the perennating bud during adverse environmental conditions as the principal features of plant adaptation to climate. On this basis he classified higher plants into five major life form classes, namely (*i*) Phanerophytes (P), (*ii*) Chamaephytes Ch, (*iii*) Hemicryptophytes (H), (*iv*) Crytophytes (Cr), and (*v*) Therophytes (Th).

Phanerophytes

These plants may be trees, shrubs or climbers. They are found mostly in tropical regions and decrease progressively from the tropics to the temperate to the polar regions. In these plants, growing buds are not well protected. They are located in upright shoots much above the ground surface. Phanerophytes are usually divided into four sublife forms depending upon the height of nature plants. These are (a) trees more than 30 metres tall, called megaphanerophytes, (b) trees between 8 and 30 metres tall, called mesophanerophytes, (c) trees between 2 and 8 metres, called microphanerophytes, and (d) shrubs less than 2 metres high, called nanophanerophytes.

Chamaephytes

In these plants the buds are located close to the ground surface or up to a maximum height of 25 cm. These commonly occur in high altitudes and latitudes, e. g. *Trifolium repens*, found in temperate North America.

Hemicryptophytes

Here the penetrating buds are protected under the surface soil. In this category, plants are usually biennial or perennial herbs and grow in cold climatic regions.

Cryptophytes

In these plants, the buds are usually buried in the soil or in bulbs and rhizomes where food is stored to withstand long periods of adverse climatic conditions. They are also called geophytes.

Therophytes

These are annula plants and produce flowers and seeds in the favourable season. They survive in adverse conditions in seed form and are found in dry, hot or cold environments.

In any plant community, the ratio of the life forms of different species in terms of their numbers or percentage is called the phytoclimatic or biological spectrum. These ratio also indicate the nature of climate prevailing in a particular region. For example, a higher percentage of chamaephytes would indicate a cold climate, whereas the occurrence of a higher percentage of therophytes would imply long dry seasons. Evidently, the occurrence of more or less similar biological spectra in different regions would suggest similar climatic conditions in those regions. Raunkiaer (1934) prepared a normal biological spectrum for the phanerogamic flora of the world and found the following percentage values of different life forms (Table 1.1).

Table 1.1: Raunkiaer's Normal Biological Spectrum for Phanerogamic Flora

Life Forms	*% Occurrence*
Phanerophytes	46
Hemicryptophytes	26
Therophytes	13
Chamaephytes	09
Cryptophytes	06

The biological spectrum of any natural ecosystem can be studied and compared with Raunkiaer's data and from this some insight can be obtained into the type of climate of the area. However, Raunkiaer's study was based on a large number of natural undisturbed habitats. But in recent times, the biological spectrum of any habitat has been affected by human activities, such as agriculture, grazing, scraping, deforestation, fire, pollution and so on. Hence, one has to be very careful while comparing different ecosystems and drawing inclusion.

Raunkiaers also classified plant communities into six types on the basis of leaf size (Table 1.2). Each leaf size, from the second type onwards, is 9 times larger than the preceding type.

According to Dansereau (1957), six plant community components are generally used for structural description besides the above classification procedures for plant communities. These components are (a) Life form, (b) plant size, (c) function, (d) leaf shape, (e) leaf texture, and (f) canopy cover. Six categories have been identified in each of these six components. Dansereau also used notations and symbols. The subgroups for life form components are T: trees; F: shrubs; H: herbs; M: bryophytic forms; E: epiphytes; L: lianas.

Table 1.2: Community Classification on the Basis of Leaf Size

Community Type	*Leaf size (mm^2)*
Leptophyll	25
Nanophyll	225
Microphyll	2,025
Mesophyll	18,225
Macrophyll	164,025
Megaphyll	>164,025

The following notations are used for the three main size groups of plants:

T = tall; m = medium; and l = low height. Four subgroups are used for functions:

D: deciduous; S: semideciduous; e: evergreen, and j: evergreen succulent or evergreen leafless. Six subgroups are identified for the leaf shape: n: needle or spine; g: graminoids; a: medium or small; h: broad; v: compound; and q: thalloid. Four subgroups are identified for leaf texture: f: filmy; z: membranous; x: sclerophylls, and k: succulent.

The canopy cover in divided into four subgroups: b = barren; I = discontinuous; p = intuffs or bunches, and c = continuous.

Frequency

Species population dispersal in a community varies due to soil conditions, seed dispersal mechanisms, grazing and other biotic factors. Raunkiaers (1934) classified the species in a community into five frequency classes (Table 1.3).

Table 1.3: Raunkiaer's Frequency Classification of Plant Communities

Class	*Species with Frequency Range (%)*
A	1–2
B	21–40
C	41–60
D	61–80
E	81–100

The frequency is expressed as a percentage and is estimated from the following relationship:

$$\textbf{\% Frequency of Species A} = \frac{\text{No. of quadrats in which species A occurs}}{\text{Total no. of quadrats examined}} \times 100$$

The relative frequency (RF) is determined by the use of the following formula:

$$\textbf{R.F. of the Species} = \frac{\text{No. of occurrence of a species}}{\text{No. of occurrence of all species}} \times 100$$

The following figure depicts the Dan-sereau's notatious and symbols for plant life form components (Figure 1.1).

LIFE FORMS		LEAF SHAPE	
T	Trees	n	Needle or spine
F	Shrubs	g	Graminoid
H	Herbs	a	Small or Medium
M	Bryolds	h	Broad
E	Epiphytes	v	Compound
L	Lianas	q	Thalloid

FUNCTION		LEAF TEXTURE	
d	Deciduous	f	Filmy
s	Semideciduous	z	Membranous
e	Evergreen	x	Sclerophyll
j	Evergreen leafless or Evergreen succulent	k	Succulent

Figure 1.1: Dansereau's Notations and Symbols for Plant Life Form Components

Since environmental conditions vary, the frequency values differ in different communities. Raunkiaers observed the following relationships between different classes: A = 53; B = 14; C = 9; D = 8 and E = 16 and generalized his findings in the law of frequency, which showed that

$$A > B > C \underset{<}{\overset{>}{=}} D < E$$

It has been accepted that this relationship occurs in undisturbed communities, or communities which disturbance is minimum. In a uniform vegetation, class E is always larger than class D and when the reverse occurs, it indicates considerable disturbances.

Density

The numerical strength of organisms in a community represents their number per unit area. The unit area may be a square metre (for smaller organisms), an acre or a hectare. Different types of samplers and sampling techniques are used to study the density of different populations in a community (Southwood, 1966). The relative density of a species is calculated as follows:

$$\textbf{RD} = \frac{\text{No. of individuals of the species in all quadrats}}{\text{No. of individuals of all species in all quadrats}} \times 100$$

Sometime the term "abundance" is used by plant ecologists which is defined as:

$$\textbf{Abundance} = \frac{\text{Total No. of individuals of a species in all quadrats}}{\text{Total no. of quadrats in which the species occurs}}$$

Cover

The above ground parts of the plant cover a certain area. If this area is demarcated by vertical projection, the area of the ground covered by the plant canopy is called canopy cover or foliage cover. In the forest community, measurements of canopy cover are obtained from different strata of vegetation. Stratification causes an increase in the canopy cover to sometimes more than 200 times the ground area. This cover is an expression of the area occupied by different species is usually expressed as a percentage. Plants with greater foliage cover can intercept more solar energy and in a forest community bring more shade, influencing the growth of understorey vegetation.

Besides the canopy cover, the basal area or basal cover of tree is important. The basal area or basal cover in trees is usually measured at breast-height (150 cm above the ground level) by the formula πr^2. For the measurement of croan cover, the diameter is taken at ground level, across the canopy perimeter passing by the main stem (Ambasht, 1988 and Misra, 1974). The relative dominance is measured as follows:

$$\textbf{Relative Dominance} = \frac{\text{Total basal area of the species in all quadrats}}{\text{Total basal area of all species in all quadrats}} \times 100$$

Quantification of Dominance and Diversity

For the quantification of diversity and comparison of species diversities between different ecosystems in various climatic conditions. It is useful and convenient to calculate an index for diversity and dominance. This index is usually calculated on the basis of ratios of the part to the whole, such as n_1/N_1 where n_1 is the number of each part and N_1 is the total value in the community.

The commonly used form of the diversity index is the Shanon and Weaver Index popularly known as Shanon Index of diversity. The commonly used dominance index is that of Simpson. If n_1 refers to number for each species in community and N to the total number in that community, then the ratio n_1/N will represent the probability function p for each component of the whole.

Shanon Diversity Index

This index is designated as H which is by definition:

$$\overline{H} = -\Sigma (n_i/N) \log e\, (n_i/N)$$

Or if $p_i = n_i/N$

Then $\overline{H} = -\Sigma\, p_i \log e\, p_i$

Since the log of fractions are negative, the minus sign is inserted in the formula to convert the sum to a positive number.

Simpson's Index of Dominance

It is designated as D, which is by definition

$D = \Sigma\, n_i/N)^2$ or $D = \Sigma\, (p_i)^2$

In the Shanon index the ratio of each component is multiplied by the log e of the ratio and then summed up while in Simpson's Index the ratio of each component is squared. Simpson's Index is an index of dominance whose maximum value is 1, which is obtained in the case of a single species (complete dominance). Near zero values are obtained when numerous species are present (no dominance), each species representing a small fraction of the total value.

The Shanon index is a diversity index. The higher the value of $\overline{H}$, the greater is the diversity. The maximum value of H can be more than 1. Total diversity depends upon (a) the number of species or number of parts, called the variety component, and (b) the evenness component or the distribution of relative abundance. Higher overall diversity occurs when the number of species and the evenness component are large (low dominance). For a comparison of Simpson's Index with Shanon's, it is necessary to scale the latter to a maximum value of 1 and minimum of zero (Odum, 1975). This is done by diversity H by the log e of S. In general nature and stable communities will have high diversity values (0.6 to 0.9 one a zero–to–one scale) and real communities or unstable communities and communities under severe stress exhibit low diversity, usually nearer zero on a zero-to-one scale. Simpson's index is weighed in favour of dominant species and Shanon's in favour of rare species.

The diversity index can be used as a good measure for studying the effect of industrial pollution, because industrial wastes and sewage almost always reduce the diversity of natural systems into which they are discharged. A measurement of diversity often provides a better index of pollution than a direct measurement of pollutants. For example, waste discharges into a river or an aquatic system at night may be in detectable by the next day but the effect of the pollutants on the living community will be there for a longer time (Wilhm and Dorris, 1968).

Similarity Index

Czechanovski (1913) developed an index called the similarity index, which is used to compare the species diversities of two systems. It is also useful for a comparison of the species diversity of an unpolluted site with a polluted one. The index of similarity is expressed as:

$$S = 100 \times \frac{2C}{(a+b)}$$

Where, a is the number of species in one site, b is the number of species in another site, C is the number of species common to both sites and S is Czechanovski's similarity index.

Although some species dominate in a community, several species grow near each other. In other words, they exhibit some associations. The inter specific association between two species can be quantified by an association index, which is calculated as follows:

If we examine 100 quadrats of the same size from a community and find that species and occurs alone in 65 quadrats and with species B in 56 quadrats, the association index of A = $^{56}/_{65}$ = 0.86.. The association of two more species in a community can be analyzed by the Chi square, and contingency table method and conclusions can be drawn. Joshi and Behera (1990) observed the inverse relationship between diversity and dominance in tropical forests of Orissa.

Importance Value Index

In the community structure, this index is used to determine the overall importance of each species. The percentage value of relative frequency, relative density and relative dominance are summed up together and this value is designated as IVI or importance value index of the species. The IVI gives an idea of the sociological structure of a species in its totality in the community but does not indicate its position separately with regard to other, aspects, such as frequency density etc. Animal ecologists use an importance value as follows (Sajise *et al.*, 1976):

$$\text{IV per cent} = \frac{Wa}{Wt} \times 100$$

Where Wa is the average dry weight of the species and w_t is the total dry weight of all species of the community of the area.

Quantitative Methods

The quantification of community structure and function depends largely upon taking periodic samples and referring the findings to the community. The sampling must be very representative of the population and community. Various sampling techniques are used, but before adopting any of them the habitat must be surveyed.

Survey methods are divided into two broad categories–(a) Extensive study of the habitat, and (b) Intensive study of the communities. The extensive study of the habitat may include the preparation of a survey map showing major ecological units like grasslands, forests, swamps, ponds, lakes and so on. The intensive study of the communities may include a survey of floristic composition ecological factors, vegetational changes etc. Floristics and faunistics include a systematic examination of the habitat, taxonomic identification, classification and listing of organisms. Vegetation sampling is usually done either by the transect or by the quadrat method. For sampling animal population from the soil, quadrats of different sizes may be required for different groups (Murphy, 1982). Many books (Simpson et al., 1960; Greig-Smith, 1964; South-Wood, 1966; Misra, 1968 and Michael, 1982) deal exclusively with quantitative methods used in the analysis of the community.

References

Ambasht, R. S. 1988. Studies on the biomass productivity of some tropical grazing lands and plantation forests. In: *Rangelands Resource and Management*, (Eds.) Singh, P. and Pathak, P.S. Indian Grassland and Fodder Res. Inst. Jhansi, India.

Dansereau, P. 1957. *Biogeography: An Ecological Perspective.* Ronald Press, New York.

Dash, M. C. 1993. *Fundamentals of Ecology.* Tata McGraw-Hill Publ. Co. Ltd., New Delhi, pp. 373.

Dash, M. C. and Guru, B. C. 1980. Distribution and seasonal variation in numbers of Testsacea (Protozoa) in some Indian soils. *Pedobiologia*, 61: 1025–1029.

Dash, M. C., Nanda, B. and Behera, N. 1980. Fungal feeding by Enchytraeidae (Oligochaeta) in tropical woodland from Orissa, India. *Oikos*, 34: 202–206.

Dash, M. C., Mishra, P.C., Mohanty, R. K. and Bhatt, N. 1981. Effect of specific conductance and temperature on crease activity in some Indian soils. *Soil Biol. Biochem.*, 13: 73–74.

Dash, M. C., Mishra, P. C., Choudhary, K. and Das, R. C. 1988. Bioaccumulation of mercury in river Ib. In: *Ecology and Pollution of Indian Rivers*, (Ed.) Trivedi, R.K. Ashish Publ. House, New Delhi, p. 293–302.

Dash, M. C. and Pradhan, G. B. 1984. Distribution and population dynamics of soil nematodes and their relationship with primary production in a tropical hill ecosystem of Sambalpur, India. *Pedobiologia*, 26: 349–359.

Dash, M. C., Satpathy, B., Behera, N. and Dei, C. 1984. Gutload and turnover of soil, plant and fungal material by *Drawida calebi*, a tropical earthworm. *Rev. Ecol. Biol. Sol.*, 21: 379–393.

Elton, C. 1927. *Animal Ecology*. McMillan Co., New York, pp. 207.

Greigh-Smith, P. 1964. *Quantitative Plant Ecology* (2nd ed.). Butterworth and Co. Publ., London, pp. 256.

Joshi, S. K. and Behera, N. 1990. Qualitative analysis of vegetation from a mixed tropical forest of Orissa, India. *Ind. Forester*, 117: 200–206.

Kar, G. K., Mishra, P. C., Dash, M. C. and Das, R. C. 1987. Pollution studies in river Ib. III. Plankton population and primary productivity. *Ind. J. Environ. Hlth.*, 29: 322–329.

Misra, K. C. 1974. *Manual of Plant Ecology*. Oxford and IBH Publ. Co., New Delhi, pp. 376.

Misra, R. 1968. *The Ecology Work Book.* Oxford and IBH Publ. Co., Kolkata, pp. 244.

Murphy, P. W. 1982. *Progress in Soil Zoology.* Butterworths, London.

Odum, H. T. 1975. Energy quality and carrying capacity of earth. *Trop. Ecol.*, 16: 1–8.

Raunkiar, C. 1934. *The Life Forms of Plants and Statistical Plant Geography*. Clarendon Press, Oxford, pp. 639.

Sajise, P. E., Orlido, N. M., Lales, J. S., Castillo, L. S. and Atabay, R. 1976. The ecology of Philippine Grasslands I. Floristic composition and community dynamics. *The Philippine Agriculturist*, 59: 317–334.

Shanon, C. E. and Weaver, W. 1963. *The Mathematical Theory of Communication*. Univ. of Ilinois Press, Urbana, pp. 117.

Simpson, E. H. 1949. Measurement of Diversity. *Nature*, pp. 688.

Simpson, G. G., Roe, A. and Lewonties, R. C. 1960. *Quantitative Zoology*. Harcourt, Brace and World Inc., New York.

Southwood, T. R. E. 1966. *Ecological Methods with Particular Reference to the Study of Insect Populations*. Methuen, London, pp. 391.

Wilhm, J. L. and Dorri's, T. C. 1968. Biological parameters for water quality criteria. *Bio. Sci.*, 18: 477–481.

Chapter 2

The Invertebrate Colonization During Decomposition of *Eichhornia crassipes* Solms in the Mouth Zone of Guareí River into Jurumirim Reservoir (São Paulo, Brazil)

R. Henry and N. de L. Stripari

Department of Zoology, Institute of Biosciences, State University of São Paulo, C.P. 510, 18618–000 Botucatu, SP, Brazil

ABSTRACT

In the mouth zone of the Guareí River into Jurumirim, litter bag experiments were used to measure decay rate (k) of *Eichhornia crassipes* and evaluate invertebrate colonization on the remaining biomass. Nitrogen, phosphorus, and polyphenol contents were also determined on decomposing water hyacinth; experiments were performed in the dry and rainy seasons to examine the influence of plant chemical composition on colonization. In both periods, a fast loss (c.a. 20 per cent) of aquatic macrophyte initial biomass was observed in the first 24 hours. Decay rate was higher in the rainy ($k = 0.0406$) than in dry ($k = 0.0138$) season. In both seasons, increased nitrogen in *E. crassipes* residual biomass was observed from the first to seventh day of the experiment; after this a decrease was seen. Only in the dry season, phosphorus and polyphenol content had negative correlation to k. There was an inverse relationship between invertebrate densities and polyphenol concentrations throughout both experimental periods. In the dry season, no functional feeding groups prevailed, but collectors dominated in the rainy season. The collectors were mainly composed of Cladocera, which predominated due to high amount of particulate matter, an indication of detritivorous food chain on remaining biomass.

Keywords: *Decomposition, Colonization, Eichhornia crassipes, Tropical river.*

Introduction

There are three processes involved in aquatic plant breakdown: leaching, fragmentation, and catabolism of the detrital components (Golley, 1983; Bianchini Junior, 2003). The first is enhanced by water flow in rivers (Hammerly *et al.*, 1989). In the other two, organisms such as bacteria, fungus, and invertebrates have an active role. A direct relationship between invertebrate number and decomposition phase has been reported by Short *et al.* (1980). An inverse association between increased invertebrate density and reduced remaining biomass of *Eichhornia azurea* in litter bag experiments has been found (Stripari and Henry, 2002). However, in a study of invertebrate colonization in detritus during *Eichhornia crassipes* decomposition in a secondary arm of the Paranã River, Poi de Neiff and Neiff (1989) showed that, during the first 17 days of the experiment, the number of organisms/g DW of detritus was similar in litter bags containing roots and leaves but, between the 30th and 50th days, there was an increase and a decrease of densities in the litter bags with leaves and roots, respectively.

Factors, such as water temperature, nutrient content (Kaushik and Hynes, 1971; Carpenter and Adams, 1979), and chemical composition of aquatic macrophytes (Godshalk and Wètzel, 1978a, b) affect breakdown velocity.

One of the chemical components in aquatic macrophytes is the polyphenols, which are readily soluble in water and rapidly released into the environment (Cameron and La Point, 1978). High loss of polyphenols in detritus of *Eichhornia azurea* during the first days of decomposition experiment has already been reported (Roland *et al.*, 1990; Stripari and Henry, 2002).

Colonization by invertebrates during water hyacinth decomposition appears to be related to polyphenol content in the remaining detritus. In the decomposition of *Eichhornia azurea,* the reduction in the polyphenol levels was followed by an increase in invertebrate densities (Stripari and Henry, 2002).

The aim of this paper was:

1. To show the qualitative and quantitative differences of invertebrate colonization during *Eichhornia crassipes* decomposition,
2. To examine the influence of chemical composition, including polyphenols, total phosphorus, and nitrogen of the remaining detritus on the colonization process and,
3. To speculate on the role of water chemical and physical factors as controlling agents in the decomposition and colonization processes.

Material and Methods

The Study Area

The Guareí is a secondary tributary of Paranapanema River, in the south region of São Paulo state, Brazil. The confluence of these two rivers makes the riverine zone of Jurumirim Reservoir, the first in a cascade sequence in the Paranapanema River (Figure 2.1). In a survey during the dry season ($n = 8$), the river water had high transparency (0.68±0.16 m SD Secchi disk readings), high dissolved oxygen (8.0±0.9 mgO_2.L^{-1}SD), low temperature (19.3±3.2°C SD), and low total phosphorus (86±4 μg L^{-1} SD). In the rainy period, there was low transparency (0.28±0.13 m SD), low dissolved oxygen (5. 7±0.5 mgO_2.L^{-1}SD), high temperature (26.5±3.7°C SD), and high total phosphorus (140±39 μg.L^{-1} SD). Comparing the two periods, water alkalinity (1.045±0.058 meq.L^{-1} SD and 0.881±0.122 meq.L^{-1} SD in dry and rainy seasons, respectively), conductivity (113±20 μS.cm^{-1} SD and 105±20 μS.cm^{-1} SD,

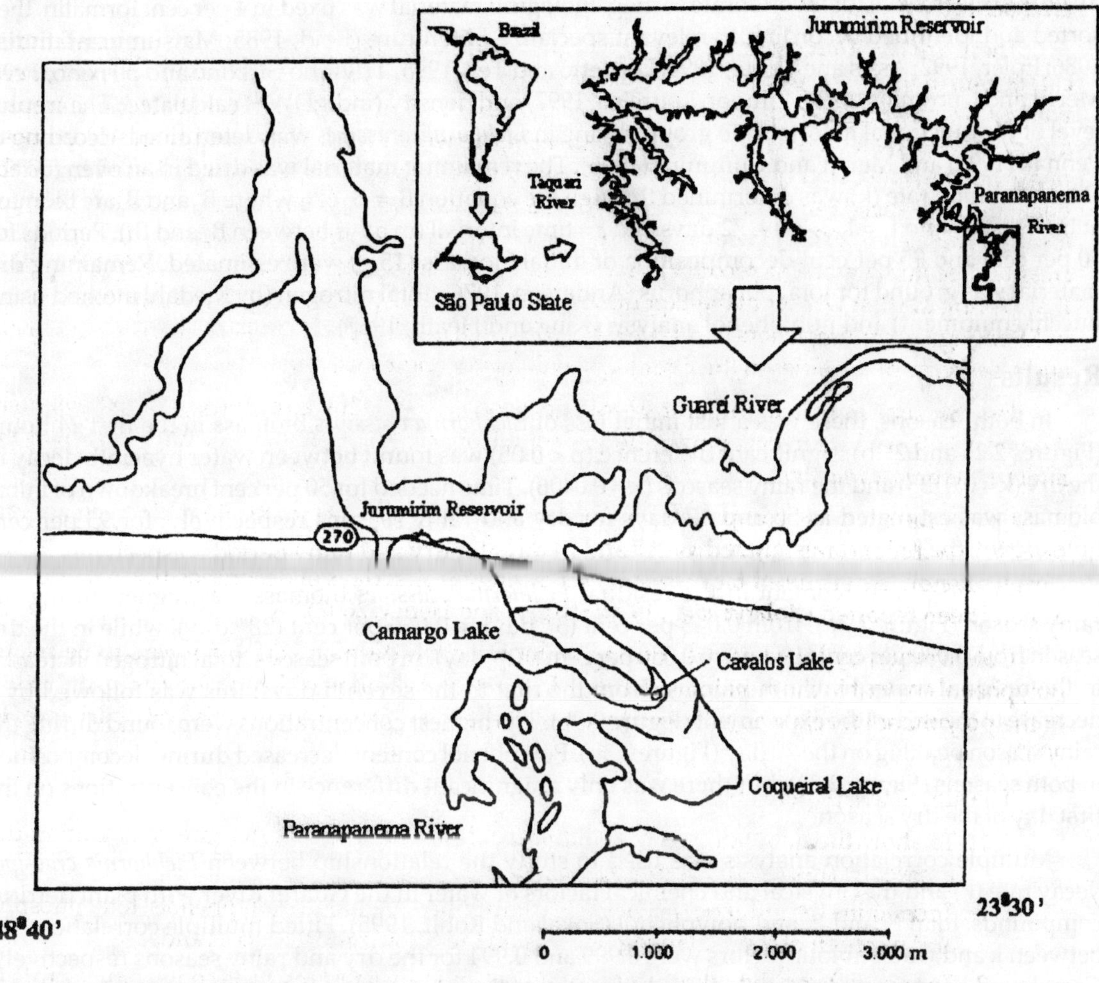

Figure 2.1: The Study Site: The Mouth Zone of Guareí River into Jurumirim Reservoir (São Paulo, Brazil)

in dry and rainy seasons respectively) and, total nitrogen (395±139 μg.L^{-1} SD and 410±169 μg.L^{-1}SD in the dry and rainy seasons, respectively) presented no significant differences.

Experimental Procedure

The litter bag method was used to evaluate *Eichhornia crassipes* decomposition rate and invertebrate colonization in two experiments, one during the dry season (from August 4 to October 10, 1997) and the other in the rainy season (from January 12 to March 9, 1998). In each experiment, 42 litter bags (21 for decomposition and 21 for colonization), measuring 15 × 20 cm with 2 mm mesh net, were incubated *"in situ"* near an *Eichhornia crassipes* stand. Plant leaf and petiole (15 g.DW) were placed in each litter bag. After 1, 3, 7, 14, 28, 56, and 78 days of incubation, 6 litter bags (three for decomposition and 3 for colonization measurements) were drawn and then removed. The remaining material was cleaned in a 0.25 mm mesh sieve for invertebrate sorting. Biological material was fixed in 4 per cent formalin, then sorted and identified according to relevant specialized literature (Reid, 1985; Matsumura-Tundisi, 1986; Epler, 1992; Assis and Costa, 1994; Lopretto and Tell, 1995; Trivinho-Strixino and Strixino, 1995; Merrit and Cummins, 1996; Elmoor-Loureiro, 1997) and density (ind.gDW^{-1}) calculated. The trophic level of the functional invertebrate groups living in *Eichhornia crassipes* was determined according to Pennak (1991) and Merrit and Cummins (1996). The remaining material was dried in an oven (60°C). Decomposition rate (k) was determined through the equation $B_t = B_o e^{-kt}$, where B_o and B_t are biomass at times t_o = O and t_1 = 1, 3, 7, 14, 72 days and t = time interval (in days between B_o and B_1). Periods for 50 per cent and 95 per cent decomposition of initial biomass (15 g) were estimated. Remaining dry material was ground for total phosphorus (Anderson, 1976), total nitrogen (by Kjedahl method using Bucchi equipment) and polyphenol analysis (King and Heath, 1967).

Results

In both seasons, there was a fast initial loss of *Eichhornia crassipes* biomass in the first 24 hours (Figures 2.2a and 2.2b); significant difference ($p < 0.05$) was found between water hyacinth decay in the dry (k = 0.0138) and the rainy seasons (k = 0.0406). Time needed for 50 per cent breakdown of initial biomass was estimated as 50 and 17 days for dry and rainy seasons respectively; for 95 per cent breakdown, the time was estimated as 217 and 74 days for dry and rainy periods respectively.

Total phosphorus content in the residual *Eichhornia crassipes* biomass was higher during the rainy season (Figure 2.3a), from 0.043 per cent (3rd day) to 0.066 per cent (28th day), while in the dry season from 0.030 per cent (1st day) to 0.050 per cent (72nd day). In both seasons, total nitrogen increased in the residual water hyacinth biomass from the first to the seventh days; this was followed by a decrease to the end of the experiment (Figure 2.3b). The highest concentrations were found during the rainy season peaking on the 7th day (Figure 2.3b). Polyphenol content decreased during decomposition in both seasons (Figure 2.3c), but there was only a significant difference in the concentrations on the first day of the dry season.

Multiple correlation analysis was used to study the relationship between *Eichhornia crassipes* decay rate (k) and the physical and chemical factors of water in the Guareí River with plant detritus compounds: total N and P, and polyphenol (Sokal and Rohlf, 1995). Fitted multiple correlations (R) between k and water abiotic factors were 0.999 and 0.994 for the dry and rainy seasons respectively. Correlation between variables was used to evaluate the relationship between each environmental factor and the detritus compounds with k (Table 2.1). During the dry season, dissolved oxygen, pH, total P, and polyphenols in detritus presented negative correlations with k, while transparency was positive. In the rainy period, no significant correlation was found.

Table 2.1: Correlation Coefficients (r) Between Decomposition Rates (k) of *Eichhornia crassipes* and Water Abiotic Factors and Detritus Compounds During the Experiments (Dry and Rainy Period)

Factors	*Dry Period*	*Rainy Period*
Temperature	0.0088	– 0.2320
Dissolved oxygen	– 0.4883	0.4342
Transparency	0.8269*	0.0260
Total P in detritus	– 0.5805*	0.0458
Total N in detritus	– 0.4544	0.3099
Polyphenols in detritus	– 0.8286*	0.3018

* Correlations with r values > 0.5 were considered as significant for interpretation.

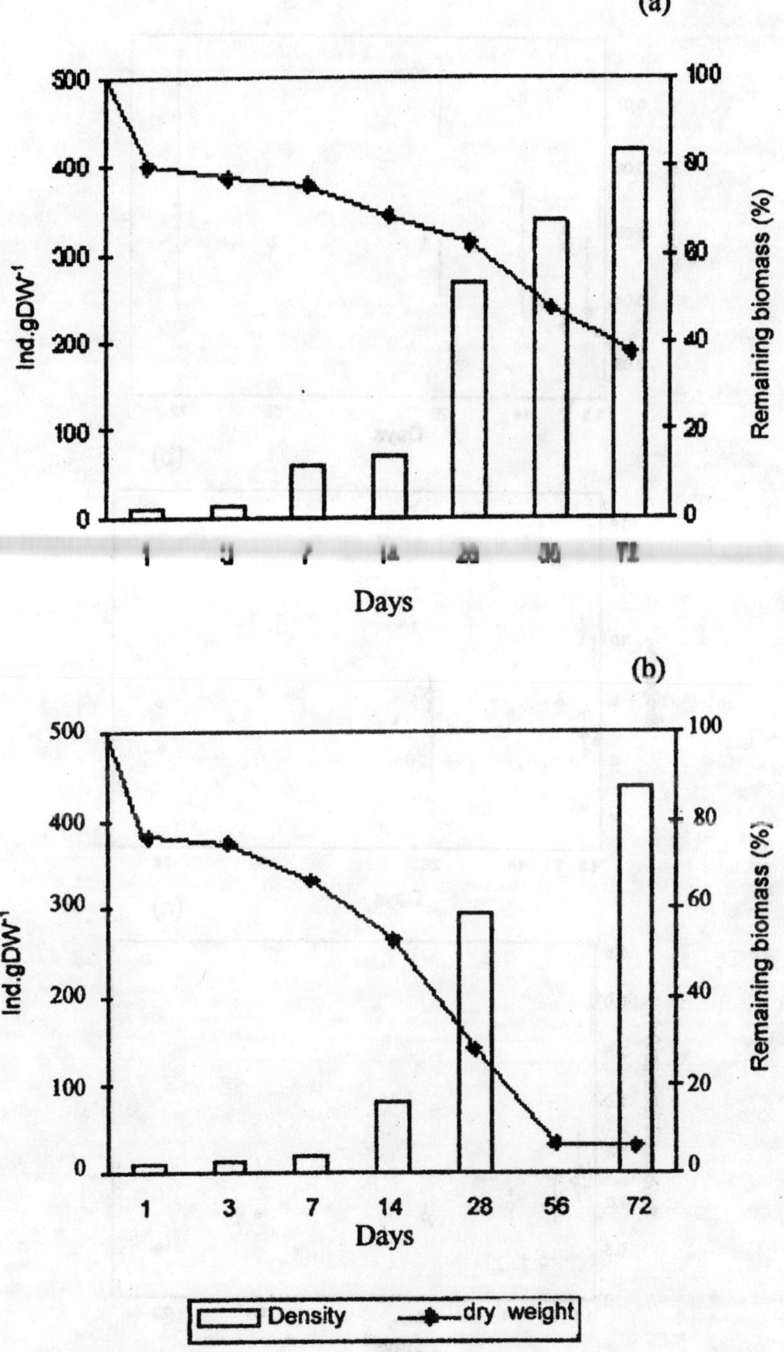

Figure 2.2: Total Invertebrate Density (Ind.gDW^{-1}) in *Eichhornia crassipes* during the Decomposition Experiment in Guareí River in the Dry (a) and Rainy (b) Period

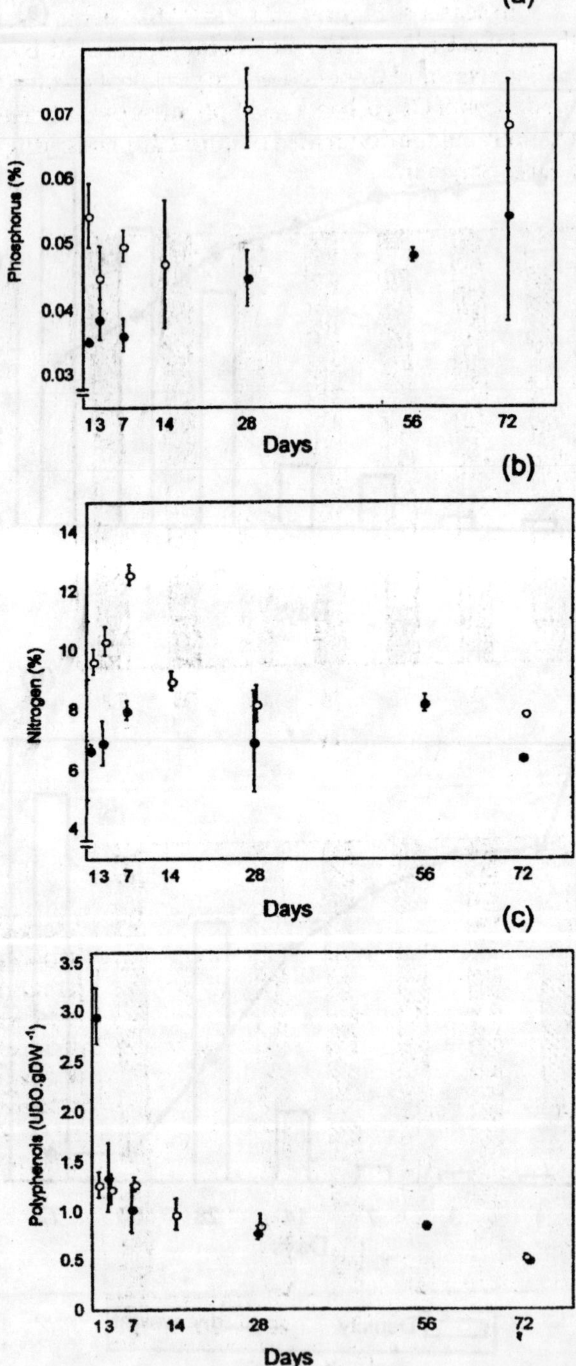

Figure 2.3: Mean ± SD of Phosphorus (a), Nitrogen (b) and Polyphenols (c) in the *Eichhornia crassipes* Remaining Biomass in the Dry (•) and Rainy (o) Period

The invertebrate density increased from the 14th day; the highest levels were found at the end of the experiment (72nd day) for both seasons (Figures 2.2a and 2.2b). On the last day of the dry season, Chironomidae presented the highest relative density, being dominated by *Dicrotendipes* sp. and *Chironomus* sp. (Figure 2.4a). Cladocera and Copepoda were prevalent on most experimental days. In the rainy season, relative abundances of Oligochaeta and Ephemeroptera increased up to the 28th day; at the end of the experiment Chironomidae dominated (Figure 2.4b), consisting mainly of *Polypedilum* sp., *Parachironomus* sp., and *Ablabesmyia* sp.

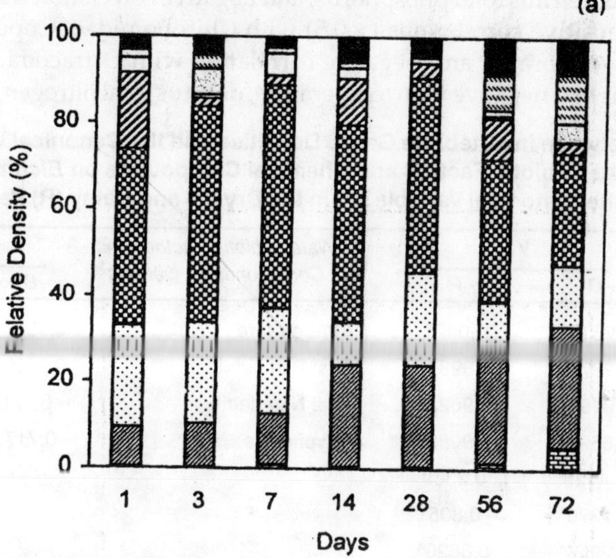

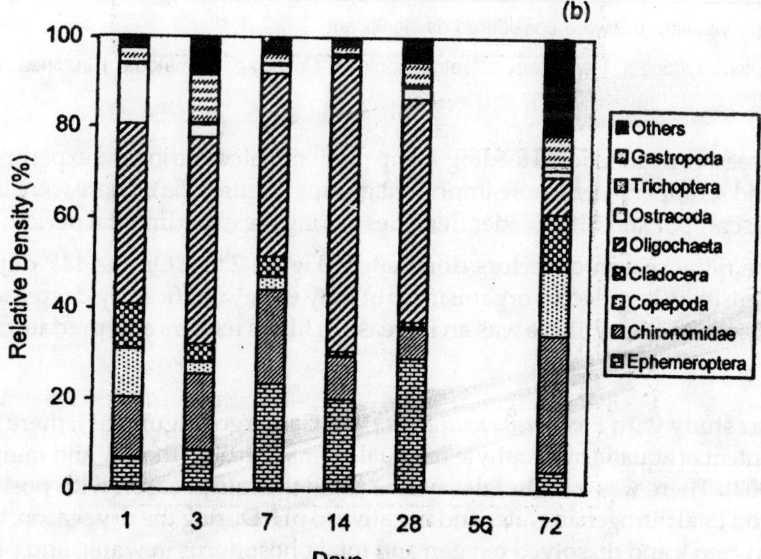

Figure 2.4: Relative Density of Invertebrates Living on *E. crassipes* Detritus in the Dry (a) and Rainy (b) Period

Canonical correlation analysis was used to evaluate the influence of environmental factors and chemical composition on density of invertebrates in remaining biomass (Manly 1994). Fitted correlations of canonical variable pairs (V1 and W1) were 0.09865 and 0.09997 for the rainy season respectively, both were significant considering $p < 0.05$. Canonical variable V1, that includes the densities of all invertebrate groups on *Eichhornia crassipes* during their breakdown, was positive (> 0.5) for all groups except Oligochaeta and Gastropoda in the dry season (Table 2.2). Canonical variable W1 corresponding to water abiotic factors and detritus chemical compounds had positive correlations (> 0.5) with temperature and detritus total phosphorus, and negative correlation with polyphenols. In the rainy season, V1 had positive correlations (> 0.5) with Chironomidae, Copepoda, Cladocera, Gastropoda, and others invertebrates, and negative correlation with Ostracoda. W1 was positive (> 0.5) with total phosphorus and negative with temperature, detritus total nitrogen, and polyphenols.

Table 2.2: Correlations Between Invertebrate Group Densities and the Canonical Variable, V1 and, Correlations Between Water Abiotic Factors and Chemical Compounds on *Eichhornia crassipes* Detritus and the Canonical Variable W1 in the Dry (D) and Rainy (R) Period

Invertebrates Groups	*V1*		*Water Abiotic Factors and Compounds of Detritus*	*W1*	
	D	R		D	R
Ephemeroptera	0.7317*	0.0421	Temperature	0.9587*	– 0.9969*
Chironomidae	0.9677*	0.8983*	Total P in detritus	0.7599*	0.8003*
Copepoda	0.9194*	0.9622*	Total N in detritus	– 0.0716	– 0.5480*
Cladocera	0.6694*	0.9088*	Polyphenols	– 0.7175*	– 0.8125*
Oligochaeta	0.4129	0.2389			
Ostracoda	0.7370*	– 0.8051*			
Gastropoda	0.2950*	0.8620*			
Others**	0.8838*	0.8708*			

* Correlations with r values > 0.5 were considered as significant.

** Others: Coleoptera, Odonata, Hemiptera, Ceratopogonidae, Culicidae, Nematoda, Hirudinea, Collembola, Bivalvia, Trichoptera.

In the dry season, no functional feeding group predominated during the experiment, but collectors, filter feeders and scrappers were more important groups (Figure 2.5a). There were increased collector and decreased scrapper and filter feeder densities during the experimental period.

During the rainy season collectors dominated (Figure 2.5b). On the 14th day, 87.7 per cent of invertebrate density was collector organisms, but they were significantly decreased on the last day (Figure 2.5b). On the 72nd day, there was an increase in filters feeders and predators (Figure 2.5b).

Discussion

In a similar study with *Eichhornia azurea* in Lake Camargo (Figure 2.1), there was a decrease in polyphenol content of aquatic macrophyte residual biomass in both the dry and rainy seasons (Stripari and Henry, 2002). There was a higher decay rate (k) in the rainy season with positive correlation to conductivity and total nitrogen in water and negative to pH. During the dry season, there was negative correlation between k and dissolved oxygen and total phosphorus in water, and positive correlation with alkalinity, water transparency and detritus total P. Despite k for *Eichhornia crassipes* being higher

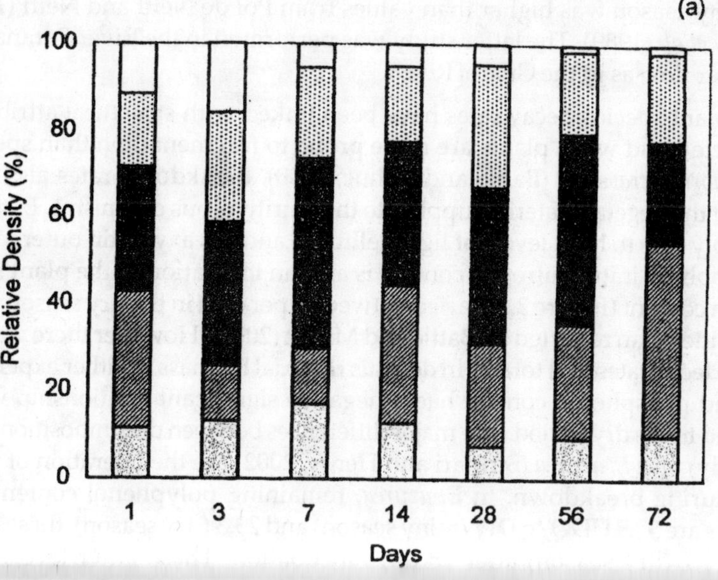

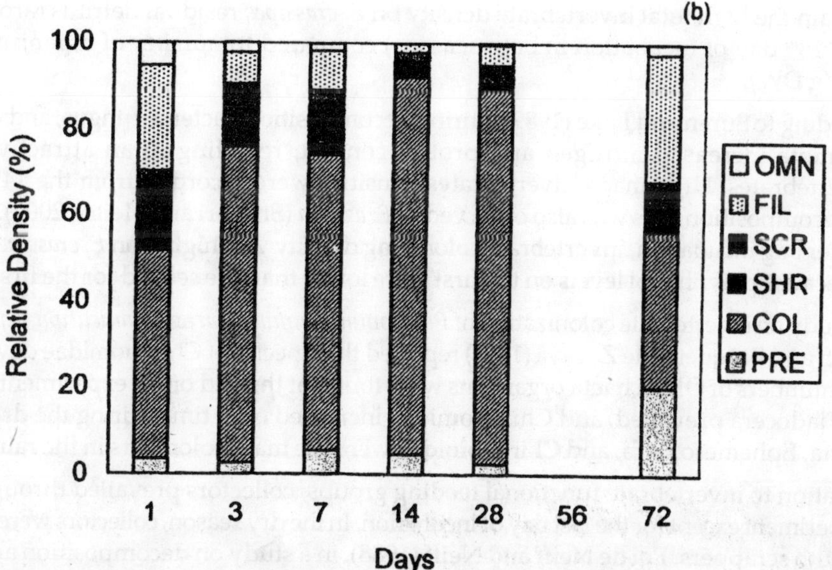

Figure 2.5: Relative Density of Functional Feeding Groups of Invertebrates Living on *E. crassipes* Detritus Durign the Dry (a) and Rainy (b) Period (OMN: Omnivorous; FIL: Filter feeders; SCR: Scrappers; SHR: Shredders; COL: Collectors; PRE: Predators).

in the rainy season, there was no significant correlation with water temperature. One of the main factors determining biomass decomposition velocity is temperature since it increases microbial activity (Howard-Williams and Davies, 1979; Esteves and Camargo, 1986; Gadelha *et al.*, 1990). *E. crassipes* decay rate in the rainy season was higher than values from Poi de Neiff and Neiff (1988; 1989) and similar to Hammerly *et al.* (1989). The latter study was performed in the River Paraná, an ecosystem with permanent water flow as in the Guareí River.

Differences in plant species decay rates have been linked with structural attributes of vegetal tissue. Highly dissected and week plants are more prone to fragmentation than species with solid blades such as *Eichhornia crassipes* (Battle and Mihuc, 2000). Breakdown rates also depend on the nutritional value that the vegetal material supplies to the detritivorous organisms. *E. crassipes* had low degradation, probably due to high levels of lignocellulose and a waxy-cutin outer layer (Battle and Mihuc, 2000). Macrophyte initial nitrogen content is also an indication of the plant's value as food. Our data on nitrogen content (Figure 2.3) varied between 6 per cent in the dry season to 10 per cent in the rainy; this was wider than reported by Battle and Mihuc (2000). However there was no significant correlation between decay rates and total N in detritus residual biomass in either experimental period. Only polyphenol and phosphorus content had a negative significant relationship with *E. crassipes* decay rate, but limited to the dry period. The major differences between decomposition experiments on *E. crassipes* (this study) and *E. azurea* (Stripari and Henry, 2002) are the liberation of polyphenols by two macrophytes during breakdown. In *E. azurea,* remaining polyphenol content after one day incubation in detritus are 32.9 UDO/g DW (rainy season) and 23.3 (dry season), these decreased to 3.9 (rainy season) and 1.7 (dry season) after 28 days incubation. These latter levels approximately correspond to initial polyphenol contents (day one) in *E. crassipes* detritus. These low concentrations could explain the high total invertebrate density on *E. crassipes* residual detritus (around 300 ind/g DW at the 28th day of incubation in both seasons) compared to number of organisms on *E. azurea* (< 100 ind/g DW).

According to Suren and Lake (1989), during decomposition bacteria, fungus, and then periphytic algae promote increased nitrogen and proteic content, resulting in an attractive increase for macroinvertebrates. High macroinvertebrates densities were recorded from the 14th day on, in *E. crassipes* decomposition; this was also observed in *E. azurea* (Stripari and Henry, 2002). In the first few days of litter bag incubation, invertebrate colonizing density was higher on *E. crassipes* than *E. azurea* detritus because polyphenol levels on the first were lower than the second for the first seven days.

In a study on invertebrate colonization of *Polygonum acuminatum* and *Panicutum elefantipes* detritus, Poi de Neiff and Bruquetas de Zozaya (1991) reported that species of Chironomidae developed quickly, and large numbers of Oligochaeta organisms were found at the end of the experiment. On *E. crassipes* detritus, Cladocera prevailed, and Chironomidae increased over time during the dry season, while Oligochaeta, Ephemeroptera, and Chironomidae were the main colonizers in the rainy season.

In relation to invertebrate functional feeding groups, collectors prevailed throughout the rainy season experiment except for the last day of incubation. In the dry season, collectors were also important, together with scrappers. Poi de Neiff and Neiff (1988), in a study on decomposition and colonization by invertebrates on *E. crassipes,* reported that collectors dominated; there were also some predators and a few shredders. The predominance of collector invertebrates is probably due to the high quantity of particulate matter, an indication of the prevalence of detritivorous food chain on the macrophyte remaining biomass (Trivinho-Strixino *et al.*, 1995). The high number of scrappers on *E. crassipes* during the dry season were represented by Cladocera-Chydoridae that scrappe the particles from the submerged macrophytes (Elmoor-Loureiro, 1997). The shredders presented higher densities in the first phase of both studied periods. According to Cummins *et al.* (1989), shredders begin to feed when microorganisms produce structural and biochemical changes in plant tissues converting them to an edible nutritional state.

There are various conditions where invertebrates can be linked to macrophytes. In both seasons, all invertebrate feeding groups can be found on the residual biomass of *E. crassipes*. This is probably due to the multifunctionality of the substrate, such as food supply and shelter.

Acknowledgements

The authors are grateful to FAPESP (Projeto temático Proc. 97/04999-8) for financial support; N.L.S. received a scholarship grant from FAPESP (Proc. 96/11943-6); to Hamilton A. Rodrigues who assisted in the field and laboratory; and Colin E. Knaggs revised the English language.

References

Anderson, J.M. 1976. An ignition method for determination of total phosphorus in lake sediments. *Water Res.*, 10: 329–331.

Assis, C.V. and J.M. Costa. 1994. Seis novas larvas do gênero *Micrathyria* Kirby e notas sobre a distribuição no Brasil (Odonata, Libellulidae). *Rev. Bras. Zool.*, 11: 195–199.

Battle, J.M. and T.B. Mihuc. 2000. Decomposition dynamics of aquatic macrophytes in the lower Atchafalaya, a large floodplain river. *Hydrobiologia*, 418: 123–136.

Bianchini Junior, I. 2003. Modelos de crescimento e decomposição de macrofitas aquaticas. In: *Ecologia e Manejo de macrófitas aquáticas*, (Eds.) Thomaz, S.M. and L.M. Bini. Eduem, Maringa, p. 85–126.

Cameron, G.M. and T.W. La Point. 1978. Effects of tanins on the decompositions of Chinese tallow leaves by terrestrial and aquatic invertebrates. *Oecologia*, 32: 349–366.

Carpenter, S.R. and M.S. Adams. 1979. Effects of nutrients and temperature on decomposition of *Myriophylum spicatum*. L. in a hard water eutrophic lake. *Limnol. Oceonogr.*, 24: 520–528.

Cummins, K.W., Wilzbach, M.A., Gates, D.M., Perry, J.B. and W.B. Taliaferro. 1989. Shredders and riparian vegetation. *Science*, 39: 24–30.

Elmoor-Loureiro, L.M.A. 1997. *Manual de identificação de cladóceros límnicos do Brasil*. Editora Universa, Brasília, p. 156.

Epler, J. H. 1992. *Identification Manual of the Larval Chironomidae* (*Diptera*) *of Florida*. Depart. Environ. Reg. Orlando, pp. 308.

Esteves, F.A. and A.F.M. Camargo. 1986. Sobre o papel das macrófitas aquaticas na estocagem de nutrientes. *Acta Limnol. Bras.*, 1: 273–298.

Gadelha, C.L.M., Watanabe, T. and A.M. Passerat-de-silans. 1990. Liberavção de nutrientes inorganicos durante o processo de decomposivção de *Ludwigia natans* (Ell.) (Dicotyledoneae: Onagraceae) e de *Salvinia auriculata* (Aubl.) (Pteridophyta: Salviniaceae). *Acta Limnol. Bras.* 3: 633–652.

Godshalk, G.L. and R.G. Wetzel. 1978a. Decomposition of aquatic angiosperms. I. Dissolved components. *Aquat. Bot.*, 5: 281–300.

Godshalk, G.L. and R.G. Wetzel. 1978b. Decomposition of aquatic angiosperms. II. Particulate components. *Aquat. Bot.*, 5: 301–307.

Golley, F.B. 1983. Decomposition. In: *Ecosystems of the World Tropical Rainforest Ecosystems: Structure and Function*, (Ed.) Golley, F.B. Elsevier, New York, p. 157–166.

Hammerly, J., Leguizamon, M., Maine, M.A., Schiver, D. and M.J. Pizarro. 1989. Decomposition rate of plant material in the Paraná Médio River (Argentina). *Hydrobiologia*, 183: 179–184.

Howard-Williams, C. and B.R. Davies. 1979. The rates of dry matter and nutrient loss from decomposing *Potamageton pectinatus* in a brackish south-temperate coasta1 lake. *Freshwater Biol.*, 9: 13–21.

Kaushik, N.K. and H.B.N. Hynes. 1971. The fate of dead leaves that fall into streams. *Arch. Hydrobiol.*, 68: 465–515.

King, H.C. and G.W. Heath. 1967. The chemical analysis of small samples of leaf material and the relationship between the disappearance and composition of leaves. *Pedobiologia*, 7: 192–197.

Lopretto, E.C. and G. Tell. 1995. *Ecossistemas de águas continentals.* Ediciones Sur, La Plata, T. 3, p.

Manly, B.F.J. 1994. *Multivariate Statistical Methods.* Chapman and Hall, London, p. 215.

Matsumura-Tundisi, T. 1986. Latitudinal distribution of Calanoida copepods in freshwater aquatic systems of Brazil. *Rev. Bras. Biol.*, 46: 527–553.

Merrit, R. and K.W. Cummins. 1996. *An Introduction to the Aquatic Insects of North America*, 3rd ed., Kendal/Hunt, Dubuque, pp. 862.

Pennak, K.W. 1991. *Freshwater Invertebrates of the United States*, 3rd ed., John Wiley and Sons, New York, pp. 628.

Poi de Neiff, A. and I.Y. Bruquetas de Zozaya. 1991. Colonizacion por invertebrados de macrófitas emergentes durante su descomposicion en el rio Paraná. *Rev. Hydrobiol. Trop.*, 24: 209–216.

Poi de Neiff, A. and J.J. Neiff. 1988. Decomposition of *Eichhornia crassipes* solms in ponds of Paraná river valley and colonization by invertebrates. *Trop. Ecol.*, 2: 79–85.

Poi de Neiff, A. and J.J. Neiff. 1989. Dry weight loss and colonization by invertebrates of *Eichhornia crassipes* litter under aerobic conditions. *Trop. Ecol.*, 3: 175–182.

Reid, J.W. 1985. Chave de identificayção e lista de refêrencias bibliográficas para as espécies continentais sulamericanas de vida livre da ordem Cyclopoida (Crustacea, Copepoda). *Bolm. Zool.*, 9: 17–143.

Roland, F., Esteves, F. A. and I. E. Santos. 1990. Decomposiyção da macrófita aquática *Eichhornia azurea* (Kunth), corn ênfase na colonizayção por bactérias epifiticas. *Acta Limnol. Bras.*, 3: 653–673.

Short, R.A., Canton, S.P. and J.W. Ward. 1980. Detrital processing and associated macroinvertebrates in a Colorado mountain stream. *Ecology*, 61: 727–732.

Sokal, R.R. and F. J. Rohlf. 1995. *Biometry: The Principles and Practice of Statistics in Biological Research.* W.H. Freeman, New York, pp. 887.

Stripari, N. de L. and R. Henry. 2002. The invertebrate colonization during decomposition of *Eichhornia azurea* Kunth in a lateral lake in the mouth zone of Paranapanema River into Jurumirim Reservoir (São Paulo, Brazil). *Braz. J. Biol.*, 62(2): 293–310.

Suren, A.M. and P.L. Lake. 1989. Edibility of fresh and decomposing macrophytes to three species of freshwater invertebrates. *Hydrobiologia*, 178: 165 –178.

Trivinho-Strixino, S. and G. Strixino. 1995. *Larvas de Chiromidae (Diptera) do Estado de São Paulo guia de identificação e diagnostico dos gêneros.* Pós Graduação em ecologia e recursos naturais, São Carlos, p. 229.

Chapter 3

Effects of Prescribed Burning on Bacterial and Fungal Communities of Top Soil in Olokemeji Forest Reserve, Nigeria

A. Akinsoji and Elizabeth Sowemimo

Department of Botany and Microbiology, University of Lagos, Lagos-Nigeria

ABSTRACT

Bacterial and fungal communities of the top soil of the three fire experimental plots in Olokemeji Forest Reserve, Nigeria were studied in August 2002. Plots A and B had been given periodic early and late dry season fire treatments for over fifty years while plot C had been fire-protected. A total of twelve bacteria species were isolated. The Gram negative non-spore forming species like *Pseudomonas, Proteus* and *Escherichia coli* were the prevalent species. Plots A and B recorded eleven species each while plot C recorded only seven species of bacteria. Bacterial density ranged between 5×10^9 and 2.8×10^{10} (cfu/g) while plot B had the highest mean density of 2.4×10^{10} cfu/g. A total of twelve fungal species were also isolated. Eleven species each were isolated from plots A and B while plots C recorded ten species. *Aspergillus* spp. were the most prevalent. Fungal density ranged between 7.8×10^8 cfu/g and 1.98×10^{10} cfu/g while the mean density of 8.3×10^9 cfu/g recorded for plot B was the highest. Bacterial and fungal densities were highest in plot B which also had the highest organic matter content. The amount of organic matter seemed to determine the rate of decomposition process which determined the microbial density. Fire seemed to have enhanced bacterial density and diversity in plots A and B.

Keywords: *Bacteria, Fire, Fungi, Nigeria, Olokemeji Forest Reserve, Soil.*

Introduction

In the tropical rain forest, where the soil is not usually nutrient-rich, the soil must continually be replenished by the active mineralization of litter or any other organic material by soil heterotrophs. The soil-plant-animal complex is maintained by the dynamic processes of nutrient mineralization and conversion; nutrient uptake and return through litter fall, animal waste and conversion, and mineralization by heterotrophs most of which are bacteria and fungi. Apart from the heterotrophs, another important factor that affects vegetation is fire. Indiscriminate fires are known to destroy vegetation while prescribed burning has been used as a management tool by foresters. Range managers have also used prescribed burning to maintain desired grasses for cattle and wildlife.

The fire experimental plots at Olokemeji Forest Reserve in SW Nigeria were set up by Forestry Research Institute of Nigeria (FRIN) primarily to monitor the effect of prescribed periodic burning on long-term vegetation development. Studies conducted several decades after the experiment commenced (Charter and Keay, 1960; FRIN unpublished reports) showed that burning in the late dry season promoted the development of savanna vegetation while burning in the early dry season was changing the vegetation to forest type but at a rate slower than the fire protected plot. It is believed that soil heterotrophs may be playing some role in mineralization of litter and ash resulting from burning of plant material on the experimental plots. Thus, this study was set up to document and compare the aerobic communities of top soil at the Olokemeji Fire Experimental Plots.

Study Area

Olokemeji Forest Reserve is located in Ogun State, SW Nigeria. It is situated on an undulating topography 90–140 m asl 35 km NE of Abeokuta (Ogun State capital city) and 32 km W of Ibadan (capital city of adjoining Oyo State) (Figure 3.1). The reserve lies on latitude 70°2′ N and longitude 3°32′ E. The vegetation is a lowland rain forest comprising a mixture of evergreen and deciduous species dominated by Moraceae, Sterculiaceae and Ulmaceae families. However, part of the reserve has been over the years converted to derived Savanna due to incessant bush burning, shifting cultivation, grazing and hunting (Udugba, 1997). The area has a mean annual rainfall of about 1400 mm, mean daily maximum temperature of 28.8°C and a mean RH of 82 per cent. The meteorological data for FRIN, Ibadan (the nearest meteorological station) for 2002 are presented in Figure 3.2. There are two marked seasons. The rainy season runs from April to November while the dry season runs from December to March with a brief cold spell of dry harmattan winds in late December and January. Other characteristics of the site have been described by Charter and Keay (1960).

Materials and Methods

Sampling was done on the three Olokemeji Fire Experimental Plots in August 2002. The plots were set up in 1929 to monitor the effects of time of burn on vegetation development. Three contiguous plots each measuring 0.1735 ha were demarcated with a 3.3 m fire trace surrounding each plot. Plot A was burned in late dry season (ca. March); plot B was burned early in the dry season (Nov/Dec) and plot C which served as the control was fire-protected. Bradley *et al.* (1992) had described late burning as burning through the overstorey consuming large woody surface fuels while early burning consumed surface fuels and understorey plants. The fire regimen was applied periodically (annually initially, but later periods between treatments increased). Five random soil samples were taken from each plot at 0–15 cm depth rising a soil cover 2.5cm in diameter. Sampling was restricted to surface soil as it has been reported by Adegeye and Ayodele (1992) that burning of vegetation has no effect on lower soil depth. Samples were collected in cellophane bags and taken in containers packed with ice cubes to the

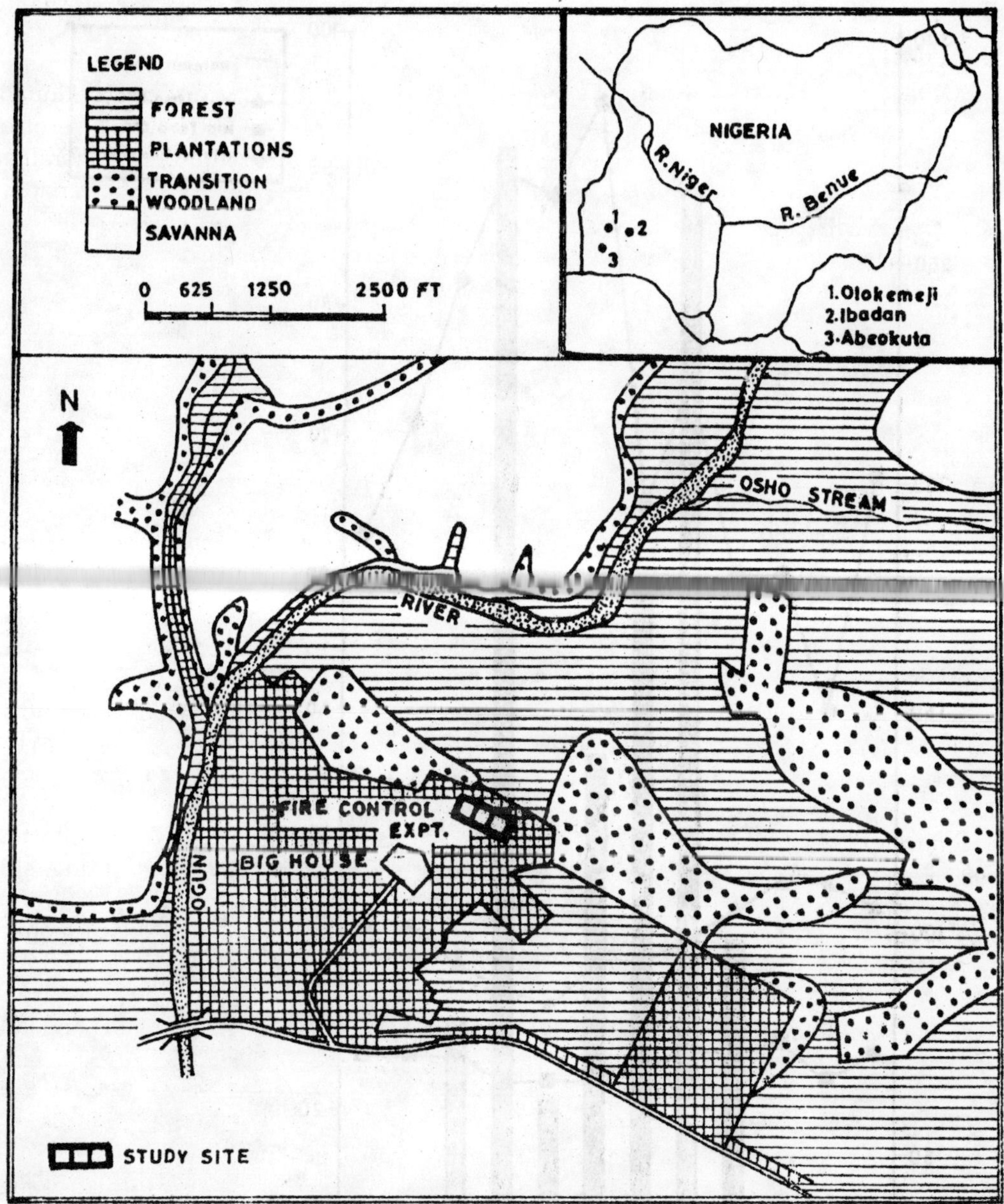

Figure 3.1: Vegetation Map of the North West Corner of Olokemeji Forest Reserve Showing Research Sites

laboratory where they were stored in a refrigerator set at 4°C and the analyses commenced within three days.

The samples were sterilized by placing in the UV room for 24 hr while the laboratory bench top was sterilized with 70 per cent ethyl alcohol. Glassware was sterilized in hot air oven at 160°C for 3 hr.

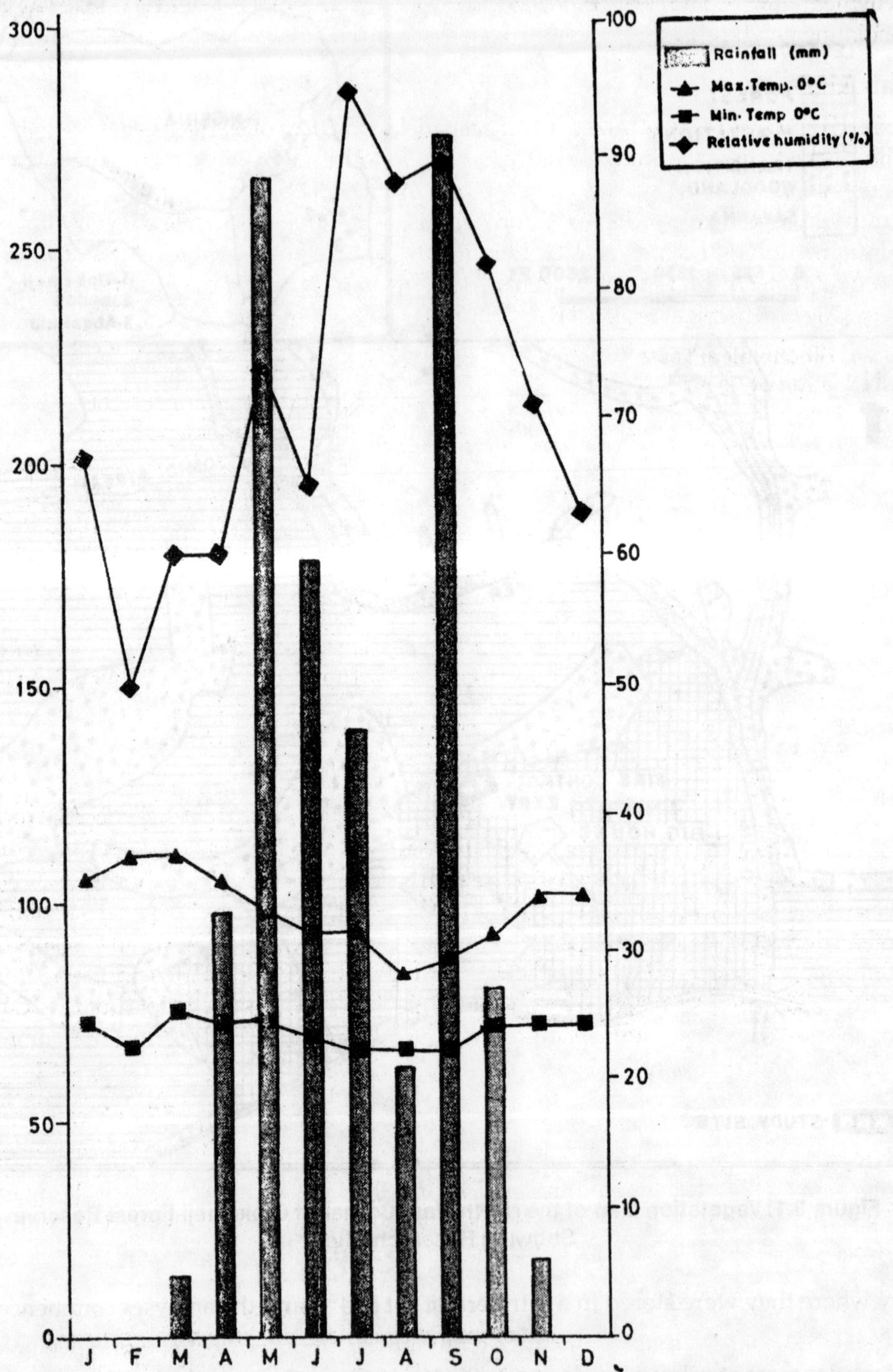

Figure 3.2: Meteorological Data for FRIN, Ibadan (2002)

The growth media and diluent were autoclaved at 121°C under 1.1 kgf/cm^2 for 20 mins. The inoculating loop, inoculating needle and the hockey stick were sterilized by dipping in ethyl alcohol and flaming.

Bacterial Analysis

1 g of each soil sample was mixed and a suspension in 10 ml of sterile distilled water was made. 1 ml of the suspension was diluted serially (ten-fold) and used in estimation of bacteria and fungi populations by the standard spread plate method of Prescot *et al.* (1999). For bacteria, plating was done on nutrient agar incubated at 37°C for 72 hr. Pure isolates of bacteria colonies that developed were maintained on agar slants at 4°C. Identification was based on cultural, morphological and microscopic characteristics. The cultural characteristics used included pigmentation, form, elevation and margin. The biochemical tests used are listed in Table 3.1.

Table 3.1: Biochemical Tests Used for Identification of Bacteria Species

Gram Stain	*Carbohydrate Utilized*
Citrate	Adonitol
H_2S	Arabinose
Lysine	Dulcitol
Ornithine	Fructose
Threonine	Galactose
Urea	Glucose
Oxidase	Glycerol
Indole production	Inositol
Acid Production	Lactose
Gelatine liquefaction	Sorbitol
	Sucrose

Fungal Analysis

The procedure used for fungi was similar to that used for bacteriological analysis but the growth medium used was potato dextrose agar containing 250 mg tetracycline hydrochloride and incubation was done at room temperature for 5 days. Identification of isolates was done according to Talbot (1978) and Deacon (1980).

Soil Analysis

Soil moisture, pH and chemical analyses were done as described by Andrews (1973) and IITA (1979).

Results

Bacterial Analysis

A total of twelve bacterial species were isolated from the soil samples collected from each of the three plots. Table 3.2 shows that eleven species each were isolated from plots A and B while only seven species were isolated from plot C. Bacterial densities ranged between 5×10^9 and 2.8×10^{10} with the average mean densities of 1.95×10^{10}, 2.44×10^{10} and 1.32×10^{10} for plots A, B and C respectively. The species composition of the different morphological types of bacteria are shown in Table 3.3. The Gram negative non-spore forming bacteria were the most common on the three plots but plot C had fewer number of species. The three plots recorded only one species of Gram positive non-spore forming bacteria *i.e., Corynebacterium* sp.

Fungal Analysis

Twelve fungal species were also isolated from the soils of the three plots. Table 3.4 shows that eleven fungal species were isolated from each of plots A and B while only ten species were isolated from plot C. Fungal densities ranged between 7.8×10^8 cfu/g and 1.98×10^{10} cfu/g while the mean densities were 1.6×10^9, 8.3×10^9 and 3.2×10^9 for plots A, B and C respectively.

Soil Analysis

Some selected characteristics of the top soil from the three plots are presented in Table 3.5. The moisture, silt content and the clay content were significantly higher in plot C than the two other plots

(P < 0.05). The nitrate and phosphate levels were also significantly higher in plot C than the two other plots. However, plot B recorded a significantly higher organic matter content than the other two plots.

Table 3.2: Bacterial Species Isolated from the Three Fire Experimental Plots at Olokemeji Forest Reserve

Isolates	*Family*	*Plot A*	*Plot B*	*Plot C*
Bacillus cereus	Bacillaceae	+	+	+
Bacillus megaterium	Bacillaceae	+	+	–
Bacillus subtilis	Bacillaceae	+	+	+
Corynebacterium sp.	Corynebacteriaceae	+	+	+
Escherichia coli	Enterobacteriaceae	+	+	+
Klebsiella aerogenes	Enterobacteriaceae	–	+	+
Micrococcus sp.	Micrococcaceae	+	+	–
Proteus vulgaris	Enterobacteriaceae	+	+	–
Pseudomonas cepacia	Pseudomonadaceae	+	+	–
Pseudomonas putida	Pseudomonadaceae	+	–	–
Staphylococcus aureus	Micrococcaceae	+	+	+
Streptococcus sp.	Micrococcaceae	+	+	+

+: Present; –: Absent; A: Late burn; B: Early burn: C: Fire-protected.

Table 3.3: Species Composition of Different Morphological Groups of Bacteria from Soil Samples of Olokemeji Fire Experimental Plots

Morphological Group	*A*	*B*	*C*
Gram +ve Non-spore forming rods	3	3	2
Gram +ve Spore forming rods	1	1	1
Gram –ve Non-spore forming rods	4	4	2
Cocci	3	3	2

A: Late burn; B: Early burn; C: Fire-protected.

Discussion

The relatively high soil moisture in plot C can be attributed to its relatively high clay and silt contents which retain water more than sand particles. Another factor is low rate of evaporation from the soil surface due to cover provided by the canopy of trees. The densities recorded for both bacteria and fungi on these plots were within the range reported by Alexander (1977). Both plots A and B had higher microbial counts than plot C but plot B recorded the highest bacterial and fungal densities. While the difference was significant for fungal populations on the three plots, the difference was not significant for bacteria. However, bacterial densities recorded on all the plots were higher than the corresponding values of fungi on all the three plots. The substrate must have been more conducive for the bacterial species. Most of the bacteria species recorded are normal inhabitants of soil (Adegeye and Ayodele, 1992, Okoh *et al.*, 2000), but a few of them such as *Bacillus, Micrococcus, Proteus* and *Pseudomonas* have been isolated from barks of trees from Olokemeji Forest Reserve by Akinsoji (1991). Some of the fungal species such as *Aspergillus flavus, A. niger* and *Rhizopus* sp. isolated from soils in these plots have also been recorded from tree barks by Akinsoji and Obadina (2004).

Table 3.4: Fungal Species Isolated from the Soil Samples of the Three Fire Experimental Plots at Olokemeji Forest Reserve

Isolates	*Family*	*Plot A*	*Plot B*	*Plot C*
Aspergillus flavus	Eurotiaceae	+	+	+
A. fumigatus	Eurotiaceae	+	+	+
A. japonicus	Eurotiaceae	+	+	+
A. niger	Eurotiaceae	+	+	+
A. oryzae	Eurotiaceae	+	+	+
A. wentii	Eurotiaceae	+	+	–
Aspergillus sp.	Eurotiaceae	+	–	+
Candida albicans	Cryptococaceae	–	+	+
Mucor sp.	Mucoraceae	+	+	+
Penicillium sp.	Eurotiaceae	+	+	+
Rhizophorus sp.	Mucoraceae	+	+	+
Saccharomyces cerevisiae	Saccharomycesaceae	+	+	+

+: Present; –: Absent; A: Late burn; B: Early burn: C: Fire-protected.

Table 3.5: Characteristics of Top Soil from the Olokemeji Fire Experimental Plots

Particle Size	*A*	*B*	*C*
Sand (per cent)	90.4	90.4	78.4
Silt (per cent)	3.4	5.4	11.4
Clay (per cent)	6.2	4.2	10.2
pH	4.92	4.90	4.96
Moisture content per cent	20.56	15.51	23.20
Organic matter per cent	2.35	3.41	1.94
Ca (per cent)	0.0024	0.0029	0.0024
Mg (per cent)	0.0016	0.0011	0.0009
Na (per cent)	0.0197	0.0232	0.0202
K (per cent)	0.0018	0.0023	0.0020
PO_4 (per cent)	0.20	0.27	0.57
NO_3 (per cent)	0.67	0.88	1.64

Source of particle size data (Adegeye and Ayodele, 1992).

The diversity of soil bacteria community increases with the degree of disturbance (Willig *et al.*, 1996). Fire, which in this study can be regarded as a form of disturbance has enhanced the diversity of bacteria species, as the plot C (fire-protected plot) recorded the least density and diversity for both bacteria and fungi.

There seems to be a relationship between the decomposition process and bacterial density in the soil. This relationship is a function of the amount of organic matter deposited in the soil. This amount is determined by the type of vegetation on the soil surface. In plot A, where burning is done at the end of the dry season, the vegetation comprises mainly grasses which when burnt adds organic matter to the soil. In plot B, the grasses are completely burned and there are trees, so the burnt grass and the litter from the trees contribute to the organic matter content while plot C only the litter contributes to the organic content. Thus, organic matter input is highest in plot B and this accounts for the higher bacterial and fungal density and diversity in plot B. This observation agrees with the report of Haack *et al.* (1995). Bils and Polishook (1994) and Cornejo *et al.* (1995) also reported that litter substrate species could be part of the factors which influence microbial density and species composition in forest litter decomposition.

Acknowledgements

We acknowledge the technical assistance of Dr. Adekunle and Dr. Nwachukwu. Tayo Onadipe typed the manuscript.

References

Adegeye, A.O. and A.I. Ayodele. 1992. The effect of fire on soil nutrient, soil microorganisms and seed microorganisms in Olokemeji Forest Reserve, Nigeria. *The Nigerian Journal of Forestry,* 22(2): 11–14.

Akinsoji, A. 1991. Studies on epiphytic flora of a tropical rainforest in south-western Nigeria II: Bark microflora. *Vegetatio.,* 92: 181–185.

Akinsoji, A. and Obadina, O. 2004. Studies on epiphytic flora of a tropical rain forest in south-western Nigeria. III: Bark Mycoflora.

Alexander, M. 1977. *Introduction to Soil Microbiology.* John Wiley and Sons. New York.

Andrews, W.A. 1973. *A Guide to the Study of Soil Ecology.* Prentice-Hall of Canada. Scarbourough, Ontario, pp. 196.

Bills, G.F. and J.D. Polishook. 1994. Abundance of microfungi in leaf litter of lowland Rainforest of Costa Rica. *Mycologia,* 8: 187–198.

Bradley, A.F., N.V. Noste and W.C. Fisher. 1992. Fire ecology of the forest habitat types of Eastern Idaho and Western Wyoming. *USDA General Technical Report INT,* 290.

Charter, J.R. and R.W.J. Keay. 1960. Assessment of Olokemeji Fire Experiment 28 years after institution. *Nigeria Forestry Information Bulletin,* 3: 32.

Cornejo, F.H.A., A. Varela and S.J. Wright. 1994. Tropical Forest litter decomposition under seasonal drought: nutrient release, fungi and bacteria. *Oikos,* 70: 183–190.

Deacon, J.W. 1980. *Introduction to Modern Mycology.* Blackwell-Scientific Publications, London, pp. 197.

Haack, S.K., H. Garchos, D.A. Odelson, L.J. Forney and M.J. Klug. 1995. Analysis of factors affecting accuracy, reproducibility and interpretation of microbial community carbon source utilization pattern. *Applied Environmental Microbiology,* 61: 1458–1468.

IITA. 1979. *Selected Methods for Soil and Plant Analysis.* Manual Series, 1: 70.

Okoh, A.I., G.O. Babalola and A.O. Olaniran. 2000. Aerobic heterotrophic bacterial and fungal communities in the top soil of Omo Biosphere Reserve in south-western Nigeria. *Biotropica*, 32(2): 208–212.

Prescot, L.M., J.P. Harley and D.A. Klein. 1999. *Microbiology*. McGraw Hill Companies Inc., USA, pp. 963.

Talbot, P.H.B. 1978. *Principles of Fungal Taxonomy*. Macmillian, London, England, pp. 252.

Whitley, L.S. and T.N. Seng. 1976. Studies on the bacterial flora of tubificid worms. *Hydrobiologia*, 48(1): 79–83.

Willig, M.R., D.L. Moorhead, S.B. Cox, and J.C. Zak. 1996. Functional diversity of soil bacteria communities in the Tabonuco Forest: Interaction of anthropogenic and natural disturbance. *Biotropica*, 28(4a): 471–483.

Chapter 4

Muga Based Ecological Farming System: An Approach to Sustainable Rural Development and Ecorestoration

L.N. Kakati*# and B.T. Kakati**

**Department of Zoology, Nagaland University, HQs. Lumami, Nagaland*

***Research Extension Centre C M E R& T I, Central Silk Board, North Lakhimpur – 787 001, Assam*

ABSTRACT

Muga culture (rearing of *Antheraea assama* and production of muga silk) which is known for tremendous employment and income generation in Assam has emerged as major concerns in recent years due to uncertain environmental condition, continuous deforestation, habitat degradation of muga host plants, insufficient quantity of quality seeds, lack of coordinated efforts in transfer of technology to grass root level. The muga based ecological farming system (MEFS) is a concept and approach for transforming the idea of conservation of biological resource and environmental protection to optimize integrated economic, ecological and social benefits to ensure sustainable rural development by means of anticipatory environmental policies. The ecofarming system ensures higher production through symbiosis of most desirable plants and animal species in space and time with gradual change from destructive land use pattern towards more stable and ecologically sound system. The present study area of Lakhimpur district of Assam is located in the foothills of Arunachal Pradesh and falls in humid subtropical region. Purposive random sampling technique using standard interview schedule was employed to generate data on economic benefit from rearing of muga silkworm under different systems. Soil analysis was made as per standard methods. Rearing performance in AMFP farm was better than the muga farm forestry, which was due to regular intercropping and proper maintenance. Importance of participatory role of women with intensive management of surplus labour has

Corresponding Author: E-mail: kakati_ln@yahoo.com.

been emphasized for not only bringing economic self reliance and sustainability but also as means of ecorestoration of degraded/unfertile land in the district.

Keywords: *Antheraea assama, Joint forest management, Intercropping, Muga farm forestry, Economic sustainability, Role of women.*

Introduction

Although rapid urban industrialization has brought about food security, employment and income generation, it equally poses great threat to resources conservation and environmental protection (Wenhua and Qingwen, 1999). The technological development and environmental protection are considered as two independent elements by the present agro industrially scientifically advance society which often sacrifices all the environmental interest for the sake of growth rate of production. In this context, ecological farming system maintains an equilibrium between consumption of biological resources and environmental protection following the principle of species symbiosis, recycling, regeneration and harmony between organisms and environment (Yan, 1986) to optimize integrated economy, ecological and social benefits for sustainable development. This ecofarming system ensures higher production through symbiosis of most desirable plants and animal species in space and time with a gradual change from degraded land use patent towards more stable and ecologically sound system. Since general crop productivity and socioeconomic conditions of the farmers are very poor, any developmental activity that is being taken up for improvement of the crop productivity and better utilization of natural resources should go a long way in improving the living standards of poor farmers (Hazra, 1999). Having realized the effect of uncertain environmental conditions on muga (*Antheraea assama*) cocoon production and habitat degradation, certain eco-friendly schemes like augmentation of muga food plants (AMFP), catalytic development scheme (CDP) muga farm forestry through social forestry programme have been implemented in Lakhimpur district of Assam to create suitable infrastructure for muga culture. Accordingly a concept of muga based ecological farming system (MEFS) has been inculcated among the muga farmers through multi cropping with muga host plants as main crop so that the beneficiaries not only ensure greater stability of production but subsistence oriented requirements from the same piece of land. While highlighting the economic benefit from the schemes and participatory management role of women in muga culture the paper projects MEFS as an approach towards sustainable rural development and ecorestoration of degraded land in Lakhimpur district of Assam.

Study Area

Lakhimpur district is located in northeast corner of Assam at 26° 48′–27° 53′ N latitude and 93° 42′–94°20′ E longitude. The district covers 2953 sq. kms (rural 2893 sq. kms and urban 60 sq. kms) with the population of 7,51,517 as per 1991 census (254 per sq. kms). The greater part of the district is alluvial plain which is formed by the Subansiri and the Brahmaputra river and its tributaries. The northern part of the district is known as foothill plains. It is characterized by absence of dry hot summer season. The highest temperature is experienced during southwest monsoon season along with abundant rains and highly humid atmosphere throughout the year. The climate of the district is more endurable than that of other plain district of Assam. The district falls in humid subtropical regions. Average winter temperature lies between 10°C and 17°C and average summer temperature lies between 25°C and 30°C. The district records average winter rainfall between 20 and 40 cm and average summer rainfall between 240 and 800 cm. Tropical rainforest, reeds, grasses and Bamboos are major natural forests available in district. A strip of reserve forest is situated in Northern Part. Mixed deciduous and mixed evergreen forest are scattered all over the district but mainly concentrated at the foothills. The eastern side of the district is conspicuous with presence of naturally grown Som (*Persia bombycina*) and Soalu (*Litsaea polyantha*) plantation, the primary food plants of muga silkworm.

Materials and Methods

AMFP and CDP Schemes

498 acres of degraded and unutilised land belonging to same nos. of poor and marginal farmers were identified in different parts of the district. The AMFP scheme was implemented during 1994–1997 with 2 m × 2 m spacing while CDP scheme was implemented during 1998–2001 with 3 m × 3 m spacing. The area was either ploughed or hoed, pits (1 ft × 1 ft × 1 ft) were dug up as per specified spacing, filled with 0.5 cft FYM and well grown seedling of muga host plant (*Persia bombycina*) transplanted in the pit during rainy season covering the base with straw mulch. Further farmers were advised to take up intercropping every year with any kind of crop *i.e.*, vegetables, leguminous plants, lentils, arum etc. as per suitability of season/area and keep a record for it. Rearing of muga silkworm was conducted after 4th year of plantation and cocoon production data were recorded.

Farm Forestry

Social forestry, North Lakhimpur implemented this scheme during 1992–1993 by transplanting 56,250 nos. of Som plants (3 m × 3 m spacing) in 50 hectares of degraded land in Dorpang area located in the foothills of Arunachal Pradesh. The same methodology was followed in plantation as mentioned above. Soil analysis was made before implementation of scheme and after 6th year of the plantation as per standard methods. To assess the participatory management role of women, economic sustainability and ecorestoration through MEFS, a survey was undertaken among 50 household engaged in muga culture and intercropping from 5 villages (10 from each village) representing whole district. A purposive random sampling technique was employed to collect data using standard interview schedule. Data on family size, land holding, occupation and participation of men, women and children in muga cultural activities were recorded and analyzed. Further 50 beneficiaries of muga farm forestry of Dorpang were also interviewed to assess the economic benefits accrued from muga silkworm rearing.

Results and Discussion

Participatory Management Role of Women in Muga Culture

The total population of selected 50 families was 290 of which 71 male, 80 female adults and 66 male and 73 female children. Out of 66 male children 25 were below 10 years and 41 were between 10–18 years of age while among 73 female children 29 were below 10 years and the rest were between 10–18 years. The population size of muga rearers did not show any regular pattern and varied greatly with the smallest family comprising of 2 children and the biggest family with 6 children. The land holding and utilization pattern reflected that on an average they had 4 acres of land of which 2 to 2.5 acre were used for paddy cultivation once in a year and the rest of the land being either highland or less fertile remained unutilized or underutilized. As the average paddy production of 11 to 13 quintals per family (equivalent to Rs. 6,200 to Rs. 7,400) was not sufficient even for six months, the manfolk looked for alternative sources of income through daily wages, masonary, carpentry, trading in weekly market etc. The men and women who helped in paddy cultivation remained idle during lean season. In fact there was always surplus of labour due to lack of sufficient cultivable land and alternative earning avenues. Although almost all female in the village were illiterate they took initiative to send their children to school.

Survey reveals that since muga host plants could bring short term and long term economic benefits, the farmers generally opted for muga plantation rather than other timber yielding plants. The involvement and participation of the villagers were must from the initial period of microplanning followed by implementation and development of the programme and subsequently to its protection. The women were the most important members in decision making, nursery raising, planting trees rearing activities, harvesting of cocoons and post cocoon technology like reeling and weaving. It was observed that women have contributed 40 per cent of the total workdays in nursery preparation and transplantation of seedling. Male participate maximum in pre rearing operation (77 per cent) like preparation of rearing shed, collection of seed etc while 23 per cent had been shared by female. 45 per cent of the work in rearing stage like brushing, of tiny worms, transfer and collection of mature worms harvesting were done by the female while there was 75 per cent contribution of female in cocoon reeling and weaving of fabrics. While 92 per cent male involved in the sale of ready cocoons, female participation was only 8 per cent. It was observed that muga culture enterprise had immense potential for economic development in the rural areas by engaging surplus female labour which contributed about 42 per cent of the work force input. It is imperative that special efforts are to be made in this regard as the women play not only active part in manual operations but also in the area of decision making.

Economic Sustainability through MEFS

Income from Muga Silkworm Rearing

A case study of 50 farmers in five villages in the district revealed the fact that adoption of proper package of practice in AMFP scheme successfully reduced the gestation period of host plant from 7 to 4 years to support rearing of muga silk worm. The rearing was conducted from 1998 to 2002 in this systematic plantation with an average utilization of 200 to 500 dfls per crop depending upon the carrying capacity of plantation area. The quantum of dfls was increased from 2 to 3 crops per year as per growth of the host plants with cocoon production ranging from 10,00,000 to 40,50,000 nos. A rearer could earn at least Rs. 6,800.00 to Rs. 31,725.00 from sale proceed of cocoons in different year (Table 4.1a).

Income Generation from Intercropping

The mixed cropping systems including intercropping indicated that different kinds of crops were grown in the same field and in the same year, the order of these crops decided by the ecological adaptation to the growing season (Wenhua and Qingwen, 1999). While implementing the AMFP and CDP schemes the farmers were initially reluctant to go for plantation of muga since in initial first 3–4 years they would not get any return till the plantation was ready for commercial exploitation. The majority of farmers had very small land holdings and relied on family labour and simple tools. They neither had capacity to take risk nor had enough land to diversify cropping. Thus the concept intercropping with muga host plants was introduced for not only ensuring greater stability of production but profit and subsistence-oriented requirement from the same piece of land (Shukla *et al.*, 1989). The rearers took up intercropping of tomato, ginger, tamarind, arum, lentil, black gram and mixed vegetables during gestation period of Som plants *i.e.*, 1994–1998 and also continued till further which brought economic stability to the rearers.

Table 4.1: Muga Silkworm Rearing Performance During 1998–2002 in AMFP (a) and Dorpang Farm Forestry (b) Plantation

(a) AMFP Plantation

Sl.No.	Year	No. of Rearers	No. of Crops	Dfls Reared (gms.)	Cost of dfls (Rs.)	Cocoons Harvested (Nos.)	Cocoon: Dfl	Sale Proceed (Rs.)	Net Income (Rs.)	Income/ Rearer	Income from Other Sources
1.	1998	50	2	20000	60000	10,00,000	50	4,00,000	3,40,000	6,800	–
2.	1999	50	2	30,000	90,000	13,50,000	45	5,73,750	4,83,750	9,675	–
3.	2000	50	3	52,500	1,83,750	24,67,500	47	10,60,925	8,77,175	17,544	–
4.	2001	50	3	60,000	2,40,000	31,20,000	52	14,04,000	11,64,000	23,280	–
5.	2002	50	3	75,000	3,37,500	40,50,000	54	19,23,750	15,86,250	31,725	–

(b) Dorpang Farm Forestry Plantation

Sl.No.	Year	No. of Rearers	No. of Crops	Dfls Reared (gms.)	Cost of dfls (Rs.)	Cocoons Harvested (Nos.)	Cocoon: Dfl	Sale Proceed (Rs.)	Net Income (Rs.)	Income/ Rearer	Income from Other Sources*
1.	1998	50	2	20000	60000	6,00,000	30	2,40,000	1,80,000	3,600	–
2.	1999	50	3	30,000	90,000	9,90,000	33	4,20,750	3,30,750	6,615	–
3.	2000	50	3	46,400	1,62,400	13,45,600	29	5,78,608	4,16,208	8,324	–
4.	2001	50	3	52,253	2,09,012	18,28,855	35	8,22,985	6,13,793	12,279	–
5.	2002	50	3	61,357	2,76,106	19,02,067	31	9,03,482	6,27,375	12,548	2000

* Sale proceed of pollarded and pruned branches of host plants.

Ginger and Tamarind

Ginger and Tamarind, which were planted in two rows along the som plantations, had been considered as popular cash crop. The crop showed active growth during summer and a lien growth period during winter and brought an additional income of Rs. 500.00 to Rs. 1200.00/acre/year to farmers.

Tomato

Between two rows of som plants, 4 rows of tomato plants with 1 spacing both side were planted during winter. The annual average income from tomato was Rs. 1,500 to 2560.00 per acre.

Vegetables

The farmers used the vacant land for cultivation of different type of seasonal vegetables like, spinach, chilly, cabbage, potato mustard, onion, garlic pea etc. for domestic consumption which amounted to Rs. 1000.00 to 1400 per acre per year.

Arum

Two pits of 1 sq.ft. were dug up between each row of som plants for plantation of arum which was considered as one of the most suitable intercrop in som plantation. Due to regular hoeing done for growth of arum, the soil became loose around som plants providing better aeration and nutrient supply. Further this practice controlled excessive growth of weeds. Farmers received an additional of income of Rs. 600.00 to Rs. 1,500.00/per acre/year.

Black Gram and Lentil

The plantation area was ploughed/hoed and crops were sown in August and harvested during December. An additional income of Rs. 1,480.00 to Rs. 2,300.00/per acre/year was received. Thus it has been observed that farmers earned an additional amount of Rs. 5,880.00 to Rs. 7,700.00 per acre during gestation period (1994–1997) and also continued to earn in later years also (Table 4.2).

Table 4.2: Income Generation from Muga Silkworm Rearing and Intercropping in One Acre of Land Under AMFP Scheme

Year	*Inter Crop (Rs.)*					*Silkworm Rearing (Rs.)*	*Total Income (Rs.)*
	Tomato	*Vegetables*	*Arum*	*Ginger/ Tamarind*	*Lentil/ Blackgram*		
1994	1500	1100	900	850	1530	–	5880
1995	1575	1200	1200	500	1480	–	5955
1996	2500	1100	1500	1000	1500	–	7600
1997	2400	1400	1500	800	1600	–	7700
1998	2000	1200	1400	1200	2300	6800	14900
1999	2560	1000	1000	1100	1500	9675	16835
2000	1500	1100	800	500	2000	17544	23444
2001	1200	1000	600	900	1950	23280	28930
2002	1500	1000	900	900	1800	31725	37825

Economic Benefits from Farm Forestry with Muga Host Plant

Social forestry is part of joint forest management system where government agencies and people share the burden and bounty together and work for growth and long stay of forests. The national forest policy, 1988 stress upon the maintenance of environmental stability and restoration of ecological balance, conservation of forest, meeting the basic need of the people and maintaining the intrinsic relationship between forest and poor people living in and around (Singh and Sinha, 2000). Farm forestry offers the opportunity for the local people to plan, establish, manage and harvest forest crop to develop forest base industry which provide gainful employment to the rural people (Romadevi and Sivaramakrishnan, 1995). Indiscriminate deforestation in Lakhimpur district and Arunachal Pradesh has adversely affected the land surface in a variety of ways like soil erosion and devastating flood which not only reduces agricultural area but also greatly threatened muga culture by habitat degradation. Hence with a broader vision to develop muga culture into an organised industry social forestry in Lakhimpur district since 1984 has brought 139 hectres of community land like Government vacant land, road sides, canal banks, wasteland, catchments area and other unutilized area through which farmers were motivated to plant trees and till 1993 raised 2,55,850.00 nos. of Som plants with spacing range from 2 m × 2 m to 4 m × 4 m. The plantations raised in Government land were managed by the department continuously for 5 years and there after handed over to a locally formed committee for further maintenance and rearing of Muga silkworm. It was estimated that a 5–9 years old tree of som or soalu could support two rearings of 3–5 dfls in one year yielding at least 500 numbers of cocoons (Das, 2000). The rearing performance in Dorpang muga farm forestry during 1998–2002 for five years showed commendable results in cocoon production and the rearers earned handsome amount by selling pollarded and pruned branches [Table 4.1(b)]. A perusal of Table 4.1(a) and (b) indicates that performance of cocoon production (dfl: cocoon ratio) is quite high in AMFP farm than the Dorpang farm forestry in different years during the study period. It may be due to the proper maintenance of the farm and regular intercropping in the farmer which enhance the quantity of foliage that sustain higher quantity of dfls.

MEFS for Ecorestoration

Intercropping as Tool for Ecorestoration

Intercropping checks the weeds and provide suitable microclimatic conditions for muga host plants. The leguminous crops like black gram, Pea and lentil have rhizobium bacteria which are wholly or largely capable of bringing about appreciable nitrogen fixation on roots. The leaves etc. which fall on the surface add the organic carbon to the soil and also enrich soil nitrogen. The organic materials added during cropping act as food for soil organisms and tend to stimulate biological changes. The growing leguminous crop exerts conserving influence on the nutrient status of the soil since it takes up soluble constituents that might otherwise be lost in drainage or by erosion, thus, functioning as a cover crop (Singh *et al.*, 2003). Because of their nitrogen fixing ability, legumes not only transfer nitrogen to the associated muga plants but also leave behind the soil fertile. This leads to decrease in use of chemical fertilizer thereby bring down the maintenance cost of plantation. Because of their early fast growth and spreading canopies the legumes cover the soil surface faster and weeds underneath. The light reaching the weeds is cut off partially or wholly causing reduced photosynthesis. This results in suppression of weed growth, causing reduction in flowering and seed production of weeds (Shukla *et al.*, 1989). Positive effect of intercropping with different crops in mulberry garden enhancing growth and leaf quantity of mulberry has also been reported by Saraswat and Gupta (1990), Singh *et al.* (2003), Tikader and Sen (1991).

Muga Farm Forestry for Ecorestoration

Dorpang area in the foot hills of Arunachal Pradesh which became barren due to complete removal of vegetation followed by heavy grazing was considered as highly degraded land as reflected on the initial status of important physico-chemical properties and available nutrient status. However reanalysis of soil status after six years of plantation indicated an appreciable improvement of soil nutrients and organic carbon over initial level which was due to afforestation and muga cultural activity (Table 4.3). The marginal drop in soil pH may be due to greater opportunity period of soils to retain water and gradually got percolated alongwith increase organic matter content in soils (Hazra, 1999). The excreta of worms and dead worms drop on the forest floor enriched the floor with nutrients that helped to grow a second stratrum of herbs and shrubs. Again the root systems of all such plants penetrated to different depths thus transporting the minerals to forest floor and loosen the soil for better retention of water and further growth of root systems. The ground vegetation stopped soil erosion and increased water retention thereby improving soil condition. It has been observed that barren lands around forest area having regular sericulture activity turned fertile. The role of afforestation and silvipasture has also been highlighted by many workers for restoring soil productivity and reducing soil and water loss through erosion (Patil, 1979; Hazra, 1989).

Table 4.3: Soil Fertility Status Before and After Afforestation of Dorpang Community Land

Soil Properties	*Soil Texture*	*Organic Carbon (%)*	*pH*	*N kg/ha*	*P kg/ha*	*K kg/ha*
Before plantation (1990)	Sandy loam	0.35	6.7	131.5	8.5	105.9
After plantation (1998)	Sandy loam	0.95	6.4	285.5	29	210.0

Conclusion

Sustainable use of limited land through MEFS brings not only rural economic development but also help ecorestoration in following ways:

1. Package of practice available for maintenance of economic block plantation can be effectively implemented to increase the quality of foliage.
2. The gestation period of muga host plant can be reduced from 7 to 4 years to take up early rearing.
3. The schemes link the environment, natural resources and socio-economic factors in development of rural area.
4. The landslide prone area can be identified and effectively controlled by Muga host plant for its commercial utilization.
5. It offers chances for employment and socio-cultural development of local people in the generation to come by the system of exploitation, utilisation and management of natural resources.
6. Establishment of true symbiosis between man and his environment leading to lasting synergy, human achievement, development and promotion of self reliance.
7. Being semidomesticated, rearing of muga silkworm becomes a part of forest ecosystem, which requires exploitation without injury to plants and requires a system of training and monitoring in forest management and sericulture. Muga culture has multidirectional action in regenerating the forest.

8. Loss of leaves is for a brief period of time as fresh leaves appear within 30 to 40 days after rearing. The newly sprouted leaves get better exposure to sunlight causing more trapping of energy and provide greenness to the area.
9. Increase vegetation stops soil erosion and increases water retention, penetrating root system transport the minerals to the floor, thereby bringing eco-restoration of degraded environment and barren land can be converted into fertile land. Further the green canopy of the plants checks the dropping of rains and protect underlying soil.
10. The rural people can utilize pollarded and pruned branches of tree as fuel.
11. The new concept encourages the local educated youth and women group for adoption of upgraded technology and inculcate self confidence, self reliance and self dependence which check urban migration for any other employment opportunity.
12. Beside improving the ecosites, afforestation helps improving the overall environmental condition. Although its effect on the air could not be quantified in terms of quality improvement, but people around realize its effect on lowering of temperature specially during summer months.

References

Das, P.K. 2000. Muga silk industry in Assam. In: *Proc. Development of Muga and Eri Silk Industries through Improved Technology*, p. 1–7.

Gaur, O.P. 1999. Integrated rural development through farm forestry on waste lands: A cooperative experience. In: *Environmental Education for Sustainable Future*, (Eds.) Desh Bandhu and R. Dyal. Indian Environmental Society, New Delhi, p. 221–227.

Hazra, C.R.1989. Forage and soil productivity under agroforestry and silvipasture system in Bundelkhand region. In: *Promotion Fuel Wood Trees*, (Eds.) Hegde, N.G., Relwani, L.L. and Ketkar V.D. Baif Publication, Pune, p. 89–98.

Hazra, C.R. 1999. Rehabilitation of degraded community grazing land through participatory resource management on watershed basis: A case study of Kharaiya Nala Watershed. In: *Environmental Education for Sustainable Future*, (Eds.) Desh Bandhu and R. Dyal. Indian Environmental Society, New Delhi, p. 155–171.

Patil, B.D. 1979.Agroforestry with accent on forage-fuel production. *Bull. IGFRI Research Area Series 1*, Jhansi, p. 1–22.

Romadevi, O.K. and Sivaramakrishnan, V.R. 1995. Future of tasar silk industry in relation to Forestry in India. *Indian Silk*, p. 35–38.

Saraswat, S.B. and Gupta, Y.K. 1990. Tomato as intercrop in mulberry garden. *Indian Silk*, p. 28–30.

Shukla, P., Misra, R.K. and Choudhury, P.C. 1989. Intercropping: A new vista in mulberry cultivation. *Indian Silk*, p. 9–11.

Singh, B.D., Singh, P.K. and Kant, R. 2003. Lentil and Moong-intercropping with mulberry. *Indian Silk*, p. 10–12.

Singh, B.M.K and Sinha, A.K. 2000. Role of sericulture on sustenance of joint forest management. *Indian Silk*, p. 14–18.

Tikadar, A and Sen, S.K. 1991. Ginger and Field pea as intercrops with mulberry. *Indian Silk*, p. 47–48.

Wenhua, L. and Qingwen, M. 1999. Integrated farming system: An approach towards sustainable agriculture in China. *Ambio*, 28(8): 655–662.

Yan, J. 1986. The main practice and types of ecological engineering for the conversion of wastewater into resources. *Rural Ecology and Environment*, 8: 40–44.

Chapter 5

Water Management and Analysis

K. Bayapu Reddy, R.V.S.S.L. Revathi** and T. Manjunatha****

**Chairman, B.O.S and Head, Department of Environmental Sciences, Acharya Nagarjuna University, Guntur – 522 510*

***Research Scholar, Department of Environmental Sciences, Acharya Nagarjuna University, Nagarjuna Nagar, Guntur – 522 510*

****Sikkim Manipal University, Indian Institute of Ecology and Environment, New Delhi*

ABSTRACT

A review of grey water reclamation as a basic concept of effective water management is the theme of the work presented. Sewage (grey water) samples from the drains of administrative blocks of Visakhapatnam Port Trust were collected and were analysed to assess the strength of the sewage. The rate of flow of sewage in peak hours was computed basing on which sedimentation tank was designed to subject the sewage for primary treatment followed by chlorination thus raising the same for sanitation purposes as a substitute to freshwater.

Keywords: Grey water, Water management.

Introduction

Most of Earth's water (97 per cent) is too salty for consumption (Gleick, 2001) only 2.5 per cent is fresh is actually unstable (Allen 2001: Earth Pulse). Growing demand for water and the declining supplies are the cause for the widening gap between demand and supply. This has created a rift between rural and urban water needs exacerbated by the trend of expanding urbanisation. With people having no interest in using water carefully, the sustainability of the water resources has itself become a question mark. Overpumping of groundwater resources (aquifers are pumped faster than they are replenished) is acute in parts of India, China, North Africa, the Middle East and USA (Ellwood, 2000).

Sustainable water management and the development of recycled water systems have accelerated with the restrictions of Australia's water industry. In Australia water reuse from STPs in 1998 was 113 GL (giga litres: a thousand million litres), being 7 per cent of STP effluent and this is expected to double in 4 years (Dillon, 2000). The innovation of recycling of water is not new. Some of the earliest beneficial uses in the USA dated from 1925 when the Grand Canyon village developed dual reticulation to conserve water to meet the demand from tourism which has since been expanded (1960) and upgraded (Gasthe EC, Okun 1996). Japan is another active user of reclaimed water. Due to serious urban water shortages, Japan began recycling wastewater within large buildings, for toilet flushing (Arano, 1996). 37 per cent of reclaimed water is being used for dual systems 23 per cent for cleaning and only 18 per cent for landscaped irrigation (Okun, 1996). The first thing is to augment available water resources. Rainwater harvesting provides a powerful and decentralised solution to the crisis. But water harvesting will not be enough if we do not learn to use water prudently and carefully that is being ignored completely. This is the fifth consecutive year of drought. In vast parts of the country without water the situation is crippling and dehumanising. So rich cities like Singapore learned to value water and began to recycle and reuse it but our urban settlements just destroys every water body in the vicinity. The traditional Japanese never discharge waste into the rivers, instead they compost it and then use it as manure.

On site systems of water recycling are often referred to as grey water or sullage reuse. Household wastewater, except sewage receives the minimal treatment of filtering or screening of gross pollutants, such as hair and other debris. This practice has been allowed in the USA for sub-surface watering, but there is no evidence of its use in Japan (Jeppessen Solky, 1994).

Materials and Methods

The study of possible "water management and analysis" of the two administrative buildings *i.e.*, administrative office building of Visakhapatnam Port Trust and Dock Labour Board building of Visakhapatnam have taken up. This includes the effective use of potable water and reusing of wastewater after treatment for different purposes. Visakhapatnam Municipal Corporation, Visakhapatnam, is supplying both the buildings with water. The water supply section of the Civil Engineering Department of Visakhapatnam Port Trust has monitored the same.

Administrative office building is the main administrative office block for Visakhapatnam Port Trust, situated in port area consisting of five floors (G floor + 4 floors). The number of employees in each floor are around 200 nos. including floating persons *i.e.*, visitors and employees. The total number of persons using the water through different ways are about 1600 nos. The total quantity of water supplied to AOB is 90,000 litres per day including for canteen and watering of garden around the building. The wastewater is being flown through surface drains and open drain and finally to the sea.

Dock Labour Board building is the main administrative block for Dock Labour Board situated in port area consisting of six floors (G floor + five floors). Total number of persons using the water are about 350 nos. including visitors. The total quantity of water supplied to this building is 20,000 litres per day including for canteen.

Each floor in both the buildings are provided with wash basins, Indian water closets (IWC), European water closets (EWC), urinals for the best use of the employees and visitors of the building. The facilities provided are tabulated in Table 5.1

Table 5.1: Details of Tap Points in AOB and DLB Buildings

	IWC	*EWC*	*Urinals*	*Wash Basins*	*Total Tap Points*
AOB	25 nos.	27 nos.	77 nos.	68 nos.	197 nos.
DLB	64 nos.	14 nos.	96 nos.	38 nos.	212 nos.

The wastewater (grey water) samples of AOB and DLB are collected in the drain carefully in the water sample bottles. The samples were then tested in the laboratory for the parameters. Since the source of supply is same for both AOB and DLB the freshwater sample has been collected from the tap point of AOB and got tested in the laboratory for the parameters.

The samples were analysed for different physico-chemical parameters *viz.*, pH, Total Hardness, Sulphates, BOD, Total Suspended Solids, Alkalinity, Chlorides, Total Solids and Total Dissolved Solids adopting the analysis procedures from "Standard methods for examination of water and wastewater 20th edition, APHA, AWWA".

Results and Discussion

The analysis results of the wastewater samples are tabulated as shown in Table 5.2 collected from AOB, DLB and freshwater respectively.

Table 5.2: Results of Water Sample Parameters (All values except pH and Conductivity are in mg/l)

Sl.No.	*Parameters*	*AOB Wastewater Sample*	*DLB Wastewater Sample*	*Freshwater Sample*
1.	pH	7.20	7.31	7.88
2.	Conductivity (micro mhos/cm)	575	886	208
3.	Chlorides	122	177	24
4.	Total solids	757	622	109
5.	Total dissolved solids	400	600	100
6.	Total suspended solids	357	22	9
7.	Total alkalinity	169	312	118
8.	Total hardness	158	139	97
9.	Sulphates	41	50	12
10	BOD	200	120	–

Table 5.3: Domestic Water Quality Standards (All values except pH are in mg/l)

Sl.No.	*Parameter*	*Standard*
1.	pH value	6.5 to 9
2.	Total solids	500 to 1000
3.	Total suspended solids	250
4.	Total dissolved solids	250 to 500
5.	Sulphates	250
6.	Chlorides	250
7.	Total hardness	75 to 115

The pH values of AOB and DLB sewage samples were 7.20 and 7.31 respectively while in freshwater it was 7.88 as evident from the Table 5.2.

The important factor influencing the strength of sewage are total solids and total dissolved solids. The result shown in the Table 5.2 indicates that comparatively low solids were present in DLB sample recorded as 622 mg/l and in AOB they were 757 mg/l. In both the samples the total suspended solids were within the domestic water quality standards but the total suspended solids were high in the sample collected at AOB in comparison with those in the DLB sample whereas dissolved solids were reported high in AOB sample. On a whole the sewage can be categorised as medium with respect to the table (Garg, 1999) shown below.

Type of Solid	*Strength of the Sewage*		
	Weak	*Medium*	*Strong*
Total Solids	400	800	1200
Total Suspended Solids	100	200	350

The hardness values were found to be 158 mg/l at AOB and 139 mg/l at DLB. Both the samples showed high levels in comparison to the domestic water quality standards.

Also as BOD is not high, other parameters like conductivity, chlorides, alkalinity, sulphates were analysed and were within the norms making the wastewater well suited for reclamation. As the sewage in the present case is not so strong, primary treatment is only required and the treated water can be effectively used with less economic cost. Basing on the total quantity of water supplied and the total quantity of water consumed, the total wastewater probably passing through drains is estimated as follows.

Water Management

Table 5.4: Surplus Water Per Day in AOB and DLB

	AOB	*DLB*
Quantity of water supplied (litres/day)	90,000	20,000
No. of persons utilising the service	1600	350
No. of litres per capita per day	45	45
Consumption of water per day		
1. For total persons at 45 LPCD	72,000 litres	15,750 litres
2. For gardening and canteen	10,000 litres	2,000 litres
Daily consumption–Total	82,000 litres	17,750 litres
Surplus water per day = quantity supplied – quantity consumed	90,000 – 82,000 = 8,000 litres	20,000 – 17,750 = 2,250 litres

Wastage of Water

Wastage of water through taps is 2.25 litres per minute (average wastage of water through taps–observed data).

Table 5.5: Rate of Wastage of Water in AOB and DLB

	AOB	*DLB*
No. of tap points	197	212
Rate of wastage of water	197 × 2.25 litres/min = 443.25 litres/min	212 × 2.25 litres/min = 477 litres/min

Volume of Sewage

This is based on the consumption of water. It is generally assumed that half the daily consumption of water occurs within 6 hours. This gives the average peak consumption. The open drains which carry wastewater should be designed for self cleansing velocity of 0.76 to 0.91 m/sec. A good average rule is to allow liguid waste from buildings at the rate of 0.03 cu.m./min per 100 persons based on a water consumption of 135 LPCD.

Average peak consumption = Daily consumption/(2×6) litres per head per hour

Rate of Flow of Liquid Waste

0.03 cu.m./min per 100 persons based on a water consumption of 135 LPCD. For water consumption of 45 LPCD = $0.03 \times 45/135 = 0.01$ cu.m./min per 100 persons.

Table 4.6: Total Flow of Liquid Waste Per Hour in AOB and DLB

	AOB	*DLB*
Average peak consumption	45/(2 × 6) = 3.75 lts per head per hr	45/(2 × 6) = 3.75 lts per head per hr
No. of employees	1600	350
Total average peak consumption	1600 × 3.75 = 6000 litres per hr	350 × 3.75 = 1312.50 litres per hr
Rate of flow of liquid waste	1600 × 0.01/100 = 0.16 m³/min	350 × 0.01/100 = 0.04 m³/min
Total flow of liquid waste	0.16 × 60 = 9.60 m³/hr	0.04 × 60 = 2.40 m³/hr
Total		**9.60 + 2.40 = 12.00 m³/hr**

Conclusions

1. The wastewater is appreciable as it is within the domestic water quality standards.
2. The wastewater which is being discharged from the institutions and office administrative buildings can be reused effectively so that demand for freshwater can be reduced.
3. As the sewage water is not very strong, primary treatment followed by chlorination is just required and is economically viable.
4. The treated water can be used for gardening, floor washings, flushing purposes, lavatory cleaning, etc.

Design of Tank for Treatment by Plain Sedimentation

Total flow of liquid waste per hour in AOB and DLB = 12.00 m^3/hr or say 15 m^3/hr.

The general working hours of both offices is from 8:00 A.M. to 6:00 P.M. (*i.e.*, a total of 10 hours). Detention period = 6–8 hours

Therefore, Total quantity of flow = $10 \times 15 = 150$ m^3/hr

Long, relatively narrow tanks are less affected by inlet and outlet disturbances and cross currents caused by breezes. Narrow tanks are more efficient since the time taken by the suspended matter to settle at the bottom is less. The velocity of flow in the tanks is kept ½ to 1 ft. per minute (Khanna, 2001).

Let us assume velocity of flow = 0.5 ft/min = 9.144 m/hr

Area of the tank = 150/(0.5 × 0.3048 × 60) = 16.40 m² or say 20 m².

But the proportion of breadth to length is generally 1 : 2, with 2 : 3 minimum and 1 : 4 maximum (Khanna, 2001). Area = length × breadth

A = l × b = 4b × b (since l = 4b)

Breadth = b = √(A/4) = √(20/4)	= 2.236 m.
Length = l = 20/2.236	= 8.944 m.
Depth of water in the tank	= 4.00 m.
Free Board	= 0.50 m
Depth of the sludge	= 0.80 m
Net water depth = 4.00 – 0.80	= 3.20 m
Therefore, Dimensions of tank	= 8.95 m × 2.24 m × 4.50 m

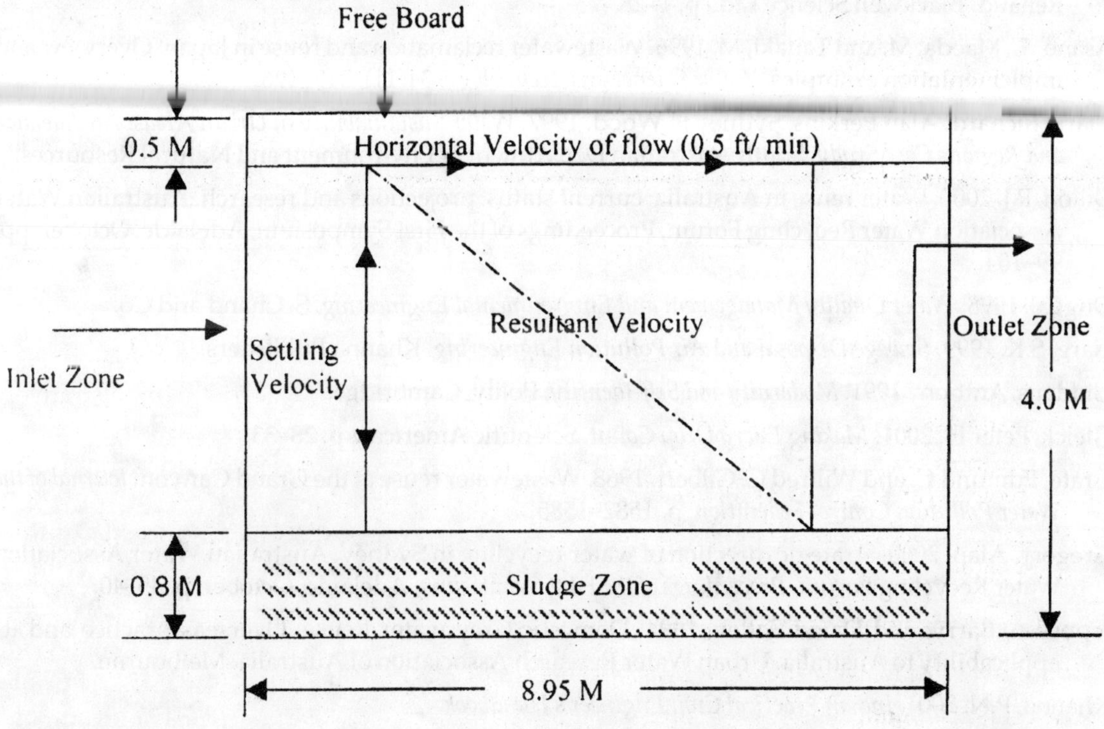

Figure 5.1

The settling velocity of organic particles (sp. gr. 1.2) of size larger than 0.06 mm is about 0.3 mm/sec or 1.2 m/hr (Garg, 1999). Therefore the particles are likely to settle to a depth of 4.0 m in about 3 hours time as against the 6–8 hours detention period. Hence it is estimated that the size of tank will be sufficient enough for the particles to be removed by the time the water drains out. This water is then stored in a separate sump for further reuse.

However as the water is proposed to be used for landscaping, floor washings, flushing purposes, lavatory cleaning, etc. bleaching powder is added. The dosage of chlorine suggested for settled sewage (fresh) is 10 mg/l. The dosage of chlorine required is calculated by the following formulae

$$\text{Dosage of chlorine} = \frac{\text{Litres of water} \times \text{dosage in mg/l}}{1{,}000{,}000} = \frac{1.00{,}000 \times 10}{1{,}000{,}000} = 1 \text{ kg}$$

The amount of bleaching powder with 30 per cent available chlorine required = 1 × 100/30 = 3.33 kg.

References

Allen, William L. (Ed.). 2001. *World of Water: Enough for All*? National Geographic, 199(4): Earth Pulse.

APHA, AWWA. *Examination of Water and Wastewater*, 20th edition.

Archuleta, E.G. 1997. National Report: USA. In: *International Water Services Association Congress*, (Ed.) P. Renaud. Blackwell Science Ltd., p. 1–28.

Asano, T., Maeda, M. and Tanaki, M. 1996. Wastewater reclamation and reuse in Japan: Overview and implementation examples. *Water Science and Technology*, 34(11): 219–226.

Clark, Richard, Alan Perkins, Sydney E. Wood. 1997. *Water Sustainability in Urban Areas: An Adelaide and Regions Case Study*. South Australian Department of Environment and Natural Resources.

Dillon, P.J. 2000. Water reuse in Australia: current status, projections and research. Australian Water Association Water Recycling Forum, Proceedings of the First Symposium, Adelaide, October, pp. 99–104.

Duggal. 1998. *Water Quality Management and Environmental Engineering*. S. Chand and Co.

Garg, S.K. 1999. *Sewage Disposal and Air Pollution Engineering*. Khanna Publishers.

Giddens, Anthony 1991. *Modernity and Self-Identity*, Polity, Cambridge.

Gleick, Peter H. 2001. *Making Every Drop Count*. Scientific American, p. 28–33.

Grate, Edmund C. and Wilfred C. Gilbert. 1968. Wastewater reuse at the Grand Canyon. *Journal of the Water Pollution Control Federation*, p. 1582–1585.

Gregory, Alan. 2000. Strategic direction of water recycling in Sydney. Australian Water Association Water Recycling Forum, *Proceedings of the First Symposium*, Adelaide, October, p. 35–40.

Jeppesen, Barry, and David Solley. 1994. Domestic Grey water Reuse: Overseas practice and its applicability to Australia. Urban Water Research Association of Australia, Melbourne.

Khanna, P.N. 2001. *Indian Practical Civil Engineer's Handbook*.

Okun, Daniel A. 1996. Distributing reclaimed water through dual systems. *Journal of American Water Works Association*, 89(11): 52–64.

Rast, Walter. 2000. Overview of the Status of Implementation of the Freshwater objectives of Agenda 21 on a Regional Basis. Sustainable Development International [online accessed, 23 February 2001] URL: http://www.sustdev.org/journals/edition.01/preview/1.053.shtm.

Santra, S.C. 2001. *Environmental Sciences*. New Central Book Agency Pvt. Ltd. Publishing.

Sharma, B.K. 2001. *An Introduction to Environmental Pollution*. Goel Publishing House.

Chapter 6

Biomonitoring Approach with Benthic Macro-invertebrates for Water Quality Assessment in a Medium Reservoir

Ch. Srinivas and Ravi Shankar Piska

Department of Zoology, University College of Science, Osmania University, Hyderabad – 500 007

ABSTRACT

The present study deals with the benthic macro-invertebrates as a tool for biomonitoring for water quality assessment in a medium reservoir, Edulabad, Ranga Reddy District, Andhra Pradesh, India. The present work carried for three years from 2000–01 to 2002–03. The water spread area is 247.31 ha. and water used for irrigation and fisheries. Polluted river Musi water enter into reservoir through a canal called Narayanarao Kathwa. Oligochaets, Hemipterans, dipterans, Coleopterans and molluscans found in the reservoir, out of which dipterans dominated. The biological data of various pollution indices like saprobic score and diversity scores calculated to quantify the water quality of the reservoir. Biomonitoring studies indicated that the reservoir was heavily polluted.

Keywords: Biomonitoring, Benthic macro-invertebrates, Water quality, Pollution, Reservoir.

Introduction

Monitoring of environmental components is an important pre-requisite for pollution control activities. It is difficult to monitor water quality only by using physico-chemical methods due to large number of pollutants and their low concentrations. To overcome this difficulty environmental scientists, all over the world, are exploring the possibility of using biomonitoring techniques in addition to physico-chemical monitoring.

Aquatic organisms have preferred habitats, defined by physical, chemical and biological features. Variation in one or more of these can lead to stress on individuals and possibly a reduction in the total number of species or organisms that are present. In extreme situations of environmental change, certain species will be unable to tolerate the changes in their environment and will disappear completely from the area concerned, either as a result of death or migration. Thus the presence or absence of certain species or family groups or the total species number of abundance, have been exploited as a means of measuring environmental degradation.

Biomonitoring is the introduction of biological variables for assessment of the structural and functional aspects of aquatic ecosystems. Through biomonitoring, the cumulative effect of all the pollutants can be determined and the overall health of the aquatic ecosystems could be properly assessed. Biomonitoring can be used as a cost-effective mean for supplementing the physico-chemical techniques. In the present study, benthic macro-invertebrates used to assess the pollution levels in the medium reservoir.

Materials and Methods

The present study conducted in a medium reservoir, Laxminarayana reservoir, Edulabad, Ranga Reddy District, Andhra Pradesh, India during three years from 2000–01 to 2002–03. For the Bio-monitoring study of the reservoir, four sites were selected and the benthic macro-invertebrates samples collected in the same sites to minimize seasonal variation. Samples collected during biologically active period (Oct-May), the collected sediment sample has been sieved for separating animals from detritus. US standard Seive No. 30 with pore size 0.6 mm used to retain organisms. The other way was to pick up organisms from sieve with forceps and placed in collection bottle.

The collected organisms preserved with 4 per cent formalin in wide mouth polyethylene bottles for further identification, verification and analysis. The criteria followed for water quality evaluation was based on the comparison of diversity score and saprobic score as per deZwart (1993) and CPCB (2002).

The bottom fauna collected at a sampling station for evaluating saprobity is randomized by gently shaking the collection bottle. The contents poured in to a flat white enamel or plastic tray with vertical and horizontal lines on the bottom. Clumps of organisms dispersed by pouring water over them. Sequential comparison Index (SCI) method involved a comparison of sequentially encountered individuals and the difference of two specimens which can easily be observed up to the species level. The diversity was the ratio of the total number of runs and the total number of organisms encountered. However, diversity alone can't indicate overall health of water body that is why the diversity of benthic animals was always compared with saprobity score.

Saprobic Score method involves inventory of the presence of local benthic macro-invertebrate forms upto species or family level. All possible families having saprobic indicator value were classified on a score–scale of 1 to 10 according to their preference for saprobic (oxygen availability) water quality. The saprobity scores of all families registered are averaged to produce BMWP (Biomonitoring Working Party). The pollution levels concluded after comparison with the help of Table 6.1 (deZwart, 1993; CPCB, 2002).

Results

The Biomonitoring water quality evaluation of Laxminarayana reservoir for three years from 2000–01 to 2002–03 indicated that this reservoir was showing D class as per CPCB Criteria (Table 6.2). The water body classified as Heavily Polluted. During present study Seven out of 24 months the water quality showed class C which was moderately polluted. This was due to the inflow of freshwater from catchment area after heavy showers. Where as 17 out of 24 months the water quality showed class D, *i.e.*, class D was dominated. The mean diversity score was 0.37 and saprobic score was 4.62. The mean scores indicate class D.

Table 6.1: Biological Criteria for Water Quality Evaluation

Sl.No.	Taxonomic Groups	Range of Saprobic Score (BMWP)	Range of Diversity Score	Water Quality Characteristics and Class
1.	Ephemeroptera, Plecoptera, Trichoptera, Hemiptera, Diptera	7 and more	0.2–1	Clean A
2.	Ephemeroptera, Plecoptera, Trichoptera, Hemiptera, Odonata, Diptera	6–7	0.5–1	Slight pollution B
3.	Ephemeroptera, Plecoptera, Trichoptera, Hemiptera, Odonata, Crustacea, Mollusca, Polychaeta, Coleoptera, Diptera, Hirudinea, Oligochaeta	5–6	0.4–0.9	Moderate pollution C
4.	Mollusca, Hemiptera, Coleoptera, Diptera, Oligochaeta	2–5	0.4 and less	Heavy pollution D
5	Diptera, Oligochaeta or no macroinvertebrates	0–2	0–0.2	Severe Pollution E

Table 6.2: Biomonitoring Water Quality Status of Lakshminarayana Reservoir from 2000–01 to 2002–03

Year	Indices and Class	Oct	Nov	Dec	Jan	Feb	Mar	Apr	May
2000–01	Diversity Index	0.5	0.38	0.4	0.4	0.35	0.39	0.32	0.4
	Saprobity Index	4.4	4.5	4.1	4.3	5.0	4.9	4.5	4.3
	Water Quality Class	C	C	C	D	D	D	D	D
2001–02	Diversity Index	0.39	0.42	0.3	0.35	0.4	0.33	0.34	0.38
	Saprobity Index	5.0	4.4	4.3	4.9	4.5	5.0	4.6	4.6
	Water Quality Class	D	C	C	D	D	D	D	D
2002–03	Diversity Index	0.31	0.28	0.33	0.29	0.36	0.45	0.39	0.4
	Saprobity Index	4.2	5.0	4.4	4.7	5.0	4.8	5.0	4.6
	Water Quality Class	C	D	C	D	D	D	D	D

Class 'C': Moderate pollution; Class 'D': Heavy pollution.

In the reservoir, oligochaeta, hemiptera, diptera, coleoptera and mollusca animals found abundantly. Apart from these animals polychaeta, ephimeroptera, plecoptera, trichoptera, odoneta and hirudenea found rarely in the reservoir (Table 6.3). Out of the above orders, diptera dominated, followed by hemiptera and coleoptera. Molluscans and oligochaets also found in good number. Based on different organisms of various groups the saprobic and diversity scores analysed to conclude the water quality of the reservoir.

Table 6.3: The Benthic Macro-invertebrates in Laxminarayana Reservoir

Orders	*Benthic Macro-invertebrates*
Hemiptera	*Notonecta glauco*
	Anisops sardea, A. nivens
	Lemnometra fluviorum L. anadyomene
	Ranetra elongata
	Laccotrephes maculates
	L. rubber
	Corixa ima
	Lithoceros indicum
Coleoptera	*Cybister limbatus, C. rugulosus*
	Hydaticus fabricii
	Lacuphilus chirensis
	Hydrophilus indicus
Diptera	Larvae of *Culex* sp., *Anapheles* sp., and *Aedes* sp.
	Chryptochiranomus sp.
	Chiranomus reparius
	Harischia sp., *Polypeedium* sp.
	Tendies sp.
Odoneta	Dragon and Damsel fly nymphs
Epimeroptera	Mav fly nymphs
Trichoptera	Caddis fly larvae
Pleoptera	Stone fly nymphs
Mollusca	*Planorbis exustus, Anisus* sp.
	Vivipora sp., *Alocinma orcula*
	Eulota scalpturita, Glessula sp.
	Pila sp., *Carbicula* sp.
Annelida (Oligochaeta)	*Tubifex tubifex,*
	Glycera sp.

Discussion

Biomonitoring is the use of biological response to assess changes in the environment due to anthropogenic causes. In the present study biomonitoring used to assess the pollution levels in the reservoir. Biological monitoring of water quality could be useful for assessing the overall health of water body because:

1. The number of chemicals to be monitored are unending.

2. Biological effects often occur at concentrations below analytical capabilities. Many of the pollutants are present in such low concentrations that the influent sensitivity is too poor to determine the micro-quantity of such pollutants.
3. The toxicants may act quite differently in mixture than individually. Such toxicants affect the ecosystem in a synergistic manner which cannot be detected by chemical analysis alone.
4. Chemical nature of the toxicant is highly dynamic in environment with time and space, whereas biological system can integrate all environmental variables are a long period of time in terms of effects, which can be easily measured and quantified.

Environmental pollutants from anthropogenic sources affect the aquatic ecosystem in a synergistic manner, which cannot be detected comprehensively by determination of selected physico-chemical parameters alone. Whereas, biological system can integrate all environmental variables over a long period of time in terms of effects, which can be easily measured and quantified.

According to CPCB (2002) during the development and validation of biomonitoring methodology, established the fact that among all the biological components of the aquatic ecosystem, benthic macro-invertebrates are best suitable for evaluation of water quality. Most of these macro-invertebrates particularly insect larvae, which are aquatic, share their biological life in freshwater, while their adults fly over for miles together in search of suitable freshwater environment for reproduction, breeding and laying eggs in suitable habitats and establish biological communities.

According to Kennish (1992) biomonitoring was a valuable assessment tool that is receiving increased use in water quality monitoring programmes of all types. deZwart *et. al* (1993) and CPCB (2002) reported that the animals such as aquatic and semi-aquatic insects like ephemeroptera, plecoptera, hemiptera, coleoptera, trichoptera, odoneta and diptera, mulluscans, annelids like oligochaetes and leaches can be used in monitoring studies to assess the pollution levels of a water body. Ramakrishnan (2003) reported that several biological communities including phytoplankton, periphyton, micro-phytobenthos and aquatic macrophytes can be used in the biomonitoring studies to assess the pollution levels of a water body. He also stated that the level of degradation of water reservoirs can be quantified by biomonitoring technology more than conventional chemical methods. Phytoplankton, periphyton and microphytobenthos used in biomonitoring by him. Cairns (1979), Gunale (1991), Pati and Sahu (1993), Acharjee *et al.* (1995), Baruah *et al.* (1997) and Sultana and Sharief (2004) reported that the planktons serve as an index of water quality in respect of pollution. They also stated that plankton population observation can be used as reliable tool for biomonitoring studies to assess the pollution status of aquatic biota. Trivedi (1999) reported that different organisms like bacteria, algae, zooplankton, benthic macro-invertebrates and fish can be used for biomonitoring studies.

Palmer (1969) presented the list of most pollution tolerant algae species. According to Hynes (1960) heavy pollution affects whole taxonomic group of macro-invertebrates. Mild pollution can bring changes in individual species. Mason (1981) reported the pollution tolerant benthic macro-invertebrates. There were *Dinocras cephalotes* (Plecoptera), *Ecdyonusus venosus* (Ephemeroptera), *Gammarus perlex* (Amphipoda), *Asellus aquatica* (Isopoda), *Chironomus reparius* (Diptera) and *Tubifex tubifex* (Oligochaeta). Trivedi (1999) reported that *Tubifex* found most abundantly near sewage outfalls and can grow in very low oxygen conditions. *Gammarus* cannot survive in low oxygen condition. Stoneflies, mayflys and caddisflies disappear with increasing pollution.

The water of Laxminarayana reservoir was highly polluted and its biodiversity of phyto and zooplankton, benthic macro-invertebrate and ichthyofauna also declined. Major fish kills also noticed during the present study. Bio-accumulation of heavy metals found in the tissues of fishes. The fish growth rates and fish yields also decreased significantly. All the above changes occurred in the reservoir due to heavy pollution, which was due to the entry of polluted river Musi water into the reservoir. River Musi water severely polluted due to sewage, effluents from industries, agriculture, etc.

Acknowledgement

We are thankful to Prof. B. Raghavendra Rao and Prof. P. Umamaheswara Reddy for their encouragement and suggestions during the work.

References

Acharjee, B.A., M. Dutta, M. Chaudhary and V. Pathak. 1995. Phytoplankton species diversity indices in Dighate Jeel, Assam, India. *Environment and Ecology*, 13(3): 660–662.

Baruah, B.K., M. Chaudhary and M. Das. 1997. Plankton as index of water quality with reference to paper mill pollution. *Poll. Res.*, 16(4): 259–263.

Cairns, J. Jr. 1979. Final summary, In: *Biological Indicators and Water Quality*, (Eds.) James, A. and L. Evison. John Wiley and Sons, New York, p. 226–227.

deZwart, D. 1995. Manual on integrated water quality evaluation Appendix 6: *Taxonomical Key for Biological Water Quality Determination*.

Gunale, V.R. 1991. Algal Communities as indicators of pollution. *J. Environ. Biol.*, p. 223–232.

Hynes, H.B.N. 1960. *The Biology of Polluted Waters*. Liverpool Univ. Press, Liverpool.

Kennish, M.J. 1992. *Ecology of Estuaries: Anthropogenic Effects*. CRC Press, Boca Raton.

Mason, C.F. 1981. *Biology of Freshwater Pollution*, Longman.

Palmer, C.M. 1969. A composite rating of algae tolerating organic pollution. *Br. Phycol. Bull.* 5: 78–82.

Ramakrishnan, N. 2003. Biomonitoring approaches for water quality assessment in two waterbodies at Tiruvannamalai, Tamil Nadu India. In: *Proceedings of the Third International Conference on Environment and Health*, (Eds.) Martin J. Bunch, V. Madha Suresh and T. Vasantha Kumaran, Chennai, India, 15–17 December.

Pati, S. and B.K. Sahu. 1993. Biomonitoring of water pollution in Rengali reservoir, Orissa. *Environment and Ecology*, 11(4): 982–984.

Sultana, M. and Sharief, D. 2004. Water pollution studies in the Double lake (Erretai eri) with relation to Phytoplankton. *J. Aqua. Biol.*, 19(1): 15–18.

Trivedy, R.K. 1999. *Encyclopedia of Environmental Pollution and Control*, Vol. 1, Enviro Media, India.

Chapter 7

Diversity of Phyto and Zooplankton with Reference to Pollution Status of Kalavam Bazaar Lake, Arcot, Vellore District

V. Indra, V. Prabakaran and R. Balachandar*

**Department of Zoology, Thiruvalluvar University, Fort Campus, Vellore – 632 004*

ABSTRACT

To study the impact of waste disposal on plankton community structure in different seasons of the Kalavam Bazaar lake. The qualitative and quantitative analysis of Phyto and Zooplankton was carried out. A total of 15 Zooplankton and 23 Phytoplankton taxa were identified from 4 sampling sites. The occurrence of diatom flora at all sites were recorded. More number of Phytoplankton species in three sites indicted the organic enrichment of this habitat. This also reveals that algae are sources for monitoring the lake water quality. The plankton communities in the discharge point indicate the nature of pollution in the lake is discussed.

Keywords: Seasonal variation, Pollution status, Zooplankton, Phytoplankton and Physico-chemical parameters.

Introduction

The Kalavam Bazaar lake in Arcot, Vellore District receives effluent from various sources of multiple nature. The addition of these materials not only influences the microflora of freshwater but also favours the development of variety of new biotic species rendering it unfit for human consumption, like fishes, crab etc., consumed by common people in and around Arcot town. The water of the Kalavam Bazaar lake is also used for irrigation, washing and drinking purpose.

Phytoplankton is the fundamental component of aquatic ecosystem as they are the major source of biologically important organic compounds, located at the base of food chain. The magnitude and dynamic of Phytoplankton population has become an essential factor to assess the general health of an aquatic ecosystem (Gaikwad *et al.*, 2004).

Phytoplanktons of freshwater ecosystem have been studied extensively in India (Bilgram's,1989; Nandan and Patel, 1992 and Thorat *et al.*, 2000). The importance of lake water has been increased now due to the provision of aquatic animals, therefore, the present study was carried out to assess the physico-chemical parameters and various forms of bacteria, Phytoplankton and Zooplankton of Kalavam Bazaar lake, based on various seasons. Such studies have also been conducted by Annon (1998) and Biswar and Konar (2000). There is no information available on the ecology of Zoo and Phytoplankton in Kalavam Bazaar lake. Hence, this study has been focused to study the Phytoplankton, Zooplankton, various forms of bacteria and characteristics of the Kalavam Bazaar lake water and its extent of pollution.

Materials and Methods

A total of 8 water samples were collected from October 2004 to March 2005 from 4 sites which comprised four each for rainy (October–December) and post monsoon period (January–March), respectively. Standard methods (APHA, 2000) were used during the collection, preservation and estimation of physico-chemical, bacteriological analysis and plankton sample analysis. The planktons were collected from all the sites, were fixed in small vials with 5 per cent formalin and few drops of glycerin was added. Samples were centrifuged and concentrated to 10ml in the laboratory. Qualitative and quantitative estimation of Phytoplankton was done with the atlas of Desikachary (1959), Baltish (1992) and Agarkar *et al.* (1994) and Zooplankton identification was done using standard keys of Edmondson (1959), Ruttner-Kolisko (1974), Ranga Reddy (1984), Sharma (1988) and APHA (2000).

Results and Discussion

The physico-chemical parameters analysed in the present study are illustrated in Table 7.1. Zooplankton was identified at different sites such as Cladocera, Copepoda, Rotifera and Ostracoda were indicated in Table 7.2. Distribution and abundance of Phytoplankton were showed in Table 7.3 and the results of bacterial analysis are expressed in Table 7.4.

The pH data of the water sample showed slightly alkaline nature during rainy season in all the stations. Whereas, during postmonsoon season all the values of the samples showed above 8. It was ranged from 7.4 to 8.5. The total dissolved solids showed more than permissible limits of (WHO 1994) *i.e.*, ranged from 1020 to 1625 mg/l during postmonsoon and 1230 mg/l to 1360 mg/l during rainy season. The BOD and COD values are quite high and indicate that inorganic and organic pollution in the lake is high. The phosphate content was more than 0.2 mg/l, may be considered as productive nature of water. Natural waters mostly contain < 0.1mg/l of phosphorous, which is permissible limit. However, excess of Soluble Reactive Phosphate (SRP) fraction of phosphate promotes eutrophication (Kodarker, 1995). The whole lake is influenced by major point source of pollutant that may be due to freshwater runoff during rainy season. More or less all the physico-chemical parameters were showed slightly higher values during rainy season than in the post monsoon season. Though the rainy season dilute the lake, the freshwater runoff may increase the concentration of organic and inorganic pollutant load. The less concentration during postmonsoon season may be due to high temperature (evaporation). The water of this lake was greenish and odour is fishy due to proliferation of Phytoplankton population *i.e.*, indication of eutrophic habitat and occasional release of water from mouth of the mixing and discharge zones.

Table 7.1: Seasonal Variations of Physico-chemical Parameters of Kalavam Bazaar Lake, Arcot, Vellore District

Parameters	*Rainy Season*				*Postmonsoon*			
(mg/l)	*S1*	*S2*	*S3*	*S4*	*S1*	*S2*	*S3*	*S4*
pH	7.8	7.4	7.5	7.4	8.61	8.54	8.41	8.5
Temp. °C	27.0	28.0	28.0	27.5	38.0	37.0	38.0	38.0
Colour	Sg	Sg	Sg	Sg	G	G	G	G
Odour	Unobjection-able	Objection-able	Fishy	Objection-able	Objection-able	Objection-able	Fishy	Objection-able
BOD	15.0	17.0	16.0	45.0	94.0	54.0	88.0	90.0
COD	110.0	2200	236.0	180.0	225.0	240.0	225.0	235.0
Total hardness	250	280	260.0	276.0	210	260.0	250.0	240.0
Total solid	1275	1280	1360.0	1230.0	1438	1625	1340	1020
DO	7.60	7.83	6.45	8.32	8.0	8.45	8.46	8.00
Organic nitrogen	0.026	0.029	0.024	0.035	0.046	0.056	0.043	0.056
Nitrate	[illegible]	[illegible]	33.45	35.47	40.37	43.86	47.44	46.7
Phosphate	2.31	2.62	2.42	3.10	1.86	1.48	1.40	1.63

Sg: Slightly green; G: Green.

Table 7.2: Seasonal Variation of Zooplankton Diversity in Kalavam Bazaar Lake, Arcot

Zooplankton	*Rainy Season*				*Postmonsoon*			
	1	*2*	*3*	*4*	*1*	*2*	*3*	*4*
Daphnia	42	78	40	31	55	64	60	68
Bosmina	13	11	7	9	8	4	11	3
Macrothrix	–	4	–	–	–	2	–	–
Moina	8	8	2	1	4	3	8	8
Ceriodaphnia	14	35	42	11	38	29	27	19
Polyhemus	12	4	9	3	2	8	1	4
Diaptomus	11	13	15	9	8	12	10	6
Cyclops	8	7	11	5	4	–	–	3
Asplanchna	31	42	33	15	18	21	20	9
Brachionus	7	11	4	9	11	6	–	–
Keratella	–	–	2	1	–	–	3	–
Halicyclops	1	7	–	3	5	1	8	–
Mesorchra	1	3	1	2	1	1	–	–
Filinia	3	–	1	1	2	–	1	3
Cypris	1	–	2	1	1	2	–	–

Table 7.3: Seasonal Variation of Phytoplankton Abundance and Diversity in Kalavam Bazaar Lake

Phytoplankton	*Rainy Period*				*Postmonsoon Period*			
	1	*2*	*3*	*4*	*1*	*2*	*3*	*4*
Spyrogyria	14	26	23	12	25	27	38	16
Volvox	22	11	14	19	29	20	31	37
Eudorina	–	6	–	4	3	1	5	1
Closterium	–	–	–	8	–	–	6	4
Hydrodictyon	4	3	1	–	1	2	–	–
Pachycladom	2	8	4	3	2	7	6	–
Cyclindrocapsa	–	1	4	1	3	1	2	2
Bumillaria	1	2	–	–	1	3	1	–
Amphora	2	1	1	–	–	–	–	1
Oscillatoria	3	5	–	–	–	–	–	1
Anabaena	18	13	21	24	–	27	–	29
Spirulina	15	17	30	12	11	–	1	15
Navicula	3	8	6	5	4	9	11	8
Melosura	–	–	4	–	1	1	3	2
Acanthes	–	–	–	1	–	–	1	–
Tabellaria	2	7	4	3	1	–	1	–
Fragillaria	2	1	2	1	1	1	–	1
Camphonema	1	1	1	–	–	–	1	–
Nitzschia	2	1	1	–	–	3	5	7
Euglena	11	8	7	–	4	–	4	2
Microcystis	3	2	5	1	4	4	2,	8
Auliscus	1	–	1	–	–	–	1	–
Coscinodiscus	3	1	5	1	3	2	1	3

Table 7.4: Total Viable Count of Various Bacteria in Kalavam Bazaar Lake

Species	*Rainy Period*				*Postmonsoon Period*			
	1	*2*	*3*	*4*	*1*	*2*	*3*	*4*
Escherichia coli	350	368	450	340	345	344	90	250
Staphylococcus aereus	132	102	130	146	157	134	231	200
Pseudomonas sp.	120	110	95	62	84	176	211	160
Micrococcus sp.	90	68	46	35	38	60	64	96
Aerobactor aerogenus	64	55	98	93	110	88	55	38
Clostridium sp.	15	22	44	38	55	37	43	63

It was also found that most of the Phytoplanktons were recorded considerably higher number in different sites, during postmonsoon season, similar findings were recorded by Chakraborty and Konar

(2000), and Gaikwad (2004). The dominant groups like Myxophyceae and Chlorophyceae were recorded. Thus, the appearance of blue green algae was an indication of organic enrichment of these habitats. According to Biswas and Konar (2000), it can be correlated with high concentration of dissolved organic load and was obviously scripted from sewage, dead body burning, damping of animal carcasses and defaecation of the lake. The appearances of tolerant taxa of Chlorophyceae also support the above explanation.

The results of the Zooplankton and bacteriological analysis are given in Tables 7.3 and 7.4. The maximum viable count was observed in rainy season than postmonsoon season. This may be due to addition of land washing and organic matter by rain. The specific bacteriological examination indicates the notable presence of *Eschericia coli,* and *Pseudomonas* sp. Sengar *et al.*, (1996) reported the presence of similar genera in different sources of water.

In the present study the abundance and diversity of Zooplankton and Phytoplankton was found to be dependent on water quality and co-existing biotic communities at given point of time. Due to organic load disturbances from various non-point sources was clearly reflected by shifting of Phytoplankton and Zooplankton at various sites, which highlights the importance of non-point and point sources to cause pollution in the Kalavam Bazaar lake, Arcot.

References

Agarkar, M.S., Goswami, H.K., Kaushick, S., Mishra, S.M., Baipai, A.K and Sharma, U.S. 1994. Biology, conservation and management of Bhojmet land 1. Upper lake ecosystem in Bhopal. *Bionature*, 14: 20–28.

Anon. 1998. The river Dhamodhar and its enrichment. *Bull. Cent. Inland Fish. Res. Inst.*, Barakpore, 79: 32.

APHA. 2000. *Manual for Water and Wastewater Analysis*, Washington, DC.

Battish, S.K. 1992. *Freshwater Zooplankton of India*, Oxford and IBH Publishing Co., New Delhi.

Bilgrams, K.S. 1989. Biomonitoring of water quality of Ganga. In: *The Ganga Project Directorate.* North Book Centre, New Delhi, p. 101–106.

Biswas, B.K and Konar, S.K. 2000. Influence of Nunia Nullah (canal) discharge on plankton abundance and diversity in the river Damodar at Narankuri (Raniganj) in West Bengal. *Indian J. Environ. and Ecoplan.*, 3: 209–217.

Chakrabotry and Konar, S.K. 2000. Studies on the hydrobiology of river Damodar from Raniganj region (West Bengal) with reference to Phytoplankton. *Indian Journal of Environ. and Ecoplan.*, 3: 485–490.

Desikachary, J. V. 1959. *Cynophyta: Monographs on Algae.* ICAR, New Delhi.

Edmondson, W.T. 1959. *Freshwater Biology*, John Wiley and Sons Inc., New York, 2nd, p. 657–663.

Gaikwad, S.R, S.R. Thorat and T.P. Chavan. 2004. Plankton diversity with reference to pollution status of river Tapi in North Maharashtra region. *J. Curr. Sci.*, 5(2): 749–754.

Kodarkar, M.S. 1995. *Conservation of Lakes*, IAAB Publi., No. 2, IAAB, Hyderabad.

Nandan, S.N and Patel, R.J. 1992. *Ecological Studies of Algae in Aquatic Ecology*, Ashish Pub House, New Delhi, p. 70–99.

Rangareddy, Y. and Radhakrishna, Y. 1984. The calanoid and cyclopoid fauna of lake Kolleru, South India. *Hydrobiol.*, 119: 27–48.

Ruttner and Kolisko, A. 1974. Plankton Rotifers. *Biennengeswasser*, 26: 1–146.

Sengar, S.G., Naqshi, A.R and Mir, G.R. 1996. Impact of floating gardens on the limnological features of Dal lake. *Poll. Res.*, 15: 217–221.

Sharma, S.S.S. 1988. New records of rotifer. *Hydrobiol.*, 160: 263–269.

Thorat, S.R. and Massarat, S. 2000. Pollution status of Salim Ali lake, Aurangabad (M.S.). *Poll. Res.*, 19: 307–309.

WHO. 1994. *Guidelines for Drinking Water Quality*. Geneva, 2: 91–96.

Chapter 8

Biochemical Changes in the Snail *Bellamya bengalensis* (Lamarck) Under Toxic Stress of Sumicidin

P.H. Rohankar and K.M. Kulkarni
Environmental Physiology Laboratory, Department of Zoology, Government Vidarbha Institute of Science and Humanities, Amravati

ABSTRACT

The glycogen, protein and lipid levels were studied in body tissues of *B. bengalensis* under pesticidal application of lethal and sublethal doses of synthetic pyrethroidal pesticide-sumicidin. All these constituents decreased significantly suggesting a rapid breakdown of glycogen, protein and lipid in energy metabolism under toxic stress.

Keywords: Pyrethroid, Glycogen, Lipid, Protein, Lethal and Sublethal concentration.

Introduction

The freshwater ecosystem is becoming increasingly polluted due to various pesticides and agrochemicals which also affect non-target organisms like fishes and snails (Holden, 1973). Study of biochemical parameters is helpful in understanding the mechanism of pesticide toxicity to organisms. The metabolites such as carbohydrates, proteins and lipids which playa major role in body construction and energy production, are adversely affected by water pollution which lead to inhibition of important enzymes, retardation of growth, longevity of organisms (Sri Nivasa Murthy, 1986).

As molluscs form an integral part of aquatic ecosystem and are good marker of aquatic pollution and few reports are available on variations in biochemical components in molluscs under pesticide stress (Kulkarni *et al.*, 1985; Rao *et al.*, 1995; Shah *et al.*, 2001; Amanulla *et al.*, 2004) the present work was undertaken.

Material and Methods

The snails were collected from Wadali tank near Amravati city. Active and healthy snails were used in study after acclimatization. These were divided into two groups as control and experimental. The test solutions prepared for biochemical experiments as LC_{50} for 96 hr (0.15 ml/l which is termed as lethal concentration and 1/3rd concentration of 96 h LC_{50} (0.0475 ml/l) which was taken as sublethal concentration of pyrethroid Sumicidin. At the end of 24, 48,72 and 96 hrs of lethal and sublethal exposure to pesticide, the required number of treated snails from each group were sacrificed by decapitation, tissues separated, weighed and analysed for biochemical' components such as glycogen, proteins and lipids. The glycogen was estimated by Anthrone method (Carrol *et al.*, 1958). Protein was estimated by Biuret method (Lowry *et al.*, 1951) and total lipids, were estimated by chloroform methanol method (Folch *et al.*, 1957) and values were expressed as mg/gm wet wt. tissues.

Results and Discussion

There is maximum reduction in glycogen level of all tissues exposed to lethal concentration as compared to tissues of the snails exposed to sublethal concentration of pyrethroids. There is decreasing trend from 24 hr to 96 hr, the reduction is maximum in glycogen level of hepatopancreas of snail. This might be due to high metabolic potency and efficiency of the gland as compared to other tissues. In general, there is significant and gradual decrease in the levels of glycogen in all tissues tested.

From the Table 8.2 it is evident that there is total decrease of lipid content of all tissues continuously at all hours, hepatopancreatic tissue exposed to lethal and sublethal concentrations show maximum decrease of lipid level than mantle, foot, gills and gonad tissue. The decrease is more significant at lethal concentration of pesticide.

Changes in protein level of body tissue are recorded in Table 8.3. Protein also show decreasing trend in body tissue with increase in concentration of pesticides and time exposure. There is maximum decrease at 96 hr exposure in all tissues of the snail. The hepatopancreas tissue show maximum decrease in trend than other tissues exposed to sumicidin.

Changes in macromolecular components like glycogen, protein and lipids are considered to be the indicators of pollution stress. In present study, marked decrease in glycogen content suggesting the possibility glycogenolysis, probably caused by a stress induced by pollutant mainly due to hypoxia. Many investigators observed similar changes (Sheela *et al.*, 1992; Vasanthi and Ramaswamy, 1987; Patil and Kulkarni, 1993; Gupta and Agrawal, 1995; Muley and Mane, 1995; Shaikh and Yeragi, 2004) decrease in glycogen content might be for high rate of energy production, at onset of various enzymatic blockage due to pesticidal toxicity (Kulkarni and Chaudhari, 1995).

Lipid is also a major source of energy after carbohydrates as it is reserved stock of food for animals. Since more energy is needed to minimize any stress condition, lipid source may be used after depletion in glycogen as immediate source (Gupta, 1987; Padmaja and Rao, 1994; Muley and Mane, 1995; Bhavan and Geraldine, 1997).

Protein is a complex organic substance of dual importance for organism as a building material and as a source of energy. Several investigators reported that pesticide interferes with protein synthesis

Table 8.1: Changes in Glycogen Content of Different Tissues of Snail *B. bengalensis* Exposed to Lethal and Sublethal Concentration of Sumicidin (mg/100 mg wet wt. tissue)

Hepatopancreas		Foot		Mantle		Gill		Gonad	
Lethal	*Sublethal*	*Lethal*	*Sublethal*	*Lethal*	*Sublethal*	*Lethal*	*Sublethal*	*Lethal*	*Sublethal*
16.11±1.70	17.18±1.67	10.93±1.74	11.00±1.91	16.25±2.13	16.43±1.26	10.11±1.0	19.13±0.93	9.11±1.19	9.14±1.12
13.24*±2.00	15.14±1.61	10.66±1.96	10.86±1.91	15.37±1.97	16.56±1.45	[illegible].15**±0.87	9.01±0.99	8.62*±1.41	9.08±1.00
(–25.62)	(–3.98)	(–2.89)	(–4.14)	(–2.26)	(–0.70)	(–2.53)	(–1.84)	(–9.41)	(–10.86)
12.48**±2.26	14.37±1.76	8.64*±2.11	9.64**±1.53	14.70±2.11	15.56±1.45	[illegible].12±1.172	8.38**±1.37	8.25*±1.45	8.69*±1.32
(–25.62)	(–16.35)	(–21.38)	(–12.36)	(–9.53)	(–5.29)	(–19.68)	(–8.21)	(–9.44)	(–4.92)
10.58±1.53	12.59*±2.05	7.47**±1.25	9.38±2.01	13.42±1.72	14.62**±1.71	7.51±1.34	8.19***±1.32	7.54**±1.51	8.24±1.13
(34.32)	(26.71)	(–32.02)	(–14.72)	(–17.41)	(–11.01)	(–25.71)	(–10.29)	(–17.23)	(–9.84)
8.49±1.71	10.72±.1.70	6.71**±1.37	8.68**±1.49	11.79*±1.78	13.46±1.40	6.39±1.28	7.47±1.05	6.84*±1.42	8.01±1.09
(–47.29)	(–37.60)	(–39.23)	(–21.09)	(–22.44)	(–18.07)	(–36.79)	(–18.18)	(–24.91)	(–12.36)

* P < 0.05, ** P < 0.01, *** P < 0.001.

Figures in parenthesis indicate percent change value.

Table 8.2: Changes in Total Lipid Content of Different Tissues of Snail *B. bengalensis* Exposed to Lethal and Sublethal Concentration of Sumicidin (mg/100 mg wet wt. tissue)

Hepatopancreas		Foot		Mantle		Gill		Gonad	
Lethal	*Sublethal*	*Lethal*	*Sublethal*	*Lethal*	*Sublethal*	*Lethal*	*Sublethal*	*Lethal*	*Sublethal*
9.06±0.97	8.65±1.29	9.18±1.17	8.86±1.10	6.15±1.14	5.91±1.17	5.58±1.28	5.52±1.35	9.10±1.07	9.07±0.99
8.11**±0.95	8.93±1.03	8.77**±1.61	8.66**±1.54	5.86*±0.93	5.71*±1.40	5.15±1.19	5.49±1.24	8.62±1.31	8.94±1.05
(–6.24)	(–1.43)	(–7.40)	(–2.59)	(–4.71)	(–3.38)	(–6.53)	(–0.54)	(–5.27)	(–1.43)
7.70±1.42	8.70***±1.31	8.06±1.25	8.36±1.31	5 40**±0.90	5.50±1.35	4.71±1.23	5.25±1.18	8.08±1.13	8.62±1.04
(–10.98)	(–3.97)	(–18.19)	(–7.67)	(–12.19)	(–6.93)	(–10.88)	(–4.89)	(–11.20)	(–4.96)
7.17±1.11	8.03±1.14	7.28±1.19	8.15±1.04	4.63±1.08	5.21±1.04	[illegible].15***±0.79	4.85*±1.16	7.57±1.38	8.35±1.33,
(–17.68)	(–11.36)	(–28.97)	(–12.64)	(–24.71)	(–11.84)	(–24.68)	(–12.13)	(–16.81)	(–7.93)
6.60±1.07	7.26***±1.33	5.59±1.34	7.71±1.12	4.37**±0.86	4.90***±0.84	[illegible].86***±0.86	4.53±1.16	7.28±1.33	8.01±0.96
(–26.01)	(–19.36)	(–39.76)	(–23.25)	(–28.94)	(–17.08)	(–29.94)	(–17.93)	(–20.00)	(–11.68)

* P < 0.05, ** P < 0.01, *** P < 0.001.

Figures in parenthesis indicate percent change value.

Table 8.3: Changes in Total Protein Content of Different Tissues of Snail *B. bengalensis* Exposed to Lethal and Sublethal Concentration of Sumicidin (mg/100 mg wet wt. Tissue)

Hepatopancreas		*Foot*		*Mantle*		*Gill*		*Gonad*	
Lethal	*Sublethal*	*Lethal*	*Sublethal*	*Lethal*	*Sublethal*	*Lethal*	*Sublethal*	*Lethal*	*Sublethal*
11.28±1.49	11.79±1.18	7.26±1.56	8.45±1.31	7.50±2.02	7.09±1.21	2.52±1.28	5.43±1.07	9.24±1.30	9.57±1.24
8.39*±1.71	11.32±1.35	7.05**±1.27	8.10**±1.43	7.33±2.25	7.04±1.07	2.38±1.15	5.33±1.02	8.37±1.05	8.53±0.83
(−25.62)	(−3.98)	(−2.89)	(−4.14)	(−2.26)	(−0.70)	(−2.53)	(−1.84)	(−9.41)	(−10.86)
8.49**±2.01	10.92±1.06	6.57**±1.25	7.60±1.01	6.56±1.33	6.32***±0.92	4.89***±0.77	5.20±0.91	7.49±1.35	7.98±1.25
(−25.62)	(−7.38)	(−9.50)	(−10.05)	(−12.53)	(−10.86)	(−11.41)	(−4.23)	(−18.93)	(−16.61)
7.24±1.58	10.58±1.10	6.35**±1.15	7.39±0.96	6.24±1.15	6.13***±0.79	4.72***±0.73	5.03±0.81	7.26±1.44	7.90±1.20
(−35.81)	(−10.26)	(−12.53)	(−12.54)	(−16.80)	(−13.54)	(−24.49)	(−7.36)	(−21.42)	(−17.45)
6.51±1.34	9.91±1.06	4.43***±1.01	6.90***±0.83	5.60±1.05	5.44***±0.82	4.55***±0.67	4.83±0.78	7.09±1.38	7.72±1.17
(−42.28)	(−15.95)	(−25.20)	(−18.34)	(−25.33)	(−23.27)	(−17.57)	(−11.04)	(−23.26)	(−19.33)

* $P < 0.05$, ** $P < 0.01$, *** $P < 0.001$.

Figures in parenthesis indicate percent change value.

and degradation resulting in alteration of equilibrium (Sivakami *et al.*, 1994; Singh and Bhati, 1994; Sinha, 1997; Rajamannar and Manohar, 1998) depletion in tissue protein may show a prior increased energy cost of homeostasis, tissue repair and detoxification during stress. The present work is in agreement with these workers.

References

Amanulla Hameed, S.V.S., Nazir Ahmed T.A. and Shah D.S.M. 2004. Effect of Butyltin toxicity on lipid metabolism in an estuarine Mussel *Sunetta scripta*. *J. Ecotoxicol. Environ. Monit.*, 14(3): 185–190.

Bhavan, P.S. and Geraldine, P. 1997. Alteration in concentration of protein, carbohydrate, glycogen, free sugar and lipid in the Prawn *Macrobrachium malcomsonii* on exposure to sublethal concentration of endosulfan. *Pestic. Biochem. Physiol.*, 58: 89–101.

Carrol, N.V., Longley, R.W. and Roe, J.H. 1950. Glycogen determination in liver and muscle by the use of anthrone reagent. *J. Biol.*, 22: 583.

Folch, J., Less, M. and Sloane-Stanley, G.H. 1957. A simple method for the isolation and purification of total lipids from animal tissues. *J. Biol. Chem.*, 26: 499–509.

Gupta, S. 1987. Effect of vegetable oil factory effluent on the lipid content in the liver of *Channa punctatus* (Bloch) studies *in vitro* incorporation of radio labelled substances. *J. Environ. Biol.*, 8: 353–357.

Gupta, R.C. and Agrawal, Sapna. 1995. Alteration in glycogen level under the stress of Azodyes in three freshwater teleost, *Channa punctatus, Channa striatus* and *Catla catla*. *Ad. Bios.*, 14(1): 97–104.

Holden, A.V. 1973. Effects of pesticides on fish. In: *Environmental Pollution by Pesticides*, (Ed.) C.A. Edwards. Plenum Press, London.

Kulkarni, A.N., Kamble, S.M. and Keshvan, R. 1985. Metabolic depression in the freshwater bivalve *Lamellidens corrianus* exposed to insecticide Hildane. *Proc. Symp. Asserr. Environ. Pollut.*, p. 225–227.

Lowery, O.H., Rosenbrough, N.T., Farr, L R. and Randall, J. 1951. Protein measurement with Folin-phenol reagent. *J. Biol. Chem.*, 193: 265–275.

Muley, D.V. and Mane, U.H. 1995. Endosulfan toxicity to freshwater mussel *Lamellidens marginalis* and pH induced changes, a biochemical approach. *Indian J. Comp. Anim. Physiol.*, 13(1): 21–26.

Padmaja, R.J. Balaparameswara Rao M. 1994. Effect of an organochlorine and three organophosphate pesticides on glucose, glycogen, lipid and protein content in tissue of the snail *Bellamya dissimilis* (Muller). *Bull. Environ. Contam. Toxicol.*, 53: 142–148.

Patil, P.S. and Kulkarni, K.M. 1993. Biochemical response of cythion exposed frog *Rana cyanophlyctis*. *Ad. Bio. Sci.*, 12: 101–108.

Rajamannar, K. and Manohar, L. 1998. Sublethal toxicity of certain pesticides on carbohydrate, protein and aminoacids in *Labeo rohita*. *J. Ecobiol.*, 10(3): 185–191.

Rao, K.R., Shejule, K.B., Sasane, S.R., Subhas, N., Patil, P.N. and Choudhari, T.R. 1995. Impact of decis on carbohydrate metabolism of the Gastropod *Indoplanorbis exustus* from Panzara river, Dhule. *Proc. Acad. Environ. Biol.*, 19(3): 215–220.

Sheela, M., Mathivanan, R. and Muniandy, S. 1992. Impact of fenvalerate on biochemical status of different tissues in the fish *Channa striatus* (Bloch). *Environ. and Ecol.*, 10(3): 547–549.

Sinha, G.M. 1997. Effect of cadmium on protein content in plasma, body muscle and liver in an Indian major carp *Labeo rohita* (Hamilton). *Environ and Ecol.*, 15(2): 275–282.

Singh, S. and Bhati, D.P.S. 1994. Evaluation of liver proteins due to stress under 2,4-D-intoxication in *Channa punctatus* (Bloch). *Bull. Environ. Contam. Toxicol.*, 53: 149–152.

Shah, P.M., Rojaramani, V., Sathik, O. and Sathiya Priya, R.C. 2001. Effect of tributyltin oxide in lipid metabolism in earthworm edible blood clam *Anadara rhombea* (Born). *J. Environ. Pollut.*, 8(1): 7–11.

Shaikh, Nisar and Veragi, S.G. 2004. Effect of Rogor 30E (organophosphate) on muscle: protein in the freshwater fish *Lepidocephalecthyes thermalis*. *J. Ecotoxicol. Environ. Monit.*, 14(3): 233–235.

Sivakami, R., Premkishore, G. and Chandran, M.R. 1994. Effect of chromium on the metabolism and the biochemical composition of selected tissues in the freshwater catfish *Myctus vittatus*. *Environ. Ecol.*, 12(2): 259–266.

Srinivasa Moorthy, K., Kasi Reddy, B., Swami, K.S. and Chetty, S.R. 1986. Dichlorovas, induced metabolic changes in tissues of freshwater mussel *Lamellidens marginalis*. *J. Environ. Biol.*, 7(2): 101–106.

Chapter 9

Air Pollution and Human Body

V. Rajendra Prasad, Y. Prasanna Kumar, P. King and V S R K Prasad*

Environmental Pollution Control Engineering Laboratory, Department of Chemical Engineering, College of Engineering, Andhra University, Visakhapatnam – 530 003, A.P., India

Introduction

Air is an essential free food of man to live. A man can stay without an eye, without hands, but not without this free food *i.e.*, Air. It is the human tendency to live a comfortable life, that's why man has started polluting the every environment in which he is living. If we go through the past human history we can easily trace out *i.e.*, point out that man, "him self" has started polluting the atmosphere in a small scale when he started lighting fire and burning fuel. Gradually, with the change of time, the human mind has also changed and it started working parallel to the time. It has led him to greater aspirations there by his exploitation of nature's stores houses at a past rate. Due to uncontrolled industrialization and urbanization and population explosion air pollution has now become a worldwide problem of indefinite magnitude. This problem "Air Pollution" has already reached to an alarming level in many highly industrialized countries like, Japan, U.S.A., also in India at some places like–Mumbai.

Continued expansion of existing industries, development of new technologies and products and increasing use of automobiles coupled with pollution growth, especially in large industrial towns are responsible for a great number and quantity of pollutants in the air. Due to this waste thrown by the industries in the atmosphere, the side effects of such activities cannot be ignored. As most of the health–effects due to air pollution start showing of after a considerable period, it is essential to anticipate the problem potential of the next 20 years and take necessary preventive measure now itself.

* Corresponding author E-mail: rpvasanta@rediffmail.com.

Composition of Air in Nature

A good composition of air contains 78.09 per cent nitrogen by volume, 20.94 per cent oxygen and the rest 0.97 per cent of the gaseous constituents of dry air includes small amounts of Carbon dioxide, Helium, Argon, Krypton, Xenon and very small amounts of other Inorganic and Organic gasses.

In the atmosphere the air is surrounded in the form of layers, and the thickness of this layer is approximately 160 km. Out of 160 km, more than half of its total weight is in the 64 km *i.e.*, just above the surface of the earth. The atmosphere is divided into two parts:

Stratosphere

It is the upper portion of the atmosphere and is an isothermal region (–50°C in polar regions and –70°C in the tropic region). In this zone clouds of vapour do not form and little convection takes place.

Troposphere

It is the lower portion of the atmosphere, and its temperature varies from the surface of the earth to the stratosphere. It contains clouds of water vapour and, air movement caused by temperature differences takes place.

Air Pollution

Actually, the presence of foreign substances in air is known as pollution. In other words dirtiness of air due to suspended or gasses particle is known as air pollution. According to the Engineers Joint Council, U.S.A. "Air pollution means the presence in the out door atmosphere of one or more contaminants such as dust fumes, gas mist (is it odour) smoke, or vapour in quantities of characteristics and of duration, such as to be injurious to human, plant or animal life or to property or which unreasonably interfere with the comfortable, enjoyment of life and property.

Sources of Air Pollutants

In all over the world stack gasses from industrial establishments are still the major sources of air pollution. The main industries letting out stack gasses include thermal power stations, steel mills, petroleum refineries, and other plants. Second source of air pollution are automobiles like cars, trains, planes which use fossil fuel in one form or the other. Third type is use of fuel for domestic purpose; refuse incineration evaporation of petroleum products etc.

Air pollution sources can be divided in two classes.

Specific Sources

Specific sources are largely industrial in nature, thus permitting their potential to pollute a community atmosphere to be readily assayed on an industry-by-industry basis. In this case, the area is fixed and is commonly a limited area relative to the community.

Multiple Sources

These are sources, which cannot be assayed practicably on source–by source basis. They are either fixed or mobile and commonly scattered through out the community. Combustion of fuel for power production, for transportation and domestic purposes, incineration of solid wastes evaporation of petroleum products are the sources that come under sources this category.

Effects of Air Pollutants on Human Body and his Environment

Man has conquered space and his technology has changed the environment, and now the environment is threatened by the dangers of pollution–a by-product of economic progress, of modern urbanization and industrialization. Today Air Pollution is the mother of numerous problems.

Since 1950, the physicians give considerable research and reports and data. It has been found that patients with Asthma, hay fever and chromic respiratory diseases suffer especially during smog period. On studying the health effects one can easily differentiate between the hazard of industrial atmospheres inside manufacturing or processing plants and those of the out side atmosphere. The health effects on worker inside industry are classified as occupational diseases. Human system can adopt itself to certain amount of air pollutants, which are being breathed into the system, and the limit, at which it becomes hazardous to the individuals, is known as maximum allowable concentration. To analyse the effects of exposure to air pollutants at lower concentrations for a longer period epidemiological and toxicological studies are carried out and found that epidemiological studies are concerned with the effects occurring in human and animal population exposed under natural conditions. Results found through above studies there is growing amount of evidence linking air pollution to respiratory conditions like chronic bronchitis, acute non-specific respiratory diseases bronchial asthma, chronic obstructive ventilatory diseases and lung cancer.

In spite of above facts the other factors affects the sensitivity of population such as age, sex general state of health, nutrition concurrent exposures, pre-existing diseases, temperature and humidity at the time of exposure. Due to air pollutants any or some of the following cause may annoyance:

1. Nose or by eye or by throat irritation and
2. By odours.

As soon as annoyance stops and danger to health begins is controversial. Eye irritation is probably the most exasperating of all the effects of air pollution. Nose and throat irritation have often been reported as effects of air pollution.

The Other Causes Due to Air Pollution

In addition to above effects on human body due to Air Pollution there are so many other causes also as listed below in brief:

1. Material Damage including soiling surface.
2. Vegetation Damage.
3. Sky-Darkening and limited visibility.
4. Damage to Domestic Animals.
5. Interference with services and production.
6. Annoyance to the senses of people and psychological.

Air Pollution Control

Air pollution can be controlled by two fundamental approaches:

1. Control of pollutant at the source so that excessive amounts are emitted to the atmosphere.
2. Control by natural dilution of the pollutant in atmosphere.

Table 9.1: Effects of Air Pollution on Human Body

Sl.No.	Air Pollutant	Typical Sources	Effects on Human Body
1.	Sulphur dioxide	Power houses sulphuric acid plants petroleum industry, oil refining, domestic use of fuel	Suffocation, Irritation of the throat and eyes, respiratory
2.	Hydrogen sulphide	Viscose rayon plants, petroleum industry, sewage treatment, Tanning industry, dry manufacture oil refining	Irritation of respiratory passages, danger of respiratory paralysis and asphyxiation
3.	Nitrogen dioxide	Acid manufacture, Automobile exhaust, explosives industry.	Irritation, Bronchitis, Oedima of lungs
4.	Hydrogen fluoride	Fertilizer industry, Chemical and aluminium industry,	Irritation, Diseases of bone (fluorosis) mottling of teeth, Respiratory disease
5.	Carcinogenic Hydrocarbons	Organic chemical industry, vehicular traffic specially automobiles	Cancer
6.	Carbonmonoxide	Fuel gasses, Exhaust emissions from automobiles mines, blast furnaces	Poisoning, Increased accident liability

Control can be made on the pollutant at the sources by:

1. Preventing the pollutant from coming into existence
2. Trapping, destroying or altering the pollutant that is emitted before it enters the atmosphere.

Out of above two, best method would be to prevent the pollution from coming into existence and if its not possible then keep the quantity to a minimum. It can be achieved as per circumstances and facilities. It can be done just by, raw material change, Process changes, Operational changes, Modification of process equipment and more efficient operation of existing equipment. The general methods used for reducing a pollutant to tolerable levels before it is emitted from the stack include.

1. Destruction of pollutants by use of fire or catalytic burners
2. Masking the pollutant
3. Countering the pollutant and
4. Collecting the pollutant from waste stream using collection equipment such as bag filters cyclone separators, scrubbers, electrostatic, precipitators etc.

The second method of controlling air pollution is to confine the contaminant at its source. Some time it's not possible in that case. By natural dilution the harmful concentration can be pollutant. Methods of attempting to accomplish natural dilution include:

1. Use of tall stacks
2. Community planning in which zoning the use of air is adopted.
3. By control of the process techniques according to meteorological conditions.

In employing tall stacks it is hoped that the discharge is high enough to disperse the pollutants into atmosphere with out reaching the ground. Thirdly manufacturing methods are curtailed or completely shutdown during period of adverse meteorological conditions.

References

____1977. *Chemical Age of India*, 28(6).

Chemical Engg. Prog. 57: 53, 146.

Chapter 10

Requirement of Dietary Vitamin E in Relation to Growth, Feed Conversion and Deficiency Symptoms for the Fingerlings of *Labeo rohita* (Hamilton)

Ashok K. Gupta

*Department of Limnology and Fisheries**

Maharana Pratap University of Agriculture and Technology, Udaipur – 313 001, India

ABSTRACT

In the present investigation vitamin E requirement was evaluated for the fingerlings of *Labeo rohita* using varying levels of vitamin E (as α-tocopherol) supplemented purified diet for a period of 60 days. Significant weight gain and absolute growth, specific growth rate and feed conversion ratio for the fingerlings were observed at 10–20 mg vitamin E/kg of diet ($p < 0.0 1$). Whereas relative efficiency of various levels of dietary yitamin E on growth, specific growth rate and feed conversion ratio for the fingerlings was estimated as 20 < 10 < 40 < 80 < 100 < 00 (Control) mg/kg diet. The dietary vitamin E deficiency symptoms observed in the fingerlings were poor growth, low conversion efficiency, low protein productive value and energy conversion efficiency, anemia, ascites and muscular dystrophy.

Keywords: Fingerlings, Labeo rohita, Vitamin E

Introduction

Fat soluble vitamins (ADEK) have been found to play an important role in maintaining normal growth and physiological functions in animals. Vitamin E (α-tocopherol) are biological antioxidants,

* Now College of Fisheries.

active in the water and lipid phases respectively of animal tissue. They protect the organism against uncontrolled oxidation by free radicals, which are generated through normal metabolism, and by oxidative drugs and pollution. Vitamin E is an essential nutrient in fish also. The demand for this nutrient has been demonstrated in relatively few fish species (Watanabe *et al.*, 1970a,b; Cowey *et al.*, 1981; Wilson *et al.*, 1984; Roem *et al.*, 1990; Hamre and Lie, 1995, Hamre *et al.*, 1997; Ruff *et al.*, 2001). There has been limited study on the dietary vitamin E requirement of Indian major carps at early stages of development. Therefore, the present study has been designed to evaluate the vitamin E requirement of the fingerlings of *Labeo rohita* in relation to growth, feed conversion and deficiency symptoms.

Materials and Methods

Modified test diet of Yone (1975) supplemented with vitamin and mineral mixture was used for vitamin E requirement of *Labeo rohita* fingerlings. In isocaloric test diet, graded levels of vitamin E (as α-tocopherol) were added separately. In each prepared diets, 7–10 per cent of water was added to prepare a thick paste. Vermicelli of diameter 1 mm size were prepared from the paste, air dried and refrigerated. Feeding trials were conducted in cemented pools of 185 lit. capacity. For each treatment three sets of experiment were run. In each replicate ten properly acclimatised healthy fingerlings of approximately similar size and length (average weight 9.16 g and length 8.5 cm) were used. Controls with vitamin E deficient diet were also run simultaneously. The fingerlings were fed at the rate of 2 per cent body weight per day in between 9.00–9.30 a.m. The ration was adjusted fortnightly in accordance with the increment in body weight. Faecal deposits were removed daily by siphoning.

Weight gains of fingerlings was recorded regularly at every fortnightly interval. The feeding trials were conducted for a period of 60 days. The data were computed and statistically analysed, *viz.*, analysis of variance and correlation coefficient (r) as per method of Snedecore and Cochran (1980) for different growth parameters such as absolute growth, specific growth rate (SGR), feed on version ratio (FCR) and gross conversion efficiency (GCE). On the basis of present observations major deficiency symptoms were also recorded and concluded.

Results

The absolute growth of rohu fingerlings was significantly high at low vitamin E levels of 20 and 10 mg/kg of diets (Table 10.2). The dietary levels of 40, 80 and 100 mg vitamin E/kg indicated poor absolute growth. At control diet, no weight gain and absolute growth were recorded in rohu fingerlings ($p > 0.01$).

The data indicate that SGR is directly related to absolute growth of fingerlings, *i.e.*, higher absolute growth resulted in higher SGR values and a positive correlation is observed at 1 per cent of significance level. Average highest SGR's of 0.253 and 0.250 g/day were observed when fingerlings of rohu fed with 20 and 10 mg of vitamin E/kg of diet respectively for a total experimental period of 60 days (Table 10.2). However, in vitamin E deficient diet, SGR of the fingerlings for 60 days was found zero ($P > 0.01$). Poor SGRs were recorded in rohu fed with 40, 80 and 100 mg vitamin E/kg of diet.

Rohu fingerlings show lower values of FCR, *i.e.*, 1.73–3.31 at all the dietary vitamin E, levels, excluding control (Table 10.2). This suggests that vitamin E levels to some extent enhances the utilization of purified diet during the feeding period. The highest GCEs, *i.e.*, 0.578 and 0.540 for the absolute growth of 15 and 15.2 g were computed at levels of 10 and 20 mg vitamin E/kg of diet respectively. The zero GCE indicate the essentiality of vitamin E in the diet for better growth of fingerlings. Obviously, it was found non-significant ($P > 0.01$) from other levels of vitamin E.

Table 10.1: Ingredient Composition of the Basal Diet Used for Vitamin E Requirement for the Fingerlings of *L. rohita* (Modified after Yone, 1975)*

Ingredients	*%*	*Ingredients*	*%*
Casein vitamin free	47.0	Gelatin	14.0
Corn oil	10.5	White dextrin	8.0
α-cellulose	2.3866	Vitamin mixture	1.10
Mineral mixture	10.0	L-phenylalanine	0.7
L-Arginine HCl	1.4	L-Cystine	0.8
L-Tryptophane	0.2	L-Histidine HCl	0.2
DL-Alanine	1.4	L-Aspartate Na	1.0
L-lysine HCl	0.5	L-Valine	0.6
Glycine	0.3		

* Proximate composition (%): Protein 54; Carbohydrate 25, Fat 10, Moisture 8, Ash 3.

Table 10.2: Growth Parameters of the Rohu Fingerlings Fed with Purified Diet Supplemented with Graded Levels of Vitamin E as α-tocopherol for a Period of 60 Days

Vitamin E (mg/kg diet)	*Initial wt. of Fish (g)*	*Final wt. of Fish (g)*	*Absolute Growth (g)*	*SGR (g)*	*FCR*	*GCE*
00 (Control)	19	19	0	0.00	0.00	0.000
10	18	33	15	0.250	1.73	0.578
20	21	36.2	15.2	0.253	1.84	0.540
40	22	32	10.0	0.166	2.95	0.338
80	20	29.3	9.3	0.155	2.98	0.334
100	26	29.0	3	0.133	3.31	0.302

SGR: Specific Growth Rate; FCR: Food Conversion Ratio; GCE: Gross Conversion Efficiency.

The gross deficiency symptoms observed in fingerlings were poor growth, poor SGR, low FCR and low GCE alongwith blood anaemia, ascites and muscular dystrophy (Table 10.3). Other dietary deficiency symptoms of vitamin E evident were low PPV and ECE leading low protein, carbohydrate and high lipid contents in the fingerlings. In vitamin E deficient diets, fingerlings also showed poor accumulation of vitamin E and high lypoxigenase activity in tissues.

Discussion

The exact requirement of α-tocopherol for any fish species may depend upon the form of vitamin E and amount and type of unsaturated fatty acids in the oil components of the ration (Woodall *et al.*,1964; Watanabe *et al.*, 1970ab). Polyunsaturated labile fish oils may invoke an increased requirement for intracellular antioxidants (Hashimoto *et al.*, 1966).

The vitamin E supplemented diets in the present study were also improved growth, SGR, FCR and GCE in the fingerlings of rohu. A significant coefficient of correlation (r) at 1 per cent was also found between weight gain and SGR; GCE and FCR and weight gain and FCR in the fingerlings fed with vitamin E supplemented diets. Whereas, vitamin E deficient diets have shown poor response for growth parameters and proved to be statistically non-significant ($P > 0.01$). Woodall *et al.* (1964) and Murai and Andrews (1974) have also reported that vitamin E deficient diets in chinook salmon and channel catfish produced poor growth and reduced food conversion with several deficiency lessions.

Table 10.3: Gross Deficiency and Hypervitaminosis Symptoms in the Fingerlings of Rohu Fed with Low and High Supplemented and Deficient Vitamin E Diet

Sl.No.	*Deficiency Symptoms*	*Sl.No.*	*Hypervitominosis Symptoms*
1.	Poor growth SGR, FCR and GCE	1.	Poor growth, SGR, FCR and GCE
2.	Low PPVs and ECEs	2.	Low mortality
3.	Low protein, carbohydrate and high lipid contents		
4.	Low vitamin E contents and high lypoxigenase activity in tissues		
5.	Blood anemia ascites muscular dystrophy and fragility in R.B.C.		

Woodall *et al.* (1964) have observed deficiency symptoms, *viz.*, poor growth, exophthalmia, ascites, anemia, clubbed gills, epicarditis and ceroid deposition in the spleen of chinook salmon. However, Dupree (1969) and Murai and Andrews (1974) have noticed vitamin E deficiency symptoms of poor growth, exudative diathesis, muscular dystrophy, depigmentation, fatty livers, anemia and atrophy of pancreatic tissue in channel catfish. Poston *et al.* (1976) fed vitamin E deficient diet to Atlantic salmon displayed deficiency symptoms of anemia, pale gills, anisocytes, poikilocytosis, exudative diathesis and muscular dystrophy. However, in the present investigation besides poor growth response, low vitamin E contents and high lypoxigenase activity in tissues, anemia, muscular dystrophy etc. have been observed in the rohu. Woodall *et al.* (1964) reported that a dietary vitamin E concentration of 5 to 30 mg α-tocopherol combined with 5 per cent herring oil provided satisfactory growth of chinook salmon and also prevented the occurrence of clinical deficiency symptoms. The present results suggest that the range of optimum dietary requirement for vitamin E as α-tocopherol for rohu fingerlings are comparatively higher despite test diet having corn oil with low levels of polyunsaturated fatty acids. Further, the present study reveals that 10 per cent corn oil in the test diet with varying, concentrations of vitamin E displayed a significant role for maintaining maximum growth, feed conversion efficiency, PPV and ECE in rohu fingerlings. At 10 per cent corn oil plus 10–20 mg vitamin E/kg diet, rohu fingerlings showed highest growth and feed conversion efficiency. The investigations of Watanabe *et al.* (1981) emphasized that diets supplemented with 50 mg α-tocopherol/kg dry diet promotes growth rate in rainbow trout fingerlings equal to those fish fed with 100, 300 or 500 mg α-tocopherol/kg of dry diet.

Cowey *et al.* (1981) have established that larger rainbow trout fingerlings (initial mean weight 10g) required 20 or 30 mg DL α-tocopherol acetate of 50 mg/kg of diet should be considered a minimum level since the requirement may vary depending on size, age, water temperature, water ionic strength, fish density, water quality etc. Further, Wilson *et al.* (1984) have also recommended that higher levels of vitamin E may be needed in practical diets that contain higher levels of polyunsaturated fatty acids or oxidized lipids. Cowey *et al.* (1984) have suggested that vitamin E requirement may increase as the water temperature decrease. Therefore, minimum dietary vitamin E requirement measured in the present study at a range of 20.6–22.4°C may be decrease if water temperature increases.

Acknowledgement

Appreciation is extended to the ICAR, New Delhi for financing the project.

References

Cowey, C.B., Adron, J.W., Walton, M.J., Murray, J., Youngsun, A., Knose, D. 1981. Tissue distribution, uptake and requirement for α-tocopherol of rainbow trout (*Salmo gairdneri*) fed diets with minimol contents of unsaturated fatty acids. *J. Nutr.*, 111: 1556–1567.

Cowey, C.B., Elizabeth, D., Tacon, A.G.J., Youngson, A., Bell, J.G., 1984. The effect of vitamin E oxidized fish oil on the nutrition of rainbow trout (*Salmo gairdneri*) grown at natural, varing water temperatures. *British J. Nutr.*, 51: 443–451.

Dupree, H.K. 1969. Vitamin E requirements of channel catfish. *U.S. Bull. Sport Fish.*, Wildlife Resour. Publ. 77: 220–221.

Hamre, K., Lie, ϕ. 1995. α-tocopherol levels of different organs of Atlantic salmon (*Salmo salar* L.): Effect of smoltification, dietary levels of n-3 polyunsaturated fatty acids and vitamin E. *Comp. Biochem. Physiol.*, 111A: 547–554.

Hamre, K., Waagbo, R., Berge, R.K., Lie, ϕ. 1997. Vitamin C and E interact in juvenile Atlantic salmon. *Free Radical Biol. Med.*, 22: 137–149.

Hashimoto, Y., Ikaichi, Y., Watanabe, T., Furukawa, A., Umezu, T. 1966. Muscle dystrophy of carp due to oxidized oil and the preventive effect of vitamin E. *Bull. Jpn. Soc. Sci. Fish*, 32: 64–69.

Murai, T. and Andrews, J.W. 1974. Interactions of dietary α tocopherol, oxidized menhaden oil and ethoxyquin on channel cat fish (*Ictalurus punctatus*). *J. Nutr.*, 104: 1416–1431.

Poston, H.A., G.F. Combs Jr., Leibovitz, L. 1976. Vitamin E and Selenium interrelation in the diet of Atlantic salmon (*Salmo salar*): Gross, histological and biochemical deficiency signs. *J. Nutr.*, 106: 892–904.

Roem, A.J., Kohler, C.C. and Stickney, R.R. 1990. Vitamin E requirements of the blue tilapia *Oreochromis aureus* (Steindachner), in relation to dietary lipid level. *Aquacult.*, 87: 155–164.

Snedecor, S.W. and Cochran, W.G. 1980. *Statistical Methods*, 7th edn., IOWA State University Press.

Watanabe, T., Takashima, F., Ogino, C. and Hibiya, T. 1970a. Requirement of young carp for α-tocopherol. *Bull. Jpn. Soc. Sci. Fish*, 36: 935–946.

Watanabe, T., Takashima, F., Ogino, C. and Hibiya, T. 1970b. Effect of α-tocopherol deficiency on carp. *Bull. Jpn. Sci. Fish*, 36: 972–976.

Watanabe, T., Takeuchi, T., Wada, M. and Uehara, R. 1981. The relationship between dietary lipid levels and α-tocopherol requirement of rainbow trout. *Bull. Jpn. Sci. Fish*, 47: 1463–1471.

Wilson, R.P., Bowser, P.R. and William, P. 1984. Dietary vitamin E requirement of fingerlings channel catfish. *J. Nutr.*, 114: 2053–2058.

Woodall, A.N., Ashley, L.M., Halver, J.E., Olcott, H.S. and VanDer Veen, J. 1964. Nutrition of salmonid fishes. XIII. The α-tocopherol requirement of chinook salmon. *J. Nutr.*, 184: 125–135.

Yone, Y. 1975. Nutrition studies on red sea bream. In: *Proceedings of the International Conference on Aquaculture Nutrition*, (Eds.) Price, K.S., Shaw, W.N., Danbert, K.S. Lewes/Rehoboth, University of Delaware, p. 34–64.

Chapter 11

Effect of Metal Poisoning on Total Body Carbohydrate in *Sphaerodema rusticum* (Belostomatidae: Hemiptera)

S. Mumtazuddin* and S. Ehteshamuddin**

**University Department of Chemistry, B.R.A. Bihar University, Muzaffarpur – 842 001, India*

***Pro Vice-Chancellor, Patna University, India*

ABSTRACT

Although many metals are required for normal physiological functions of animals in very low concentrations, altered physiological functions result when one or more of these reach sufficiently high concentrations in body cells. In the present investigation the effect of three metal compounds, *viz.*, copper sulphate, cobalt nitrate and Mohr's salt on the total body carbohydrate of an aquatic insect *Sphaerodema rusticum* has been studied. It was observed that both the male and the female insects showed decline in the total body carbohydrate concentrations on the treatment with all these three metal compounds. The decline became increasingly significant with the increase in the concentrations of the compounds, being most dramatic for total body carbohydrate in the case of copper sulphate. It is assumed that metal ions in high concentrations act as inhibitors to different enzyme systems responsible for carbohydrate syntheses.

Keywords: Sphaerodema rusticum, Total body carbohydrate, Metal poisoning.

Introduction

Many metals are required for normal physiological function of mammals, but only in very low concentrations. These include Copper, Iron, Zinc, Manganese, Cobalt, Selenium and Chromium (Harper *et al.*, 1977). It is assumed that most of these also probably have similar functions in fish and other aquatic animals. However, altered physiological functions result when one or more of these reach sufficiently high concentrations in body cells both in aquatic animals and mammals. Metals enter in the aquatic environment by a variety of industrial effluents and old mines. Acid precipitation also causes leaching of metals from surrounding soils (Norton, 1982).

In fact, aquatic environment is the ultimate sink for the pollutant and any compound either produced on the industrial scale or present in terrestrial environment is likely to reach there, sooner or later.

The study of physiological changes as means for understanding the responses of aquatic animals, like insects, fish, etc. to the stress of various types of pollutants is of great help in studying the nature and the extent of pollution. Eisler *et al.* (1979) prepared bibliographies on the biological effects of metals in aquatic environment.

The present investigation is an attempt to study the effects of three metal compounds, *viz.*, copper sulphate, cobalt nitrate and Mohr's salt on the total body carbohydrate of an aquatic insect *Sphaerodema rusticum*.

Investigation of body carbohydrate has been undertaken as it, alongwith protein and lipid, forms the principal class of organic compounds that are found in insect body. In general, carbohydrate metabolism in insects is similar to that found in vertebrates. However, insects, unlike vertebrates, possess an exoskeleton, chitin, which is rich in aminopolysaccharide. In addition, they also contain a disaccharide, trehalose, which acts as the storage form of glucose. Specific feature of insect carbohydrate metabolism, therefore, centres round the synthesis and hydrolysis of these compounds. Many workers have extensively studied carbohydrate metabolism in various insects (Chang *et al.*, 1964; Islam and Roy, 1981; Pant, 1985; Sharma and Ehteshamuddin, 1992).

Materials and Methods

The insects (*S. rusticum*) were treated separately in different batches by the following metal compounds in different concentrations.

Copper Sulphate ($CuSO_4 . 5H_2O$)

The LC_{50} of the compound for the insects was worked out to be one molar concentration (1 M) and as such ten different concentrations of the compound were taken from 0.1 M to 1 M. In each concentration of the solution, 30 to 50 insects were kept and biochemical assays were performed after 24 hours of each treatment.

Cobalt Nitrate [$CO(NO_3)_2$]

The LC_{50} for the insects in this case was found to be 0.055 M and 30 to 50 insects were placed in nine different concentrations ranging from 0.047 M to 0.055 M and the biochemical assays were performed in each case after 24 hours of treatment.

Mohr's Salt [$FeSO_4 . (NH_4)_2SO_4 . 6H_2O$]

The LC_{50} of Mohr's salt for the insects was found to be 0.1 M. Hence, ten different concentrations of the compound ranging from 0.01 M to 0.1 M were taken and in each 30 to 50 insects were placed for 24 hours for biochemical assays.

Quantitative estimation of total body carbohydrate of treated and control (untreated) insects was done by Anthrone reagent method (Carroll *et al.*, 1956) using Erma colorimeter. Mean and S.D. for each set of reading was calculated. Students-t-test, where necessary, was applied in order to find level of significance between two sets of readings.

Results and Discussion

Total Body Carbohydrate Concentration of 'Control' Insects

In the untreated male the total body carbohydrate concentration was found to be 173±5 S.D. mg/gm of the body weight and was a little higher, *i.e.*, 191±7 S.D. mg/gm of the body weight in the female insect. The adult female insect had greater body carbohydrate concentration than that of the male insect owing to greater metabolic demand for egg maturation.

Total Body Carbohydrate Concentration of 'Treated' Insects

Treatment with Copper Sulphate

On exposure to 0.1 M concentration of $CuSO_4$ itself the body carbohydrate concentration of both male and female insects showed a significant ($P < 0.01$) decline, maintaining steadily the same on treatment with 0.2 and 0.3 M concentrations. In 0.4 M concentration both male and female insects showed a very significant ($P < 0.001$) decline in concentration maintaining a gradual decline in every higher concentration of the treatment, reaching the lowest level, 15±1 S.D. mg/gm of body weight in male and 10±1 S.D. mg/gm of body weight in female insect at 1 M concentration of $CuSO_4$. The results are shown in Table 11.1.

Table 11.1: Body Carbohydrate Concentration in Insects on Treatment with Copper Sulphate

Concentration of Copper Sulphate (molar)	*Body Carbohydrate Concentration (mg/gm body weight*±S.D.)*	
	Male	*Female*
0.1	140±4	145±5
0.2	120±5	130±7
0.3	100±9	105±8
0.4	45±5	55±4
0.5	35±2	52±3
0.6	30±4	40±4
0.7	23±2	29±4
0.8	20±1	20±2
0.9	16±1	18±1
1.0**	15±1	10±1

*Mean wt. of insect:

Male insect: 250 mg±9 S.D. Female insect: 265 mg± 10 S.D.

**LC_{50}

Body carbohydrate in 'control':

Male insect: 173±5 Female insect: 191±7

Treatment with Cobalt Nitrate

In 0.047 M concentration of cobalt nitrate both male and female insects showed significant ($P < 0.05$) decline in body carbohydrate concentration maintaining a steady decline in concentration in every higher concentration of cobalt nitrate, reaching the lowest level (90±2 S.D. mg/gm of body weight in male and 88±3 S.D. mg/gm of body weight in female) in 0.055 M concentration of the compound. The results are shown in Table 11.2.

Table 11.2: Body Carbohydrate Concentration in Insects on Treatment with Cobalt Nitrate

Concentration of Cobalt Nitrate (molar)	*Body Carbohydrate Concentration (mg/gm body weight*±S.D.)*	
	Male	*Female*
0.047	155±6	158±8
0.048	145±5	148±6
0.049	130±4	136±5
0.050	125±5	125±6
0.051	120±6	115±4
0.052	110±5	115±5
0.053	105±4	102±6
0.054	100±5	95±4
0.055**	90±2	88±3

*Mean wt. of insect:

Male insect: 250 mg±9 S.D. Female insect: 265 mg± 10 S.D.

**LC_{50}

Body carbohydrate in 'control':

Male insect: 173±5 Female insect: 191±7

Treatment with Mohr's Salt

When exposed to 0.1 M concentration of Mohr's salt both male and female insects showed a significant ($P < 0.05$) decline in body carbohydrate concentration maintaining a steady decline upto 0.04 M concentration. However, in 0.05 M concentration both male and female insects showed a more significant ($P < 0.01$) decline in body carbohydrate concentration maintaining a steady declining trend upto 0.09 M concentration. In 0.1 M concentration of Mohr's salt the decline in body carbohydrate concentration of both male and female insects was highly significant ($P < 0.001$). The results are shown in Table 11.3.

The results clearly indicate that all the three compounds adversely affect the carbohydrate synthesising machinery of the insect, the effect increasing gradually from lower to higher concentrations in the case of cobalt nitrate and Mohr's salt but the effect becoming sharp from 0.4 M concentration onwards in the case of copper sulphate reaching a very low level in 1 M concentration of the compound. This is in accordance with the findings of the earlier workers on other insects (Singh, 1986; Islam and Roy, 1989; Shunmugavelu, 1993). Recently, some work has been done on the effect of organophosphorus insecticides on carbohydrate metabolism in some insects (Rekha *et al.*, 2000 a&b).

Some of the metal ions are known to act as inhibitors to different enzyme systems. In the present investigation also, it is assumed that copper, cobalt and iron in more than required concentrations act as inhibitors to various enzyme systems involved in the synthesis of carbohydrate.

Table 11.3: Body Carbohydrate Concentration in Insects on Treatment with Mohr's Salt

Concentration of Mohr's Salt (molar)	*Body Carbohydrate Concentration (mg/gm body weight*±S.D.)*	
	Male	*Female*
0.01	165±3	170±5
0.02	162±4	165±7
0.03	155±5	160±6
0.04	140±4	140±6
0.05	130±5	132±4
0.06	125±6	130±5
0.07	120±5	122±5
0.08	120±7	120±4
0.09	110±5	110±6
0.10**	92±6	85±5

*Mean wt. of insect:

Male insect: 250 mg±9 S.D. Female insect: 265 mg± 10 S.D.

**LC_{50}

Body carbohydrate in 'control':

Male insect: 173±5 Female insect: 191±7

Since the treatment with copper sulphate showed a greater decline in the concentration of carbohydrate as compared to those with cobalt nitrate and Mohr's salt, it is inferred that copper has a more potent role in enzyme system than the other two metals.

It may thus be concluded that all the three metal compounds pollute the aquatic environment and even in low concentrations affect the physiology of the insect by inhibiting the synthesis of essential biochemical components like protein (Mumtazuddin and Ehteshamuddin, 2000), ascorbic acid (Mumtazuddin and Ehteshamuddin, 2005) and carbohydrate, thus affecting the production and growth of the insect. As the insect is an important constituent of the aquatic ecosystem in maintaining a normal functional life of pond, lake, etc., the decline in the growth and production of the insect would ultimately affect the normal life of a pond. Moreover, similar adverse effect on the physiology of fishes and other aquatic animals is also expected under the influence of these chemicals.

References

Carroll, N.V., Longley, R.W. and Roe, J.H. 1956. The determination of glycogen in liver and muscles by use of anthrone reagent. *J. Biol. Chem.*, 220: 583–593.

Chang, C.K., Liu, F. and Fang, H. 1964. Studies on the metabolism of Eri silk worm (*Samia ricini*) during metamorphosis (II). The properties of trehalose and its effect on metabolism. *Acta Ent. Sin.*, 13: 494–502.

Eisler, R., Rossoll, R.M. and Gaboury, G.A. 1979. Fourth annotated bibliography on biological effects of metals in aquatic environments. U.S. Environmental Protection Agency. Res. Rep. EPA-600/3-79084, Washington, D.C.

Harper, H.A., Rodwell, V. and Mayer, P.A. 1977. *Review of Physiological Chemistry*, 16th edition. Lange, Los Altos, Calif.

Islam, A. and Roy, S. 1981. Quantitative changes of carbohydrates, lipids and protein during the post-embryonic development and metamorphosis of *Tribolium castaneum* and *T. comfusum* (Coleoptera: Tenebrionidae: Insecta), *Proc. Indian Nat. Sci. Acad.*, B47: 313–320.

Islam, A. and Roy, S. 1983. Effect of cadmium chloride on the quantitative variation of carbohydrate, protein, amino acids and cholesterol in *Chrysocoris stolli. Curr. Sci.*, 52: 25.

Mumtazuddin, S. and Ehteshamuddin, S. 2000. Effect of metal poisoning on total body protein in *Sphaerodema rusticum* (Belostomatidae: Hemiptera). *Res. J. Chem. Environ.*, 4(4): 49–51.

Mumtazuddin, S. and Ehteshamuddin, S. 2005. Effect of metal poisoning on total body ascorbic acid in *Sphaerodema rusticum* (Belostomatidae: Hemiptera). *Indian J. Environ. and Ecoplan.*, 10(1): 83–86.

Norton, S.A. 1982. The effect of acidification on the chemistry of ground and surface water. In: *Acid Rain Fisheries*, (Ed.) R.E. Johnsonz. American Fisheries Society, Betbesda, Md, p. 93.

Pant, E.G. 1985. Dynamics of the total protein, lipid, carbohydrate and glycogen content in organs of the clorado beetle (*Leptinotarsa decelineata*) Stud. Univ. Bolyai. Biol., 30: 55–61.

Rekha, Misbahuddin, S. and Ehteshamuddin, S. 2000a. Effect of the systematic insecticide phosphamidon on LDH activity in the hemolymph of sap feeding insect pest *Aspongopus janus* and *Dysdercus cingulatus. Environ. Ecol.*, 18(2): 317–319.

Rekha, Shailesh, Sah, U.N. and Ehteshamuddin, S. 2000b. Effect of phosphamidon on the hemolymph glucose concentration in heteropteran insects *Dysdercus cingulatus* and *Aspongopus janus. Environ. Ecol.*, 18(4): 863–864.

Shunmugavelu, M. 1993. Effect of heavy metals on the biochemistry of *Mylabris pustulata, Environ. Ecol.*, 11: 386–390.

Singh, G.J. 1986. Haemolymph carbohydrate and lipid metabolization in *Locusta migratoria* in relation to the progress of poisoning following bicresmethrin treatment. *Pestic Biochem. Physiol.*, 25(2): 264–269.

Sharma, S. and Ehteshamuddin, S. 1992. Carbohydrate concentration during development and maturation in *Sphaerodema rusticum* (Belostomatidae, Hemiptera). *Comp. Physiol. Ecol.*, 17(3): 112–116.

Chapter 12

A Model Approach for the Water Quality: A Case Study of River Cauvery

A.G. Nataraj, K.L. Prakash,** R.K. Somashekar** and N. Manamohan Rao***

**Department of Civil Engineering, BIT, Bangalore – 560 004*

***Department of Environmental Science, Bangalore University, Bangalore – 560 056*

ABSTRACT

The present study aims at develop a model concept and approach, which gives an acceptable elucidation relating to problem of water quality and basin management of river Cauvery. The study embodies analytical data collected between two monitoring stations (Upstream represented by Krishnarajasagar Dam and the downstream represented by Dasanapura Gauging station). Both point and non-point sources of pollution are transforming the river into a hostile aquatic ecosystem conveniently, due to increased human activities in the river basin. A strong correlation between discharge and the conservative water quality parameters was observed in river Cauvery. Application of ANFIS for developing a generalized model for the river reach yielded good results. River water quality of both dry and monsoon seasons was comparable. The pollutant load contributed by base flow to the river is insignificant both in dry and wet seasons.

Keywords: River Cauvery, Water quality, ANFIS model, Base flow.

Introduction

The rivers are considered to be the lifeline for most of the developing countries for drinking water needs. Due to improper and inadequate treatment facilities in a country like India, huge volumes of wastewater is being poured into the rivers and lakes from various sources like towns/cities, industries and indiscriminate use of fertilizers and pesticides in the irrigated lands which has significantly contributed to the non-point source of pollution. Integrating the database from both sources of pollution and with the use of mathematical models and modern computer tools it has become an easy technique to study the dynamics of the river system that in turn offers inexhaustible opportunities of integration, analysis and evaluation of complex water quality problems which will ultimately guide the planners, policy makers and decision makers to take quick steps to mitigate pollution problems and manage the water systems in a more scientific and comprehensive manner. A comprehensive and complete application of water chemistry and hydrodynamics related to river and lake water quality modelling is evident. Many water quality models are mere extension of equation, developed in 1925 by Streeter and Phelp for predicting BOD and DO in flowing streams. Although such models require considerable data and often demand significant computer time, they offer the promise of a better understanding of water quality mechanics and have the potential for promoting more efficient policies for water quality management (Viessman and Hammer, 1985). Watersheds are often characterized by taking an inventory of land uses and multiplying the areas of each use by the appropriate loading factor (Hartigan *et al.*, 1983). The direct inference approach can be extended beyond simple loading factors by the use of regression models, which have land use characteristics as independent variables and stream water quality parameters as dependent variables (Haith, 1976).

The use of statistical analysis of land use and water quality data bring out unambiguously the impact of non-point source on lake water quality (Sekhar and Raj, 1995). Statistical procedures such as correlation and regression analysis and mass balance provide a methodology for including land use and non-point source considerations in water quality management programs (Sekhar and Umamahesh, 2000).

Modelling on national and regional scale and regional mass balance studies revealed that for nitrogen, diffusive sources were more important point sources both at regional and National scale (Unk, 1991). Lorin, *et.al* (1986) designed a computer-aided methodology to determine the water quality and ecological changes due to non-point source pollution. The management problem is formulated as a fuzzy optimization problem and solved by fuzzy programming. Chang *et al.* (1997) combined the gray and fuzzy programming methods in an interactive multi-objective framework for water pollution control in a river basin.

The models are found to be very useful tool where insufficient and unclassified data require to be handled. The indirect method utilizes, however, the upstream and downstream water quality data for modelling. This approach is based on the water quality and flow, while, the mass balance studies provide rational method of separation of point source and non-point source contribution to the river and the relationship between dissolved solids and discharge in streams.

The river basin under study encounters numerous sources of point as well as non-point sources pollution from domestic, industrial and agricultural activities. Contribution from these sources is significantly affecting the water quality of the river. Given this background, a thorough study is crucial for the water quality management of the River Cauvery.

Study Area

The river Cauvery is one of the major perennial rivers of the south India and the river stretch under investigation is located between 12°00″ N and 12°45″ N and 76°00″ E and 76°15″ E. The upstream of monitoring station is Krishnarahjasagar Dam near Mysore and downstream is Dasanapura near Kollegal (Central Water Commission Gauging Station) in Karnataka. The river flows through Karnataka, Tamil Nadu and Kerala and wrap up about 11954 sq. km. The river basin in Karnataka state lies at an altitude of 600 to 1000 m above MSL and is called Mysore plateau and the basin is

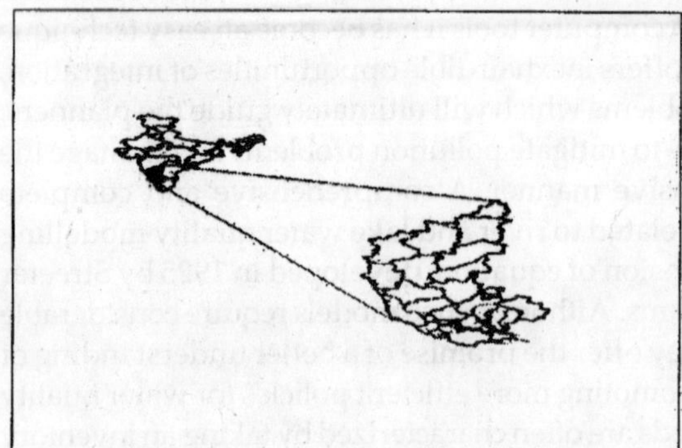

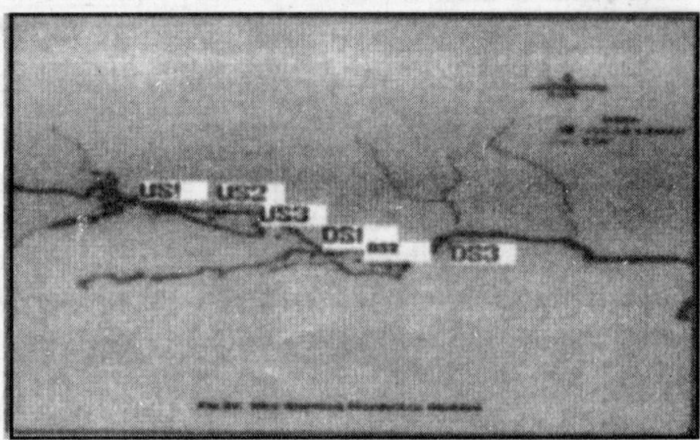

Figure 12.1: Map Showing Location and Sampling Stations

underlain by variety of rocks, but most predominant rocks occurring are magmatic gneisses of grandioritic to tonalitic composition. The three major soil types, *viz.*, laterite soils, red sandy loam soil, and dark brown, sandy loam to sandy clay loam soils. The land use in the catchment is for the varied activities; agriculture and livestock and also for the urban and industries development.

The average annual rainfall of the Western Ghats is very high and in Mysore plateau of the Cauvery basin receive inadequate rainfall. The mean relative humidity in the western part, mean humidity is higher in the months of South-west monsoon, being in the range of 70 to 80 per cent in general and about 100 per cent at the extreme western end. The mean seasonal flow at upstream of KRS Dam were recorded as 564.3, 104.8, 10 and 25.9 million m^3 whereas at Kollegal gauging station (downstream) 1324.9, 936.1, 212.8 and 159.3 million m^3 were recorded respectively for SW Monsoon, NE monsoon, Winter and Summer (1970 to 1985). Cauvery basin, due to large scale abstraction of waste, and insignificant flows during the lean months, it suffers from both point and non-point pollution source and also contributed from major towns located on the riverbank (Sreerangapattana, Bannur, T. Narisipura and Kollegal) which directly dispose the wastewater into the river.

Materials and Methods

The upstream of monitoring station is Krishnarahjasagar Dam near Mysore and downstream is Dasanapura near Kollegal (Central Water Commission Gauging Station) in Karnataka. In between these two stations, sub-stations were considered to study the water quality variations (Figure 12.2).

Table 12.1: Grouping of Upstream and Downstream Sampling stations

Sl.No.	*Upstream Sampling Station*	*Downstream Sampling Station*
1.	Krishnarajasagara Dam (US_1)	Monitoring Station 4-(DS_4)
2.	Monitoring Station 2-(US_2)	Monitoring Station 5-(DS_5)
3.	Monitoring Station 3-(US_3)	Dasanapura gauging station 6-(DS_6)

The water quality measurements available at the beginning of the study consisted of samples with temporal and spatial distribution spread over four years. Additional measurements were also performed for three more years on a monthly basis, covering different seasons, from August 1999 to May 2002.

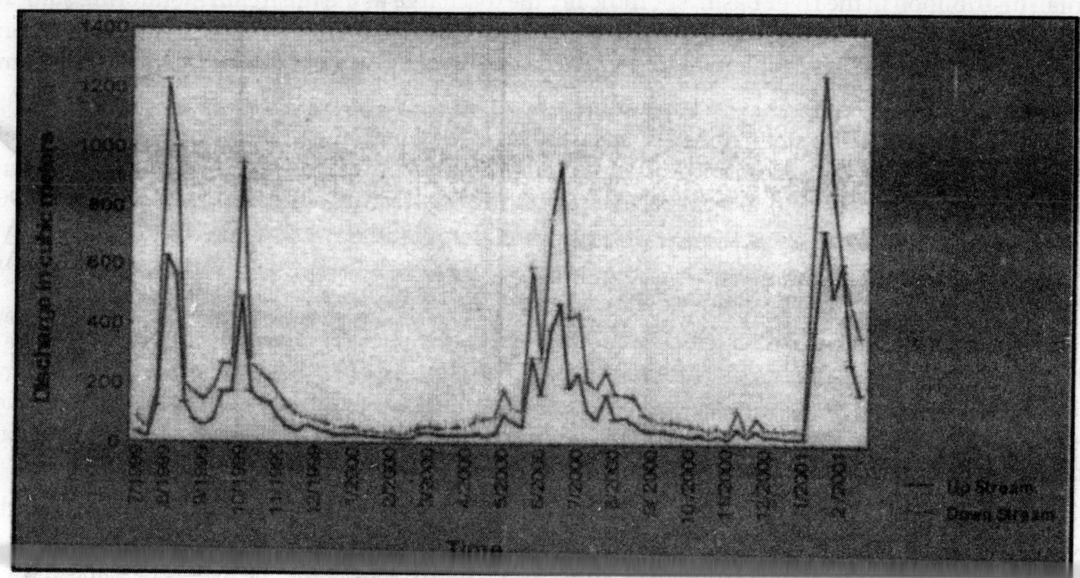

Figure 12.2: Hydrograph of River Cauvery at Upstream (KRS Dam) and Downstream (Dasanapura Gauging Station)

Representative water samples were collected at various depths including surface (ISO 5667-1 Part 1, 1980; ISO 5667-1 Part 2, 1991 and ISO 5667-1, Part 6, 1990). Samples were transported (freezer condition) to the laboratory and analysed for different physico-chemical parameters and standard methods (APHA, 1995) were followed during the analysis. The river discharge data was collected from the monitoring station SD_6 has a permanent gauging station where Central Water Commission continuously monitors river discharge

ANFIS model has been used to predict the pollutant loads in a river during the wet season. The discharges during the present time interval and the previous time interval are used as input variables to model the pollutant loads. The training of the ANFIS model is done using the Fuzzy Logic Tool Box of MATLAB software.

Results and Discussion

The river hydrology was studied in both upstream and downstream stations. The hydrograph pattern of the river makes obvious that the rain primarily recharges the river during wet season showing high discharge relative to the dry weather conditions (Figure 12.2), thus showing a characteristic rain fed riverine system.

The time series analysis revealed that the highest peak was attained during the first 10 days of August 1999 and July 2001. The mean maximum discharge in the upstream ranged between 600 to 750 m^3/sec and the downstream discharge varied between 1000 to 1250 m^3/sec. A gradual decrease in the flow of the river following monsoon to winter is evident. During winter season, the mean maximum flow was 269 m^3/sec, which is further reduced and approached a mean minimum flow of 55.66 m^3/sec. As the summer looms, the river attains a low discharge with mean minimum flow of 52 m^3/sec, which finally reaches to a lowest discharge of 41 m^3/sec. Further due to large variations in rainfall distribution in the river basin, sporadically the discharge at the upstream monitoring station marginally equals the downstream; this may be due to the presence of dams in sub-basin. In addition to the overall seasonal trends, the river hydrograph is typified by quick rises and falls in response to individual rainfall events.

The maximum base flow component was identified as the point at which the slope of the curve on plotting discharge versus concentration of conservative parameters of water quality (Chandrashekar, 2000). The estimated contribution from base flow in the downstream monitoring stations was 70 to 90 m^3/sec. The separation of base flow was done using conservative parameters like conductivity, chlorides, bicarbonates and sulphates wherein all these parameters showed more or less a similar base flow trend.

The base flow, which is the contribution from the groundwater, is estimated by plotting dissolved solids *vs.* river discharge. A similar technique was followed by Albek (1999) to separate the base flow and surface flow components using concentration distribution of chlorides. The base flow in the downstream of river Cauvery is estimated between 70 and 90 m^3/sec. In the present study the estimation is based on the concentration distribution of all conservative parameters of the river water.

During lean season the mean velocity of the river has varied between 0.5 to 0.8 m^3/sec, which takes almost 20 to 23 hours to cover the stretch under study. The changes in the concentrations of the conservative pollutants considered for the study during this short time are negligible and hence neglected. However during monsoon flows, the time of travel is considerably reduced which nullifies the changes in concentrations of the pollutants during travel time, *i.e.*, Pollutants approach quickly the downstream stations.

Application ANFIS for developing a generalized model for the river reach under study yielded good results. The correlation coefficients in the range of 0.6022 to 0.8655 RMSE (Table 12.1) indicate suitability of the models to the study area. The models productively explained the variations in loads for different flow conditions in the river. The peak flow and low flow conditions indicated the influence of previous flow and the load due to delayed effect. The models are verified using linear regression. Agreement between model predictions and observed measurements are within the uncertainty of data and are analyzed at 95 per cent confidence limit. The relation between concentration during wet and dry season served the purpose of separation of base flow and surface runoff contributions to the river. The pollutant loads contributed by base flow to the river is insignificant both in dry and wet season.

The study clearly brings out large input inventory of nutrients especially nitrate and phosphates inputs to the Cauvery river basin. This could be detrimental to water quality in the long run. The reduction in total non-point sources loads could be achieved through integrated effort, including the reduction in fertilizer application, improved irrigation practices to reduce amount of water used and retiring of agricultural fields immediately adjacent to the river, there by reducing the magnitude of both surface and subsurface return flows. Further, encouraging the development of a healthy riparian corridor would produce some degree of non point source nutrient removal. Such a practice should be popularized among the farmers through awareness campaign for achieving integrity of river water quality. Because the pollutants are conservative in nature, it is a general conclusion that, the pollutant will be cumulative downstream. Inspite of the possibility of some retention in bed sediments as a result of adsorption, the above conclusion can be regarded as reasonable when dissolved pollutants are considered. The estimated loads of dissolved pollutants provide a good reference for the relative contributions of the point and non point sources to the river. A considerable non-point contribution from the river basin as established by the present study does emphasize the need for control of non-point sources of pollution. Hence, in the Cauvery river basin, regulating and controlling both point and non-point sources can lead to the goals of good water quality management.

Table 12.1: Training and Testing of ANFIS Model for Present and Previous Discharges and Concentrations

Parameter	*Rule*	*Discharge (X_1)*	*Previous Discharge (X_2)*	*r (Training)*	*r (Testing)*	*RMSE (Training)*	*RMSE (Testing)*
Alkalinity	1	High	High				
	2	High	Low				
	3	Low	Low	0.8655	0.6952	0.0334	0.2365
	4	Low	High				
Chloride	1	High	High				
	2	High	Low				
	3	Low	Low	0.7896	0.6882	0.02569	0.1587
	4	Low	High				
Calcium	1	High	High				
	2	High	Low				
	3	Low	Low	0.8023	0.7369	0.03698	0.2568
	4	Low	High				
Magnesium	1	High	High				
	2	High	Low				
	3	Low	Low	0.7635	0.6541	0.07562	0.6581
	4	Low	High				
Sodium	1	High	High				
	2	High	Low				
	3	Low	Low	0.8014	0.7568	0.03694	0.6081
	4	Low	High				
Sulphate	1	High	High				
	2	High	Low				
	3	Low	Low	0.6984	0.5893	0.02568	0.4567
	4	Low	High				
Nitrate	1	High	High				
	2	High	Low				
	3	Low	Low	0.6022	0.5102	0.04682	0.1268
	4	Low	High				
Phosphate	1	High	High				
	2	High	Low				
	3	Low	Low	0.6481	0.4982	0.0452	0.1542
	4	Low	High				
Silicate	1	High	High				
	2	High	Low				
	3	Low	Low	0.7025	0.6823	0.0365	0.3651
	4	Low	High				

Recommendations for possible future development in the most significant directions are, paired sampling, continuous monitoring and sediment analysis between the monitoring stations. Representative sampling, which is a major problem especially in large rivers, needs to be emphasized. Regular frequency of sampling should depend on the importance of the sampling point in the river system. Representative sampling at different flow conditions seems to be more important than a high frequency of sampling.

References

Albek, E. 1999. Identification of the different sources of chlorides in streams by regression analysis using chloride-discharge relationship. *Wat. Env. Res.*, 71(7): 1310–1319.

Chandrashekar, M. and Uma Mahesh, N.V. 2000. Immission Approach for Modelling Dissolved Pollutants in a River for Separation of Point and Non-point Loads. Water and Environment Division, Department of Civil Engineering, Regional Engineering College, Warangal, Andhra Pradesh. *Ph.D. Thesis*, p. 69–72.

Chang, N., H.W. Chen, D.G. Shaw, and Yang, C.H. 1997. Water Pollution Control in River by Interactive Fuzzy Interval Multiobjective Programming. *J. Env. Eng.*, 123(12): 1208–1216.

Haith, D.A. and Dougherty, J.V. 1976. Non-Point Source Pollution from Agricultural Runoff. *J. Env. Engg. Divi., ASCE*, (102) (EE5): 1055–1069.

Hartigan, L.P., T.F. Quasebarth and Elisabath, S. 1983. Calibration of NPS Model Loading Factors. *J. Env. Engg. Div.*, 109(6): 1258–1272.

ISO 5667–1. 1980. Water quality–Sampling–Part 1: *Guidance on the Design of Sampling Programmes.*

ISO 5667–1. 1990. Water quality–Sampling–Part 6: *Guidance on Sampling of Rivers and Streams.*

ISO 5667–1. 1991. Water quality–Sampling–Part 2: *Guidance on Sampling Techniques.*

Lorin, E.R., Richard, R.H and Brain, W.M. 1986. *Non Point Source Pollution Monitoring Program Design.* ASCE Spring Convention held in Seatle, Washington, p. 335 –350.

Sekhar, M.C and Raj, P.A. 1995. Landuse-Water Quality Modelling. *Wat. Sci. Tech.*, 31(8): 383–386.

Streeter, H.W. and Phelp, E.B. 1969. A Study of Pollution of Natural Purification of the Ohio River. Public Health Bulletin 146, Public Health Service, University of Washington, Seattle.

Viessman, W. Jr. and Hammer, M.J. 1985. *Water Supply and Pollution Control.* Harper and Row Publishers, New York.

Somashekar, R.K., Rameshaiah, V. and A. Chethana Suvarana. 2000. Groundwater chemistry of Channapatna Taluk (Bangalore rural district): Regression and cluster analysis. *Journal of Env. Pollution*, 7(2): 101–109.

Chapter 13

Impact of Tourist Influx on the Courtallam Water Quality Index

G. Gitanjali and A. Kumaresan
ICAR Centre, Sri Parasakthi College for Women, Courtallam – 627 802, Tamil Nadu

ABSTRACT

Water Quality Index (WQI) of Courtallam water was studied after analysing nine hydrochemical parameters. Water samples were collected in triplicate at five sites. Samples collected in main, five and old Courtallam falls before and after anthrogpogenic activity and after were designated as site I and II respectively. Down stream samples were collected at Melagaram, Yanaipalam, Tenkasi outer and they were referred as site III, IV and V respectively. Dissolved Oxygen, biochemical oxygen demand, pH, total alkalinity, calcium, magnesium, sulphate, chloride and nitrate were analysed in the samples collected from all the five sites. WQI calculated as per the method of Kaur *et al.* (2001) giving appropriate unit weights and ratings. The water quality was monitored for five year period considering the tourist influx and cumulative rainfall. The WQI decreased from site I to V and the reduction was drastic from site III onwards. Though the rainfall and flow of water in the falls were more during the season (June to September), the WQI decreased on account of heavy tourist influx. Hence, proper surveillance is needed to maintain the quality of Courtallam water.

Keywords: Water quality index, Courtallam water, Tourist influx, Hydrochemical quality.

Introduction

Water is essential for survival but today about 200 million people in India do not have access to safe drinking water. Most of our water sources are polluted with untreated or partially treated wastes from industry, domestic sewage and fertilizer / pesticide runoff from agricultural fields. The quantity and quality of available water is decreasing and aggravating the already serious situation. When water quality deteriorates, water loses its economic value. Tourism and environment are intricately and inseparably interlinked with many positive relationships as well as several irreversible negative impacts. Tourists are attracted to destinations of high environmental quality. Courtallam is one of the most popular tourist spot in South India known for its water falls and ancient Siva temple (Kurtallanathar). Courtallam falls water is used as drinking water in most villages around. The total population of residents of Courtallam is around 3982 in a total area of 7.56 km^2. According to the report given by Tamil Nadu Tourism Development Corporation, the tourist inflow is approximately 50,000 per day during season (June to September). Nearly 2 lakh Iyyappa devotees visit Courtallam in the months of December and January. Tourism can cause large increase in the number of temporary residents of a tourist area and resulting surge of garbage wastes. Tourists often use far more packaged and disposable goods than average resident and the local disposal system frequently can not handle all the extra trash. Increasing number of tourists year after year with simultaneous reduction in rain fall accelerates the pollution load of Courtallam water thereby reduces the water quality considerably. Therefore, a detailed study was undertaken to assess the water quality index in five sites in relation to tourist influx for a period of five years.

Materials and Methods

Water samples were collected from main, five and old Courtallam falls where 90 per cent of the tourist take bath. Water samples collected before human activities were pooled together and designated as site I. Samples collected after anthropogenic activities were referred as site II. Samples collected at Melagaram, Yanaipalam (entry point of Tenkasi Town) and Tenkasi (exit point of Tenkasi Town) were indicated as site III, IV and V.

The samples were collected in triplicate and five such sampling was carried out in one year period. The study was continued for a period of five years to compare the water quality index year after year. The hydrochemical parameters namely pH, alkalinity, dissolved oxygen, biochemical oxygen demand, calcium, magnesium, chloride, sulphate and nitrate were determined in all samples as per the methods shown in APHA (1985). Water quality index was calculated as per the calculation of Kaur *et al.* (2001).

Results and Discussion

Water Quality Index

Water quality index (WQI) may be defined as the rating that reflects the composite influence of a number of water quality factors on overall quality of water. It is one of the most effective ways to communicate information on water quality trends to policy makers to shape sound public policy and implement the water quality improvement programmes efficiently. Water quality index first formulated by Horton (1965) and then by Ott (1978). The mathematical designs given by Horton (1965) and Ott (1978) were reexamined and modified by Kaur *et al.* (2001).

Weighting and Rating Scale

The word weighting implies relative significance of each of the factor in overall water quality and is dependent on the permissible levels in drinking water as suggested by ICMR. Factors having low permissible limits can harm the quality of water to a large extent even on a slight increase. Such factors have high weightings. On the other hand, factors which have higher permissible limit are less harmful and have low weightings. Each chemical factor has been assigned a rating value (Vr) to calculate WQI. The values fall between 0 to 100. The rating Vr = 0 implies that chemical factor exceeds standard maximum permissible limits and water is highly polluted while Vr = 100 implies that chemical factor has the most desirable value and water is absolutely clean.

To calculate water quality index, the products of unit weights (Wi) and rating (Vr) of nine chemical parameters were summated as per Kaur *et al.* (2001).

$$WQI = Wi + Vr$$

where,

$$Wi \times Vr = Wi(DO) \times Vr(DO) + Wi(BOD) \times Vr(BOD) + Wi(pH) \times Vr(PH) + Wi(T\,Alk) \times Vr(T\,Alk) + Wi(Ca) \times Vr(Ca) + Wi(Mg) \times Vr(Mg) + Wi(SO_4) \times Vr(SO_4) + Wi(Cl) + Vr(Cl) + Wi(NO_3) \times Vr(NO_3)$$

Table 13.1: Classification of Water Quality (According to Kaur *et al.*, 2001)

WQI	*Water Quality*	*Designation*
100	Absolutely clean	Class I
80–99.99	Slightly polluted	
60–79.99	Moderately polluted	Class II
40–59.99	Excessively polluted	Class III
0–39.99	Severely polluted	

Maximum water quality index was observed in site I which gradually decreased in site II, III, IV and V (Table 13.1). Considerable variation in the WQI between site I and II was noticed. Such variation also existed between other sites namely III, IV and V. Minimum WQI was recorded in site V which was not different from samples of sites III and IV. Therefore, there was significant difference ($P < 0.01$) in the water quality of samples collected at various sites. Though the quality of water decreased between sites I to V, no significant difference was recorded between the periods of study (Table 13.2 and Figure 13.1).

Table 13.2: Water Quality Index of Courtallam Water in Different Sites and Periods

Period of Study	*Sampling Sites*				
	Site I *n = 15*	*Site II* *n = 15*	*Site III* *n = 5*	*Site IV* *n = 5*	*Site V* *n = 5*
First year	68.58	60.69	51.93	52.65	52.23
Second year	70.69	61.54	59.41	52.65	46.29
Third year	80.59	67.47	65.68	65.68	52.98
Fourth year	68.15	67.89	58.79	58.35	58.35
Fifth year	85.21	67.47	59.63	46.29	46.51

Impact of tourists influx on water quality index during season and off-season are shown in Figure 13.2. The water quality index decreased in site I, II and III due to heavy tourist influx during season. Impact of tourists influx on water quality index during season and off-season for various sites

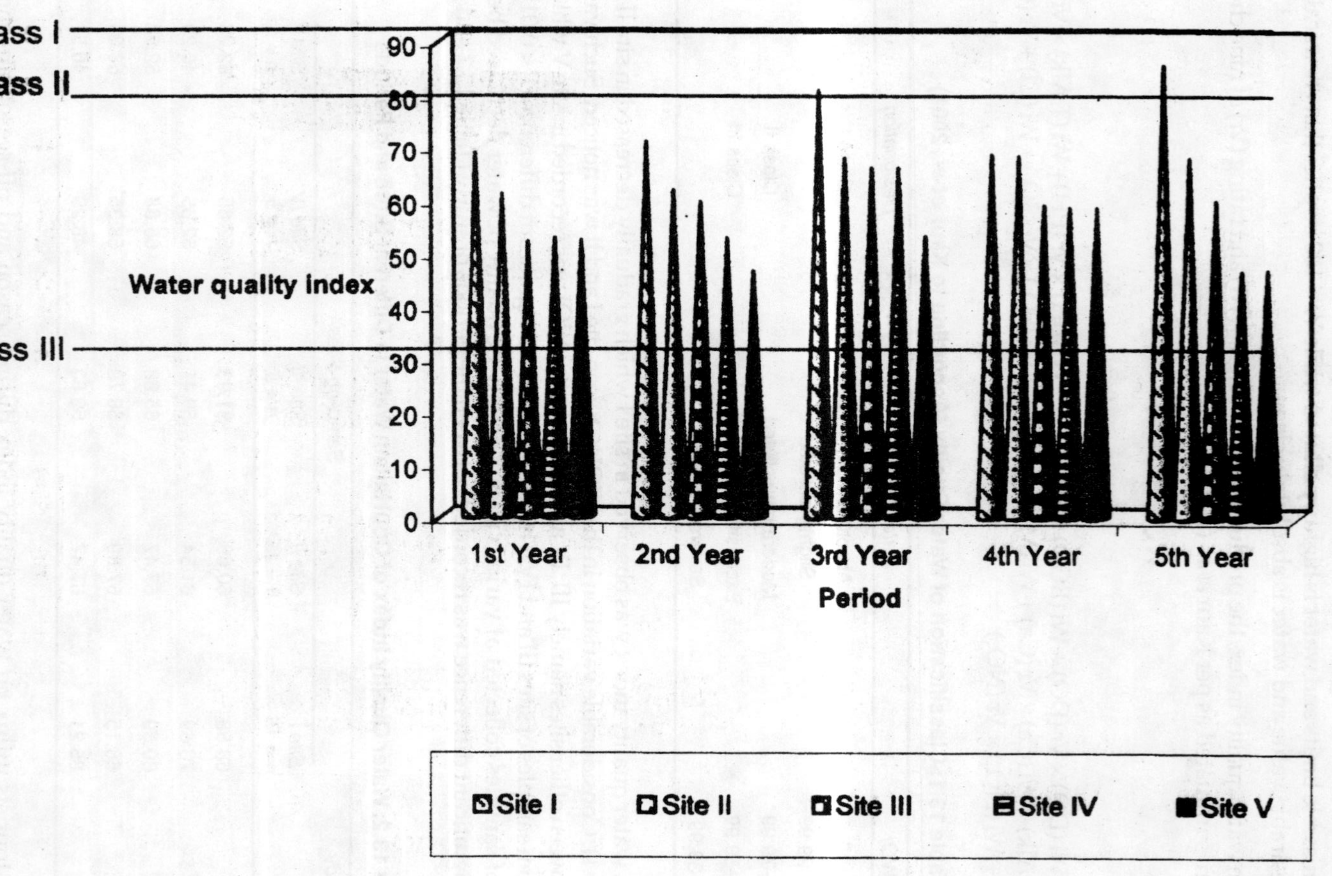

Figure 13.1: Courtallam Water Quality Index Calculated for Various Sites During the Study Period

are shown in Figure 13.2. Cumulative rainfall during various seasons for the period of study is given in Figure 13.3. Normally highest rainfall is recorded in the months of July, August, September and October of every year. January to April is particularly rainless but in May and June fair amount of rain may occur. More than half of the annual rainfall is brought by the south-west monsoon while the rest occurs in October and November. Comparing the rainfall of five year period, maximum was recorded in first two years which reduced to one half in subsequent three year period (Figure 13.3).

Total, floating and accommodated tourists arriving Courtallam and predicted numbers for future years are presented in Figure 13.4. Tremendous increase in tourist influx is observed during the season (June to September) of every year.

Impact of Tourists Influx on Water Quality Index

Travel and tourism creates jobs and wealth and has tremendous potential to contribute to economically, environmentally and socially sustainable development in both developed countries and emerging nations. If travel and tourism are managed badly, they can have a detrimental effect and can damage the small fragile environment. Because so many tourists are attracted to natural, unspoiled destinations, the environmental degradation caused by tourism can hurt the tourism industry. Tourism produces a variety of unwanted byproducts that are disposed, intentionally and unintentionally to the environment and modify the same, they are collectively called as the "externalities". Thus,

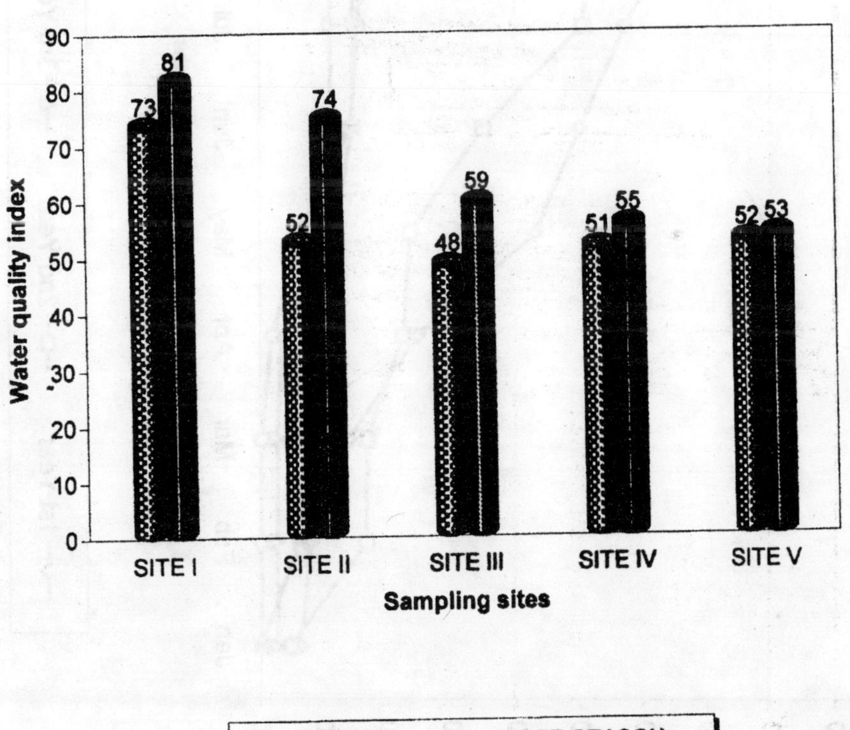

Figure 13.2: Impact of Tourist Influx on Courtallam Water Quality Index During Season and Off-Season

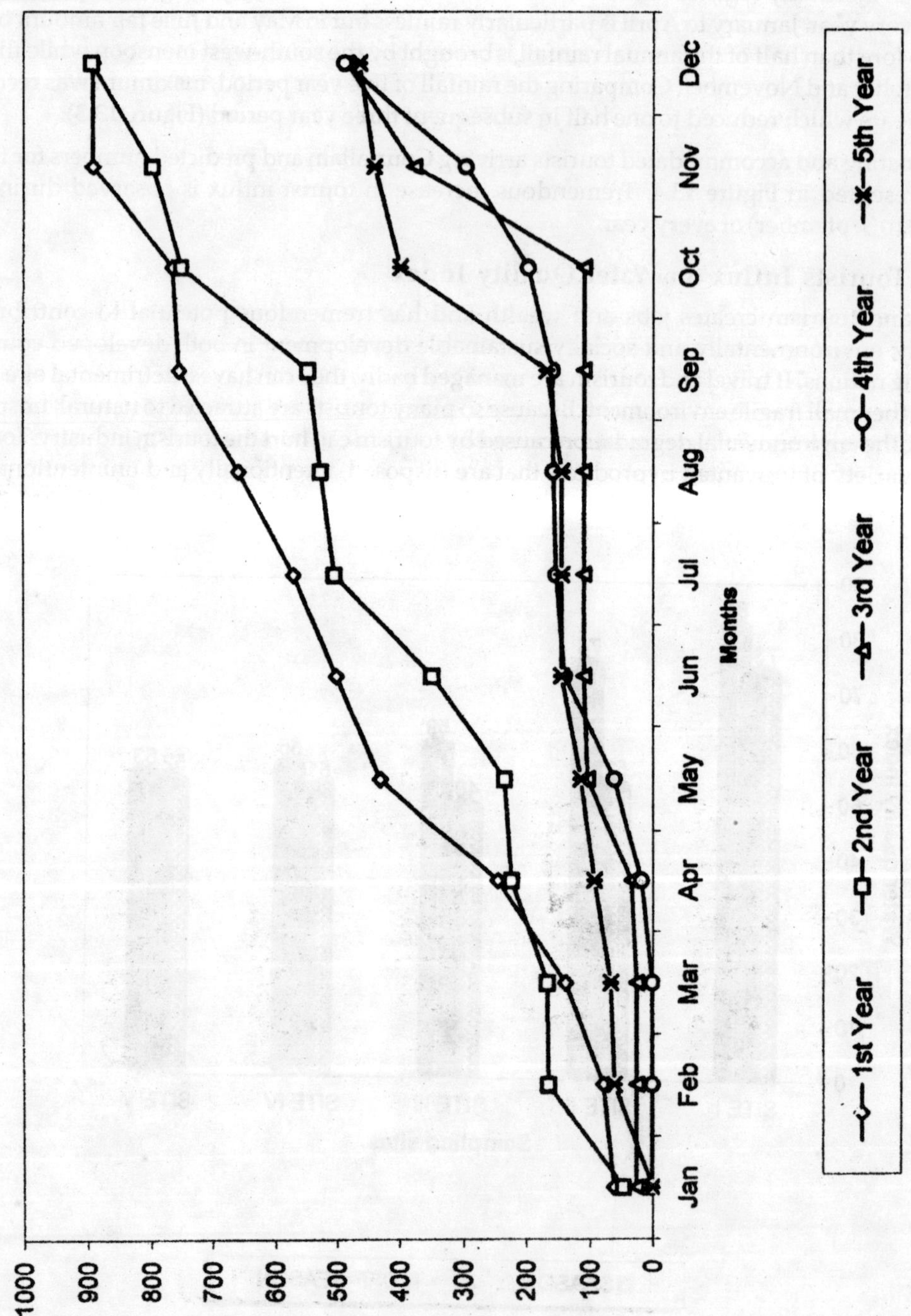

Figure 13.3: Cumulative Rainfall of Courtallam During the Study Period

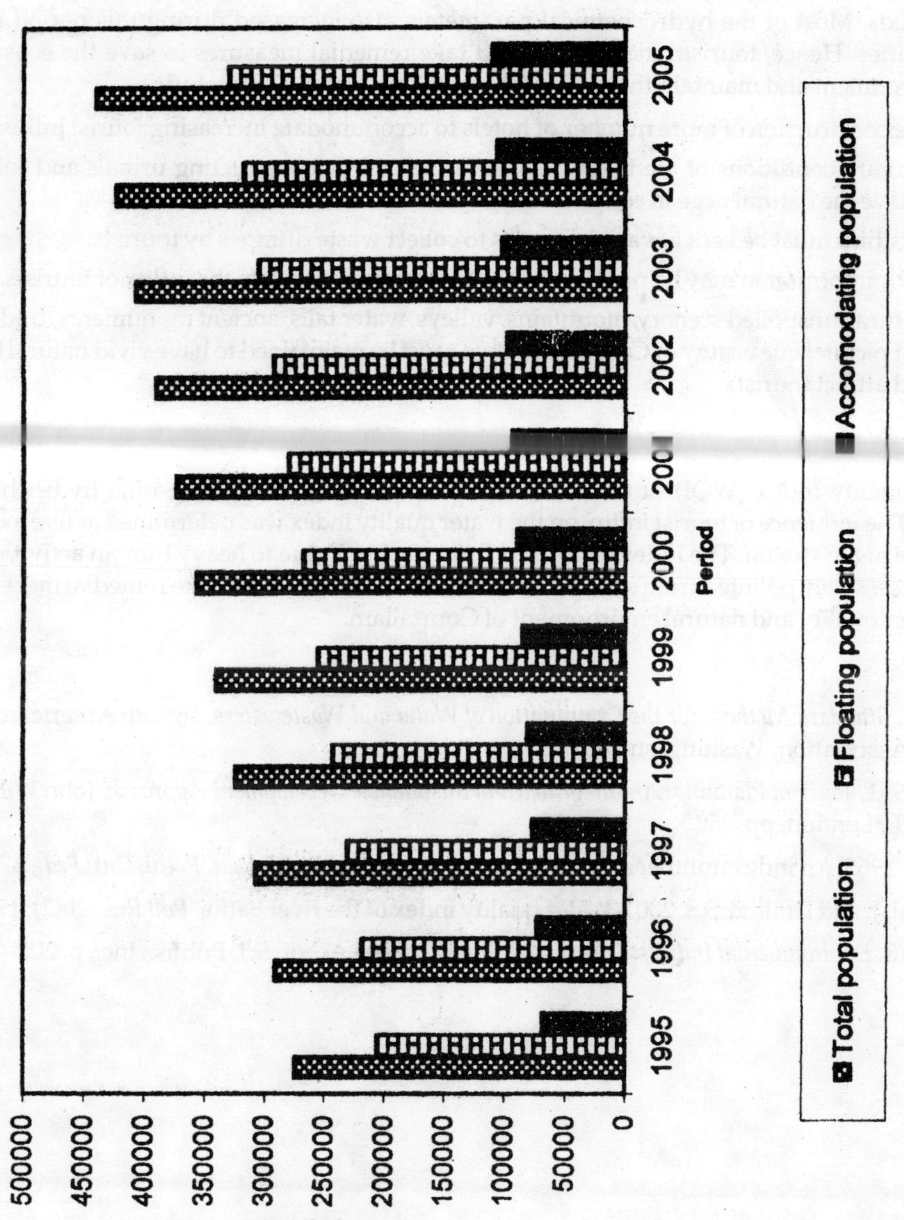

Figure 13.4: Recorded and Predicted Tourist Inflow to Courtallam

sustainable tourism depends in many important ways on the proper handling of the relationship between tourism and the environment (Inskeep, 1991).

Tourism industry should take appropriate measures to protect the environment and the health of local people. There is no facility available to receive the large number of tourists and pilgrims during various periods. Most of the hydrochemical parameters also increased during this period due to human activities. Hence, tourism industry should take remedial measures to save the ecosystem, natural environment and maintain the quality of courtallam water. This include

1. The construction of more number of hotels to accommodate increasing tourist influx.
2. Sanitary conditions of the town must be increased by constructing urinals and toilets to relieve the natural urge in confined environments.
3. Dustbins must be kept at various points to collect waste dumpes by tourists.
4. Prior information may be published in newspapers to regulate the influx of tourists.
5. Natural, unspoiled scenery, mountains, valleys, water falls, ancient monuments, traditional and picturesque beauty of Courtallam town must be maintained to have vivid natural beauty and attract tourists.

Conclusion

Water Quality Index (WQI) of Courtallam water was calculated using nine hydrochemical parameters. The influence of tourist influx on the water quality index was determined at five locations on the course of the stream. The index decreased from site I to V due to heavy human activities. The water was excessively polluted from site III onwards. The authorities must take remedial measures to save the water quality and natural environment of Courtallam.

References

APHA. 1985. *Standard Methods for the Examination of Water and Wastewaters*, 16th ed. American Public Health Association, Washington DC.

Inskeep, E. 1991. *Tourism Planning: An Integrated and Sustainable Development Approach*. John Wiley and Sons Ltd., London, pp. 252.

Horton, R.R. 1965. An index number system for rating water quality. *J. Wat. Pollu. Cnt. Fed.*, 37: 300.

Kaur, H., Syal, J. and Dhillon, S.S. 2001. Water quality index of the river Satluj. *Poll Res.*, 20(2): 199–204.

Ott, W.R. 1978. *Environmental Indices: Theory and Practice*. Ann. Arbor. Sci. Publis., Inc., p. 202–213.

Chapter 14

Water Quality Index for Ground Water Affected with Bicycle Manufacturing Industrial Wastes: An Environmental Quality Audit

Vineeta Shukla, Sharda Abusaria**, Monika Dhankhar* and K.V. Sastry**

**Department of Biosciences, Maharshi Dayanand University, Rohtak – 124 001, Haryana*

***Lecturer, Department of Zoology, A.I.J.H.M. College, Rohtak*

ABSTRACT

The present study is intended to calculate water quality index for ground water at Sonepat in Haryana State of India from various, parameters, *viz.*, pH, dissolved oxygen, total hardness, chlorides, total alkalinity, biological oxygen demand, sulphates, iron, chromium, cyanide and MPN coliforms to ascertain the quality of water for human consumption and to assess the impact of bicycle manufacturing industrial wastes on ground water quality. The high values of water quality indices for ground water samples from various sources located in the industrial premises and vicinity of the industry indicate the influence of industrial wastes on ground water quality and make it unsuitable for drinking purposes. The trend of variation in water quality index values collected around and inside the industry was irregular while for control samples, water quality index values were well below the permissible limit throughout the study period and did not vary widely. The importance of water quality index and pollution control has also been discussed and emphasized.

Keywords: Ground water, Bicycle manufacturing industrial wastes, Water quality index, Pollution control.

Introduction

Environment is our biological capital. It makes up a huge, enormously complex living machine on which our total productivity depends. Water is one of the basic natural resources needed for survival of life. Nobel laureate A. Szent-Gyorgy, physician and biochemist, called water the "matrix of life". Many organisms can live without air, but none without water. The use of this under ground fluid for agriculture and drinking purposes has a long history. But ground water is safe only as long as it is free from disease causing organisms and chemicals that are deleterious to human health. Unfortunately, in recent years industrial growth has created severe environmental pollution, and degradation has not spared even ground water aquifer system. A number of places in the country face serious ground water contamination problem due to indiscriminate dumping of toxic chemicals by polluting industries. Water soluble chemicals are not retained by soil matrix. they percolate through porous medium and enter the ground water. Environmental protection and mankind's survival are closely linked. Ground water once polluted cannot be treated, therefore, the effect is long lasting and problem is of vital concern. Hence, there is need for protection and management of ground water quality. Accurate and timely information on water quality is essential to frame a sound public policy and to implement the water quality improvement programs.

Water quality index is the most efficient tool of environmental audit process to convey the overall water quality of a water resource. It is a rating of water quality parameters by a single numerical expression reflecting the complete' influence of water quality parameters. Unfortunately in India water quality index determination has been much neglected. Few attempts have been made to calculate water quality for river waters. Studies on ground water quality index are scanty.

Sonepat in Haryana State is known for bicycle manufacturing industry whose product is available globally. But the industrial effluents and sludge generated as byproduct may cause great environmental stress locally. Spent waters of different operations such as electroplating and heat treatment are normally stored inside the industrial premises before discharge, For the past more than forty five years, surrounding land has been receiving untreated/partially treated effluents of the industry. Sludge is being dumped within the premises and in low lying areas located near the industry. The dumping and holding of such chemical wastes are likely to affect the ground water quality in these areas. So far, no detailed investigation on the effect of these industrial wastes on ground water quality has been carried out. The present study has, therefore, been undertaken to calculate the water quality index for ground water in and around the industry to measure the extent of damage to water quality.

Material and Methods

In the industry and surrounding regions ground water is the only source of drinking water and is being tapped by tube wells. Four sampling sites were selected for the present investigation. Details of these sites are given in Table 14.1. In the formulation of water quality index, the importance of various parameters depends upon the intended use of water. In this study eleven water quality parameters of ground water are studied from the point of view of suitability of water from different sources at Sonepat for human consumption. For this purpose monthly samples of ground water from selected sites were collected and analysed as per Standard Methods during March, 1999 to February, 2000.

Based on the ground water analysis data, water quality index was calculated for each sample by a simple method[4]. The 'standards' (permissible values) of various parameters for drinking water, recommended by the ISI are used for comparison. For dissolved oxygen and BOD, European Economic Community (EEC) Standard and United Public Health Service (UPHS) have been considered respectively[4].

Table 14.1: Particulars of Ground Water Sampling Sites

Reference Number	*Name and Location*	*Description*	*Age (Yrs)*	*Depth (M)*
GW1	Tubewell No. 3 in bicycle industry premises in contaminated area	20M from effluents bodies	40 120	115 up to May, 1999. June onwards 120 Closed in August
GW2	Tubewell No. 7 in bicycle industry premises in contaminated area	Earlier 30M, now 1000M from sludge dumping area and effluents bodies	8	105
GW3	Sujan Singh Tubewell in Agricultural field contaminated with industrial effluents	Near cyanide sludge dumping area of industry and about 1000M from effluents flow channel	8	75
GW4	Public Health Tubewell in unpolluted area	Beyond the influence of industrial wastes and it served as control ground water	5	83

Calculation of Subindex (SI)

To calculate the Water Quality Index, first of all subindex (SI) i for the ith parameter was calculated which was the product of the quality rating (qi) and the unit weight (wi) of the ith parameter *i.e.*

$$(S)\, i = qi\, wi \qquad ....(1)$$

Calculation of Quality Rating (qi)

The quality rating, qi for the ith water quality parameter was obtained from the equation:

$$qi = 100\,(Vi/Si) \qquad ....(2)$$

where,

Vi: Value of the ith parameter at a given sampling station; and

Si: Recommended standard for the ith parameter.

Equation (2) ensures that qi = 0 when pollutant (the ith parameter) is absent in the water, while qi = 100 if the value of this parameter is just equal to its permissible value for drinking water.

Quality ratings for pH, dissolved oxygen and MPN coliforms required special handling.

The permissible range of pH for drinking water is 6.5 to 8.5. Therefore, the quality rating qpH was calculated from the relation:

$$qpH = 100[(V_{pH} \sim 6.5)/(8.5–6.5)] \qquad ....(3)$$

where,

V_{pH}: Value of pH; and

Symbol "~" means simply the numerical differences between V_{pH} and 6.5 ignoring the algebraic sign.

The situation is slightly complicated in the case of dissolved oxygen, since, in contrast to other pollutants, the quality of water is enhanced if it contains more dissolved oxygen. Therefore quality rating qDO was calculated from the relation.

$$qDO = 100\left[\frac{14.6}{14.6-5.0} - \frac{V_{DO}}{14.6-5.0}\right] \qquad \text{.............(4)}$$

where,

V_{DO}: Value of dissolved oxygen.

14.6 is the solubility of oxygen (mg/l) in distilled water at 0°C; and 14.0 is the standard (mg/l) for drinking water.

In case of MPN coliforms, the permissible value for the drinking water is 1/100 ml, while their actual number in untreated water may be several thousands/100 ml. Therefore to get a convenient number for quality rating qMPN was calculated from the relation

$$qMPN = 100\left[\frac{Log_{10} V_{MPN}^{\;+1}}{Log_{10} S_{MPN}^{\;+1}}\right] \qquad \text{.............(5)}$$

where,

V_{MPN}: MPN count/100 ml; and

S_{MPN}: MPN standard for drinking water.

Calculation of Unit Weight (wi)

$$wi = k/Si \qquad \text{.............(6)}$$

where,

wi: Unit weight;

Si: Standard for the ith parameter (i = 1, 2.............11); and

k: Constant of proportionality which was determined from the condition

$$\sum_{i=1}^{i=11} wi = k\sum_{i=1}^{i=11}\left(\frac{1}{si}\right) = 1 \qquad \text{.............(7)}$$

Final Value of Water Quality Index

Finally the water quality index was obtained by aggregating these eleven subindices (SI) i linearly

$$W.Q.I. = \sum_{i=1}^{i=11} (qiwi) \qquad (8)$$

Results and Discussion

Unit weights assigned to water quality parameters and a sample calculation of overall water quality index from the subindices for ground water samples from different sources have been shown in Table 14.2. Water quality indices for one year study period are depicted in Table 14.3. On extensive comparison with control sample to document the impact of industrial pollution on ground, water quality present study reveals an alarming feature that in 100 per cent samples from affected area, quality of ground water is much altered with respected to natural and local background indicating pollution status of ground water and is not fit for human consumption at any time period of the year due to non–conforming water quality index with drinking water limit (100). A critical study of Table 14.2 indicates that for GW1, GW2 and GW3 samples the main dominating contribution to water quality index comes from the existence of un-natural toxic chemicals–chromium and cyanide–which are used in bicycle manufacturing industry. It is further evidenced by the fact that the more harmful a given pollutant is, the smaller is its permissible value for drinking water. So the, unit weights for various water quality parameters are assumed to be inversely proportional to the recommended standards for the corresponding parameters. Maximum value of the unit weight for water quality parameters in the present study was in the order of chromium and cyanide > iron > MPN coliforms > BOD and dissolved oxygen> pH > total alkalinity and sulphates > chlorides> total hardness, consequently due to high values of chromium and cyanide subindices WQI increased drastically. It may thus be, inferred that higher water quality indices for ground water samples from different sources–in and around the bicycle manufacturing industry–may be due to leaching of chemicals from industrial wastes through soil into ground water. Further, WQI for GW1, GW2 and GW3 samples varied depending upon the location of sampling site and variation was sizeable. Very high value of WQI for GW1 indicates the influence of proximity of effluents tanks. There was a wide difference in the values of WQI between GW1 and other two samples coinciding with levels of pollutants. Amongst above three samples, GW1 water exhibited a maximum value of WQI in the month of April which is about 39 times greater than the limit for drinking water. However, GW2 and GW3 samples also exhibited deteriorated status of water quality. On the other hand chromium and cyanide were not the constituents of WQI for GW4 sample (control) coinciding with their nil quality rating values. WQI for GW4 was quite less than other three samples and remained below the limit for drinking water. The variability in WQI among different ground water samples was measured statistically by calculating the coefficient of variation. It confirmed well marked variation in WQI for ground water samples from different sources (Table 14.3).

Table 14.2: Water Quality Index of Ground Water Samples (March)

Water Quality Parameters	*Drinking Water Standard (Si)*	*Unit Weight (wt)*	*Parameter Sub-index (qiwi)*			
			GW1	*GW2*	*GW3*	*GW4*
pH	6.5–8.5	0.00262	0.1179	0.1048	0.1179	0.1048
Dissolved oxygen	5.0 mg/l	0.00446	0.3836	0.3791	0.4371	0.4371
Total hardness	300.0 mg/l	0.00007	0.0104	0.0238	0.0149	0.0141
Chlorides	250.0 mg/l	0.00008	0.0028	0.0214	0.0071	0.0053
Total alkalinity	200.0 mg/l	0.00011	0.0157	0.0092	0.0109	0.0144
BOD	5.0 mg/l	0.00446	1.6484	0.8028	0.0347	0.2382
Sulphates	200.0 mg/l	0.00011	0.0037	0.0049	0.0037	0.0028
Iron	0.3 mg/l	0.7430	0.2194	3.4178	3.2194	2.4764
Total chromium	0.05 mg/l	0.44580	3584.2320	53.4960	53.4960	0.000
Cyanide	0.05 mg/l	0.44580	71.3280	53.4960	53.4960	0.0000
MPN coliforms	1/100 mg/l	0.02229	0.0000	0.0000	0.0000	0.0000
Water quality index			3660.9619	111.7528	111.8377	3.2931

Table 14.3: Monthly Variation in Water Quality Index of Ground Water Samples

Months	Water Quality Index (WQI)			
	GW1	GW2	GW3	GW4
Mar	3360.96	111.75	111.84	3.29
Apr	3911.12	111.02	120.73	3.43
May	3633.90	114.32	142.60	3.07
Jun	36354.88	110.46	123.77	4.28
Jul	3654.14*	110.69	148.53	9.48
Aug		119.45	113.44	7.43
Sep		132.22	131.19	3.18
Oct		141.64	139.60	3.19
Nov		129.55	122.18	3.74
Dec		119.80	130.35	3.42
Jan		129.87	121.90	3.73
Feb		138.14	130.91	3.42
Average	3639.20	122.41	128.09	4.30

* GWI (Tube well No.3) was closed by the industry from August. 1999 CV (%) = 182.64.

Another striking feature emerging from the present investigation is that, different sources do not synchronize in monthly WQI variation. This may be attributed to the fact that an index significantly increases or decreases when some of the critical parameters exceed or become less than the specified permissible value. Over a period of about half a century enormous quantity of pollutants has been added to the soil which may act as a bank of pollutants. Thus, chemical characteristics of percolating effluents, and nature and status of soil may be the possible reasons for irregular variations in WQI for affected sample? However, WQI for control samples indicates uniformly good water quality, reflecting pollution free status of ground water of area not yet contaminated, without any irregular variation. But if the bicycle industry is allowed to dump its wastes without proper treatment the possibility of spread of pollution of ground water in this area also cannot be ruled out.

It is clearly brought out by the findings of this study that the ground water environment of bicycle manufacturing industrial area has already become grossly polluted. Operational phase of the industry established beyond doubt the potential threat to ground water. It is unfortunate that groundwater, a gift of nature, turns out to be a man-made curse. Good quality ground water, as of now is a rare and scarce commodity in the area. Therefore, there is an urgent need for strict enforcement of Environment (Protection) Act, 1986. Violation of these rules can be disastrous and can have far reaching consequences. Present study also clearly confirms the utility of WQI as a proper indicator for conveying the idea of the overall quality of water with a single number. This type of information is helpful for public to understand the quality of drinking water and may have an administrative use in the field of water quality management.

References

APHA. 1985. *Standard Methods for the Examination of Water and Wastewater*. American Public Health Association, Washington, D.C.

Horton, R.K. 1965. An index number system for rating water quality. *J. Water Poll. Cant. Fed.*, 37(3): 300–306.

ISI. 1991. *Indian Standard Specification for Drinking Water* (IS: 10500), Indian Standards Institution, New Delhi.

Jeevan Rao, K. and Shantaram, M.V. 1995. Ground water pollution from refuse dumps at Hyderabad. *Indian J. Environ. Hlth.*, 37(3): 197–204.

Mariappan, P., Vasudevan, T. and Yegnaraman, V. 1999. Surveillance of ground water quality in Thiruppathur Block of Sivagangai district. *Ind. J. Environ. Prot.*, 19(4): 250–254.

Singh, A.P. and Ghosh, S.K. 1999. Water quality index for river Yamuna. *Poll. Res.*, 18(4): 435–439.

Singh, K.P. and Parwana, B.K. 1999. Ground water pollution due to industrial wastewater in Punjab State and strategies for its control. *Ind. J. Environ. Prot.*, 19(4): 241–244.

Tiwari, T.N., Das, S.C. and Bose, P.K. 1986. Water quality index for the river Jhelum in Kashmir and its seasonal variation. *Poll. Res.*, 5(1): 1–5.

Venkata Mohan, S., Rama Mohan, S. and Jayaraman Reddy, S. 1998. Determination of water quality index (GWQI) of ground water: An application. *J. Environ. Poll.*, 5(3): 219–223.

Chapter 15

Zooplankton Diversity in the Chennai Coast, Tamil Nadu

V. Indra and R. Ramanibai*

Department of Zoology, University of Madras, Guindy Campus, Chennai – 25

ABSTRACT

Zooplankton community constitute an important component in the faunal composition of the water body. Distribution of major groups of Zooplankton population were conditioned by various behavioural and physiological adaptations of the plankton population to everchanging hydrographical condition. Zooplankton density was maximum during monsoon and minimum during premonsoon season. A variety of Zooplankton taxa numbering 28 were identified from 4 major groups, the cladocera, copepoda, ostracoda and rotifera. Among the 4 groups copepoda was one of the major dominating group of the micro zooplankton fauna is represented by 16 species under 3 sub groups. After the group copepoda, the rotifera was represented by 8 species with most common species Brachionus to rare species like Trichocera. The most striking feature of the group cladocera was represented by one single species of Daphnia. The ostracods were represented by 3 species. Dominant, common, rare taxa were also isolated under each group and campared between the different zones of the coast. Relative occurrence showed distinct variations with copepoda and rotifera groups.

Introduction

Diversity refers to the range of variation or differences among some set of entities; zooplankton diversity thus refers to variety within their community. The members of the zooplankton community are important for their role in the tropic dynamics and in energy transfer in the aquatic ecosystem.

* Present Address: Thiruvalluvar University, Fort Campus, Vellore – 4.

They provide food for fishes in fresh and marine and playa major role in the fish production in the Bay of Bengal. The present study an intensive survey of zooplankton diversity from 4 different habitat of Chennai coast have been made and the collected data have been analysed to ascertain the distribution pattern in the marine ecosystem. In spite of their great importance, very little work has been done on this group from freshwater wetlands of West Bengal (Venkataraman *et al.*, 2000) Kakinada Bay (Chandramohan *et al.*, 1999).

Distribution of Zooplankton and their variation at different zones of a water body is known to be influenced by physico-chemical parameters of water. Zooplankton study provides a relevant and convenient point of focus for research on the mechanism of eutrophication and its adverse impact on an aquatic ecosystem. Zooplankton constitutes a vital link in food chain and its productivity depends on water quality at a given time. The present study reveals some aspects of Zooplanktonic dynamics and explains with physico-chemical characteristics of water.

The present investigation was conducted in the near shore coastal zone of the Chennai coast which extends between Ennore backwater on the northern side and the Kovalam backwater on the southern side. There are 8 major inlet through which the inland water bodies communicate with the Bay of Bengal. The study area was classified in to 4 zones *viz.*, (*i*) Riverine Zone (*ii*) Estrarine Zone (*iii*) Backwater Zone and (*iv*) Near Shore Zone. The nearshore area was also divided in to southern and central and northern zone. The plankton samples were collected by passing 100 litres of surface water through a 63µm plankton net. The samples were preserved in 5 per cent formalin at the sampling site itself. Enumeration of the organisms were performed in a two step protocol. Slow moving, generally abundent organisms, such as rotifers, nauplius were counted first. Counts were carried out using sub samples (or) total samples. The sub samples were then divided in to 10–12 drops on a multidepression microscope slide. During the second step, fast swimmers like copepods and cladocerans were enumerated. When rotifer counting was completed, the sample was first fixed with 5 per cent formalin. Counts of Zooplankton forms were affected on fine successive sub samples of the entire sample. Planktons were examined under a light microscope at a magnification of 10 × 10. Systemic identification was done at the level of the highest taxa using standard keys (Edmoson, 1959; Ranga Reddy and Radhakrishna, 1984; Sharma, 1988; APHA, 1995).

The most abundant Zooplankton taxa collected from the surface samples of all the 21 sampling stations were presented in the Table 15.1. Totally 28 Zooplankton taxa were identified from 4 major groups. The cladocera, Copepoda, Ostracoda and Rotifera. Among the four groups enumerated in the study area, the bulk of the biomass was accounted distinctly by few species and their occurrence is given in the (Table 15.1). Dominant, common and rare taxa were isolated under each group and compared between the different zones of the coast.

Plankton exhibit a pronounced seasonality between zones and the seasonal peaks of Zooplankton biomass varied depending on the sampling locations. Classical investigations of Zooplankton have concentrated on describing the dominant species and their seasonal abundance. Highest biomass at the nearshore coastal stations (S.C.N) was observed during monsoon seasons, while lowest biomass during summer and premonsoon seasons. Copepods were much more abundant than the other three almost all occasions. Out of a total 15 copepod species were present in monsoon season and 12 species were present in the summer season. Comparatively lower number of species was noticed in the riverine zone during monsoon season. This dominance was decreased up to single genus namely *Eucalanus* in the river adyar and *Bryocamptus* in the river cooum.

Table 15.1: The List of Zooplankters Identified in the Chennai Coast

Group	*Sl.No.*	*Name of the Organism*
Cladocera	1.	Daphnia
		Calanoida
	2.	Diaptomus sp.
Copepoda	3.	Eucalanus
	4.	Epischura
	5.	Eurytemora
	6.	Nauplii
	7.	Oithona
		Cyclopoida
	8.	Cyclops
	9.	Eucyclops
	10.	Halicyclops
	11.	Heterocyclops
	12.	Mesocyclops
	13.	Paracyclops
		Harpacticoida
	14.	Bryocamptus
	15.	Mesochra
	16.	Paradactylopodia
	17.	Stenocaris
Ostracoda	18.	Cypris sp.
	19.	Candona sp.
	20.	Limnocythere sp.
Rotifera	21.	Brachionus
	22.	Filinia
	23.	Keratella
	24.	Lecane
	25.	Monostyla
	26.	Notholca
	27.	Platyias
	28.	Trichocera

Ointhona was not present during pre and postmonsoon seasons in the coastal zones. In the estuarine (16 Nos) and in the backwater zones (12 Nos) the number of copeped taxa enumerated was quiet similar to that of near shorecoastal zone. An overall reduction in the number of copepod taxa was found during pre and postmonsoon seasons. An increased number of organisms were noticed during monsoon and summer season.

Rotifera

The second most abundant taxa was the rotifera and the species enumerated frequently on all the six zones were the *Brachionus, Keratella* and *Filinia*. More or less in all the five zones except in the riverine zone recorded up to seven species. In the river coovum only the *Brachionus* sp. was present throughout the four season. In the Adyar river 3 Taxa were present (*Brachionus, Filinia* and *Keratella*). *Filinia* was absent during summer and monsoon season and was replaced by *Karatella*.

Ostracoda

Much reduced abundance of ostracoda taxa was recorded and composed of cypris. *Candona* and *limnocythere*. All the taxa were present during monsoon in all the five zones except in the riverine zone. In the river Adyar Candona and Cypris were present in monsoon season and this density and number was reduced to the single species (Cypris) during post monsoon and summer season, and nil species was recorded during premonsoon season. In the river Coovum the ostracod population was completely altered in all the four seasons.

Cladocera

This showed high density during monsoon (Daphnia SP) season, and was absent during pre, postmonsoon and summer in the coastal zone. In the river coovum daphnia was present in the monsoon, pre and postmonsoon season. Except during summer it was also present in the estuarine zones. In the backwater zone it was absent during premonsoon season.

On the whole the planktonic abundance was more during monsoon season at all the six zones and reduction was noticed during preland postmonsoon seasons. A small increase in the planktonic abundance was noticed during summer season. This fluctuations in the abundance of Zooplankton was clearly depicted by two groups namely the copepods and rotifers.

Zooplankton

Zooplankton community constitute an important component in the, faunal composition of the water body. Plankton data vary on a wide range of spatial and temporal scales and describing this variability is an important problem in plankton ecology. Distribution of major groups of zooplankton population were conditioned by various behavioural and physiological adaptations of the plankton populations to ever changing hydrographical conditions. Tropical waters are, in general, poorly studied, but the planktonic communities surrounding India have been focus of extensive study over the past decade.

Copepods have generally been considered the most important metazoan, secondary producers in pelagic marine ecosystems, both in terms of abundance and biomass. It has been stated that copepods are the most plentiful multicellular animals on earth (Wiebe *et al.*, 1992) out numbering the insects which have more species but fewer individuals. Copepods are important in fisheries, maintaining regenerated primary productions (Madhupratao, 1999). The plankton populations were dominated by the copepods in Chennai coast which represented an average of 60 to 65 per cent of total zooplankton. The biomass value and total counts varied significantly between seasons. The present study is in conformity with the works carried out in the coast and estuarine systems of temperate regions. Copepoda dominated the zooplankton population in the Mandovi-Zuari estuarine system in Goa, west coast of India constituting 66.2 per cent of the total counts (Vadmavathi *et al.*, 1996). In the Kakinada Bay, east coast of India copepods were represented by 75–80 per cent of total zooplankton (Chandramohan *et al.*, 1999). Higher levels of abundance were relatively more during March to July than those observed from August to February in Kakinada Bay, east coast of India. Here, copepoda exhibited a decrease in density in low saline waters and many species totally disappeared during flood period. This agrees with the present study where low percentage of copepods present in the riverine zone. In Bahuda river estuary, east coast of India registered a maximum of 40 species of copepods and their peak was found during premonsoon season. The premonsoon and monsoon seasons were characterized by maximum and minimum diversity indices respectively (Sujata Mishra and Panigrahy, 1996).

In contrast to the seasonality, where an overall composition of copepods were more or less same throughout the year in the Arabian sea and small filter feeding forms and carnivores dominate (Madhupratao, 1999). Copepod population contributed 72.7 per cent and 79.6 per cent to the total zooplankton count respectively for Versova and Mahim transects of the nearshore waters of Bombay (Gajbhiye *et al.*, 1991). Copepods were abundant and 80 species were identified from coastal waters of Goa, west coast of India (Goswami, 1985) where a minimum copepod diversity was obtained during the month of October. The plankton data presented here, clearly demonstrated that, the abundance of zooplankton peaked during monsoon and decreased during pre and postmonsoon seasons. Furthermore, the abundance of dominant taxa, the copepods also decreased during pre and postmonsoon seasons. In the Nethravati–Gurupur estuary, the copepods and copepodites together formed the bulk of the plankton population. Except for great reduction during peak monsoon months, they were available in sizeable numbers throughout the year in this estuary (Bhat and Gupta, 1983).

The most diverse group within the zooplankton in the lower swan estuary in South West Australia was the copepoda, which comprised of 21-species, the majority of which were holoplanktonic forms. Copepod nauplii were by far the most abundant taxon, contributing 32.9 per cent to the total number of zooplankton in the Wilson Islet of western Australia. River zooplankton are often characteristically dominated by small forms such as rotifers, bosminids and juvenile copepods (Shiel *et al.*, 1982; Thorpe *et al.*, 1994). The biomass of river zooplankton can be substantially higher in the tidal freshwater portion than that in the non-tidal portion. On contrary to this, where as in the case of the river Coovum and Adyar the riverine zone of the study area did not register significant zooplankton biomass of either copepoda or the rotifer fauna. The probable reasons could be the poor water quality and the effect of discharge of wastewaters of the city through these rivers.

Rotifera enjoy a cosmopolitan distribution which is not limited to continents. Rotifers are important components of planktonic community seasonality. Furthermore, rotifers influence various interactions within the microbial food web which occur at several trophic levels (Arndt, 1993). Detailed studies dealing only with rotifers in estuarine environments are rare (Edborge and Tawari, 1987; Edborge, 1994; Green, 1995). The Indian freshwater rotifer fauna is well studied with reports from different parts of the Indian sub continent starting with are available.

According to there are only approximately 50 strictly marine forms of rotifers available most of which are littoral. This may be one of the reason for the sparse literature available on the occurrence of Indian marine rotifer fauna. Reports on rotifer fauna of east and west coast of India is negligible. The rare available report is from a coastal mangrove, east coast. Studies from Pichavaram mangrove, south east coast of India also confirms that *Brachionus* is dominant with more species and abundance (Govindasamy and Kannan, 1991). The earlier view that rotifers are entirely cosmopolitan based on accumulated information on their geographic distribution pattern is likely (Dumant, 1983).

The cladocerans are filter feeders and their association to phytoplankton especially diatoms is fairly well known. Ostracods inhabit a wide variety of environment mostly free-living forms (Edmoson, 1959). When compared with the other two groups, cladocerans and ostracods contribute much low percentage composition during the present study. The cladoceran and the ostracod standing stocks represent 15.19 per cent and 0.21 per cent in the coastal waters of Goa, west coast of India (Goswami, 1985). The faunal composition of Mandovi and Zuari estuarine system of Goa, west coast of India also revealed low percentage composition (1.8 per cent) of cladocera (Padmavathi and Goswami, 1996). The planktonic biomass and counts were also, poor in this estuary during monsoon period (June–September). The two groups were seen occasionally and in few numbers in the plankton hauls off Kakinada Bay (Chandramohan, *et al.*, 1999). Large cladocerans such as Daphina have been recorded in many rivers including the river Murray in Australia (Shiel *et al.*, 1982). Daphnia, has reached a maximum density of 0.7 animals l^{-1} in the Nile river, 0.8 animals l^{-1} the Ohio river (Thope *et al.*, 1994) and 2.98 animals l^{-1} is the lower river Rhine (Van Dijk, 1995). But there was no record of Daphina in Hawkesbury Nepean during 1996 (Kobayashi *et al.*, 1998).

The zooplankton of temperate Bay and estuaries were typically higher in abundance and lower in diversity than zooplankton of adjacent neritic waters. The distribution of various plankton groups was also influenced by rainfall to some extent during monsoon period. This study indicates that the zooplankton in the coastal waters of Chennai is dominated by the copepods in both number of species as well as abundance followed by rotifers. The maximum number of copepods are found in the nearshore waters, than in other zones of the coast and this has been the general observation from Indian coast (Madhupratap, 1999; Goswami, 1985; Bhat and Gupta, 1983; Chandramohan *et al.*, 1999). However, the number of species of copepods recorded in this study are only 16 in number of which half of them constitute much less in abundance.

The decline in the major planktonic communities has been predicted by (Sivasamy, 1990) in his report on plankton in relation to costal pollution at Ennore, Madras coast. Though his studies using two way correlation between the pollutants and planktonic fauna yielded statistically no significant relations, he has predicted more deterioration of water quality especially in northern coastal waters, considering the impact of urbanisation and industrial development. The present study has confirmed the trends of decline.

Gajbhiye *et al.*, 1991 have compared the data on copepod population in the creek system collected during 1990 with samples collected during 1979–1980. Though a consistent pattern could not be observed at different stations, they could record the influence of deteriorating water quality due to industrial and urban wash off from Greater Bombay on the incidence of copepod population. It is even suggested that the copepod *Acartia spinicanda*, a predominant species in the heavily sewage polluted areas of the coast as an indicator organism whose presence and abundance can indicate pollution load in the coastal waters.

The cladocerans prefer low saline conditions, unlike copepods which are truly marine, feed generally on diatoms and therefore associated with fluctuations of diatom flora. The diatom population in the Chennai coastal waters are very much reduced in diversity and abundance as reported by since diatoms are most sensitive group to pollution load. In the present study, cladocerans are impoverished and is represented by only one species (Rahukumar and Anil, 2003) established facts that nutrient availability largely determines the diversity of p. imary producers.

It appears that the Chennai coastal area has more of physically controlled than biologically accommodated communities and are low in diversity of zooplankton fauna unlike tropical/subtropical marine, environment. In the present study the sampling in the nearshore waters have been done in 50 m zone that are likely to be influenced more by monsoon inflows. The high variability of water quality, levels and zooplankton fluctuation may have been due these causes also.

References

APHA. 1995. American Public Health Association, Washington D.C.

Arndt. 1993. *Hydrobiologia*, 225/256: 231–246.

Bhat, B.V. and Gupta, T.R.C. 1983. *Ind. J. Mari. Sci.*, 12: 36–42.

Chandramohan, P., Raman, A.V. and Srinivas, N. 1999. *Ind. J. Mari. Sci.*, 28(2): 192–197.

Dumant, H.J. 1983. *Hydrobiol.*, 104: 19–30.

Edborge, A.B.M. 1994. *Hydrobiologia*, 272: 95–104.

Edborge, A.B.N. and Tawari, P. 1987. *J. Plank. Reas.*, 9: 1–13.

Edmoson, W.T. 1959. John Wiley and Sons, Inc. New York, 2nd Edn., p. 657–663.

Gajbhiye, S.N. Rosamma Stephan, Vijayalakshmi, R., Nair and Desai, B.N. 1991. *Ind. J. Mari. Sci.*, 20: 187–194.

Goswami, S.C. 1985. *Ind. J. Mari. Sci.*, 14: 177–180.

Govindasamy, C. and Kannan, L. 1991. *Hydrobiol approach. Mahasagar*, 24: 39–45.

Green, 1995. *J. Hydrobiologia*, 313/314: 47–56.

Kobayashi, T., Shiel, R.J., Gibbs, P. and Dixon, P.I. 1998. *Hydrobiol.*, 377: 133–145.

Madhupratao, M. 1999. *Ind. J. Mar. Sci.*, 28: 146–149.

Padmavathi, G. and Goswami, S.C. 1996. *Ind. J. Mari. Sci.*, 25: 268–273.

Rahukumar, S. and Anil, A.C. 2003. *Curr. Sci.*, 84(7): 884–891.

Ranga Reddy, Y. and Radhakrishna, Y. 1984. *Hydrobiol.*, 119: 27–48.

Segers, H., Sharma, S.S.S., Kattasery, F.K. and Nayar, C.K.G. 1994. *Hydrobiologia*, 287: 251–258.

Sharma, S.S.S. 1988. *Hydrobiol.*, 160: 263–269.

Shiel, R.J., Walker, K.F. and Williams, V.W.D. 1982. *Aust. J. Mari. Sci.*, 33: 301–327.

Sivasamy, S.N. 1990. *Ind. J. Mari. Sci.*, 19: 115–119.

Sujata Mishra and Panigrahy, R.C. 1996. *Ind. J. Mari. Sci.*, 25: 98–102.

Thorpe, J.H., Black, A.R., Haag, K.H. and Wehr, J.D. 1994. *J. Fish. Aqua. Sci.*, 51: 1634–1643.

Van Dijk, G.M. and Vanzantan, B. 1995. *Hydrobiol.*, 304: 29–38.

Venkataraman, K., Das, S.R. and Nandi, N.C. 2000. *J. Aquatic. Biol*, 15: 19–25.

Wiebe, P.H., Davis, C.S. and Greene, C.H. 1992. *Oceanus*, 35: 100–106.

Chapter 16

The Diversity and Seasonality of Soil Protozoans in Gir Protected Area

Pragna Parikh, Rushita Adhikari and Kiran Ahir

Department of Zoology, Faculty of Science, The Maharaja Sayajirao University of Baroda, Vadodara – 390 002, Gujarat, India

ABSTRACT

The ecology of protozoans is very complex and they are a group of ubiquitous organisms found in soil as well as in water. Ciliates, flagellates and various types of amoeba are common in soil and boggy habitats. The present study comprised of estimating the diversity of these microfauna in soil with a special emphasis on its seasonal distribution in Gir PA. The percentage of occurrences of protozoans in three different seasons was calculated. From the observations, we could conclude that the Euciliata is the only class which showed a marked seasonal variation. Its occurrence was highest in summer, moderate in monsoon and minimum in winter season. The Filosa and Lobosa exhibited less variations. Moreover, parts of Gir PA which are fully protected areas without any disturbances did not represent any specific trends in the monthly variation of the protozoan distribution. This could be attributed to the variations in the physico-chemical properties of the soil which harboured this community of protozoans. Thus, the properties of soil as well as environmental factors, alongwith the interference of anthropogenic activities, play a significant role in determining the occurrence and distribution of soil protozoa.

Keywords: Protozoa, Diversity, Seasonal distribution, Euciliata, Filosa, Lobosa, Gir PA.

Introduction

'The soil is the placenta of life', as quoted by Farb (1959), vividly suggests that although soil did not give birth to the teeming world of organisms, it is indeed the incubator for the living world. A handful of soil contains a million of organisms, ranging from ultramicroscopic to macroscopic in size. These include variety of viruses, bacteria, protozoans, nematodes, earthworms, arthropods, etc. These soil organisms play an integrated role to increase the fertility of soil they inhabit.

Protozoans comprise the soil microfauna alongwith bacteria, viruses, etc. The ecology of protozoa is very complex. They are a group of ubiquitous organisms found in soil as well as in water. Ciliates, flagellates and various types of amoeba are common in soil and boggy habitats. Since many protozoans exploit bacteria as a food source, they form a part of the decomposers' food web in nature. Recent research suggests that protozoans may stimulate the rate of decomposition by bacteria, and thus enhance the recycling of minerals. There is also evidence that some protozoans secrete a substance that promotes the growth of bacteria, thereby improving the overall soil structure.

The present study comprised of estimating the diversity of these microfauna in soil with a special emphasis on its seasonal distribution in Gir PA.

Study Area

The current study was carried out in the Gir Protected Area in Gujarat. The Gir PA is situated between 20º 40′N and 20° 50′N latitudes, and 70° 50′E and 71° 50′E longitudes. Total area of PA is 1412.1 sq.km. National Park area is 258.7 sq.km. Sanctuary area is 1153.4 sq.km. The Gir PA ecosystem falls in the semi-arid biogeographic zone.

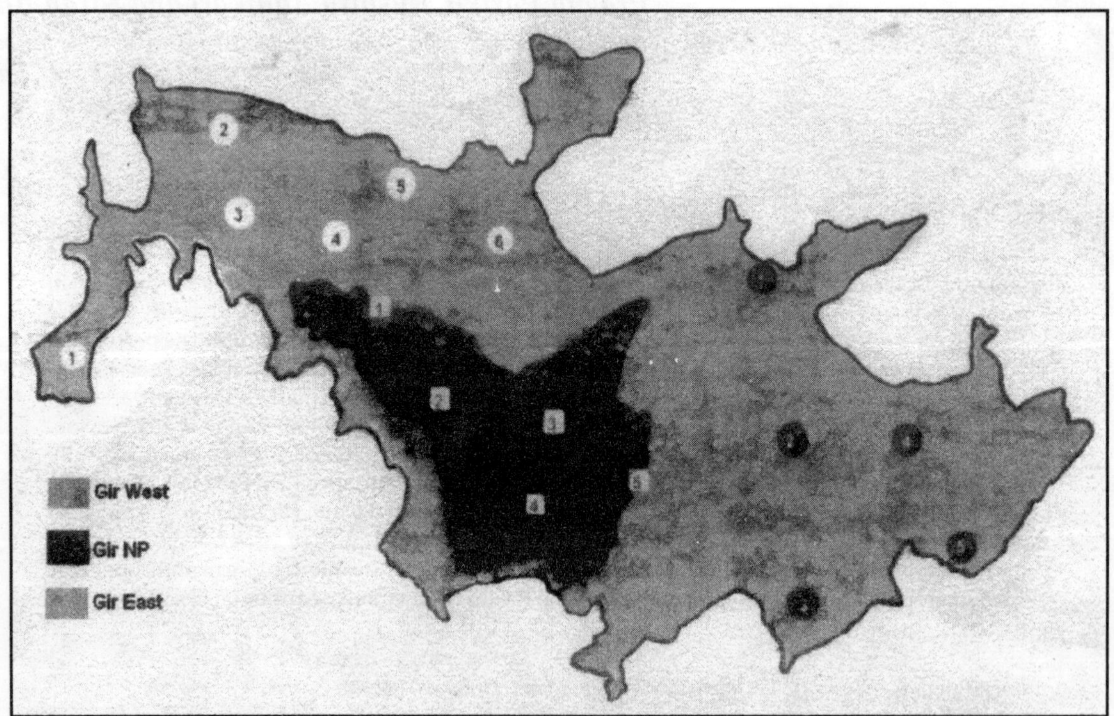

Figure 16.1: Map of Gir PA Showing Selected Study Sites and Sub-sites

Materials and Methods

As the map above shows, the Gir PA was primarily divided into three main zones: Gir West, Gir NP and Gir East. The whole of Gir PA was categorized into 16 different sub-zones as per the vegetation type. The study of protozoan fauna was categorized into edaphic habitat and ground habitat. Standard soil sampler cylinder was used for collection of soil samples. Soil upto a depth of 30cm was obtained for analysis of protozoans. Soil macrofauna was separated by passing the litter and soil separately through sieves with standard mesh sizes. Selected soil samples were specifically weighed and kept for further processing. The Berlese funnel was modified to suit the purpose of separating out the microfauna from the weighed soil samples. These samples were then taken for microscopic observations and identification of protozoans present therein.

Results and Discussion

Analysis of the protozoan diversity in Gir PA showed the presence of about 90 species which belonged to 9 orders, 4 classes and 29 families. All the classes of protozoans, namely Lobosa, Filosa, Actinopoda and Euciliata, showed variation in their seasonal distribution during monsoon, winter and summer seasons. In class Lobosa, all 12 families showed seasonal occurrence. But the Arcellidae, Centropyxidae, Difflugidae and Euglyphidae families were monsoon dominant. In class Euciliata, among the 11 families, 3 families had distribution in all the seasons. Whereas, the other 8 families showed seasonal effect. Thus, Euciliata emerged as the most diverse class with respect to seasonal distribution. The occurrence of Euciliata is highest during summer, moderate in monsoon and decreased in winter.

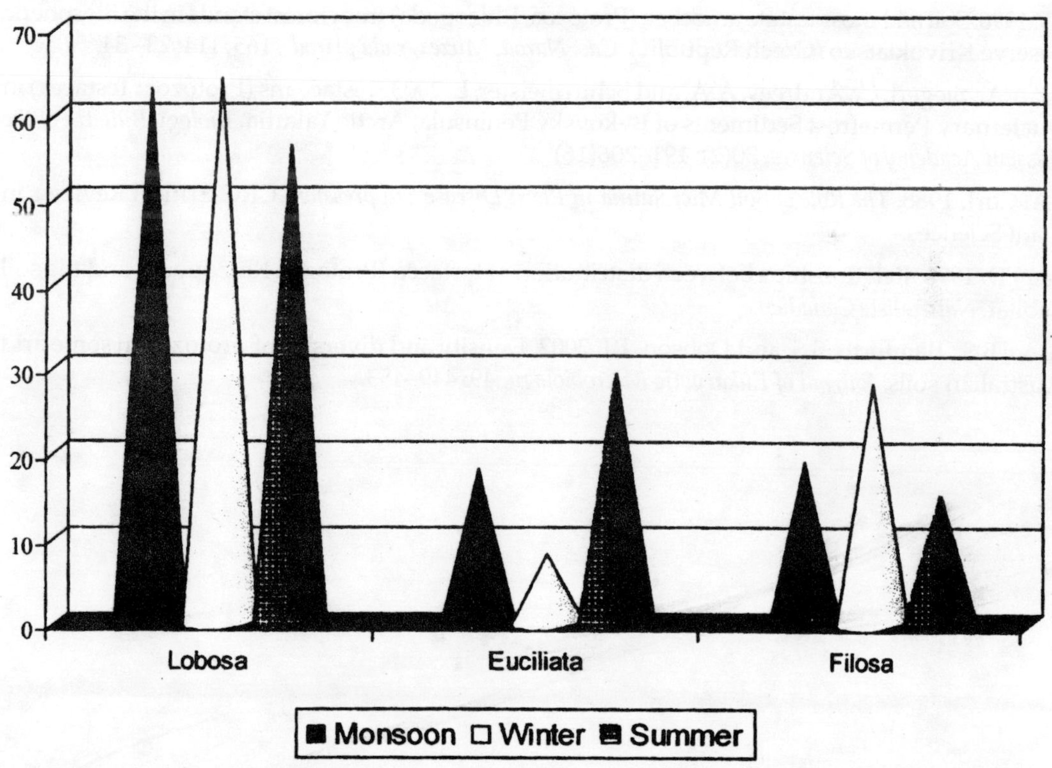

Figure 16.2: Graph Depicting the Seasonal Distribution of Lobosa, Euciliata and Filosa

Taking into consideration the vegetation type in different zones of Gir PA, occurrence of protozoans represents distinct increasing trend in Gir West. However, remaining parts of Gir PA *viz.* National Park, which is fully protected and conserved area without any sort of disturbances, and Gir East, do not represent any specific trend in monthly variation of protozoan distribution.

The protozoans have a selective influence on the composition of the bacterial population of the soil. However, the study still needs further detailed investigations wherein all physico-chemical factors should be taken into consideration. Thus, the physical environment, forest composition and the amount and kind of humus strongly affect the number and diversity of microfauna like the protozoan communities.

Acknowledgment

We are thankful to Department of Forests-Wildlife Division, Sasan Gir, Gujarat, as part of the project was sponsored under World Bank financed Ecodevelopment Project.

References

Albert Westphal and Heinz Muhlpfordt. *Protozoa.*

Anderson, R.V., E.T. Elliot, J.F. McClellan, D.C. Colemen, C.V. Cole and W.H. Hunt. 1978. *Trophic Interactions in Soils as they Affect Energy and Nutrient Dynamics.*

Balik, V. 1996. Soil and moss testate amoebae (Protozoa, Rhizopoda) from forest stands in the Biospheric Reserve Krivoklatsko (Czech Republic). *Cas. Narod. Muzea, rada prirod.*, 165, 114: 23–34.

Bobrov, A.A., Siegert, C., Andreev, A.A. and Schirrmeister, L. 2003. Testaceans (Protozoa: Testacea) in Quaternary Permafrost Sediments of Bykovsky Peninsula, Arctic Yakutia. *Biology Bulletin of the Russian Academy of Sciences*, 30(2): 191–206(16).

Elroy A. Curl. 1988. *The Role of Soil Microfauna in Plant Disease Suppression.* CRC Critical Reviews in Plant Sciences.

Lousier, J.D. 1975. Relationships between distribution of testacea (Protozoa Rhizopoda) and the soil habitat. *Naturaliste Canadien.*

Robinson, B.S., Bamforth, S.S. and Dobson, P.J. 2002. Density and diversity of protozoa in some arid Australian soils. *Journal of Eukaryotic Microbiology*, 49: 449–453.

Chapter 17

Investigation on Sub Surface Water Quality of Tarikere Taluk with Special Reference to Physico-Chemical Characteristics

K. Harish Babu and E.T. Puttaiah

Department of P.G. Studies and Research in Environmental Science, Kuvempu University, Shankaraghatta –577 451, Shimoga Distt., Karnataka

ABSTRACT

The research study was carried out in Tarikere taluk, Karnataka (India). This article mainly address the physico-chemical concentration of 56 groundwater sample during September 2004. The results of all the findings are discussed in details which reflect the present status of the groundwater quality of the study area.

Keywords: Groundwater quality analysis, Tarikere taluk.

Introduction

Rural India relies a mainly on groundwater for drinking and agriculture practices. Villages once relied on sources like wells, lakes, ponds and streams for their water needs. Contamination of most surface water sources has rendered them unfit for consumption. Also increase in water demand by an increasing population has necessitated resource to tapping groundwater. Unsustainable withdrawal of groundwater has led to the spectra of depleting the problem of water scarcity.

Every human society, be it rural or urban, industrially or technologically advanced, disposal of waste exceeds the limit of natural scavenging or removal process, they are bound to effect the normal functioning of the ecosystems and consequently they bear an adverse effect on the water resources (Miller, 1984).

Supplying inhabitants with safe and clean drinking water is one of the most common problems in developing countries like India, especially in arid and semi-arid regions. Hazardous pathogenic germs and anthropogenic substances do not only contaminate the available water quality but also geogenic substances is adversely affect the water supply of many regions.

The groundwater of Tarikere taluk had many threats such as anthropogenic activities, quality deterioration by agricultural activities and over exploitation. Persistence of continuous drought condition during 2001–2004 had declined the amount of groundwater in the region. With this background an effort has been made to know the quality of groundwater in this region.

Methodology

Study area

Tarikere taluk is located between 75°45′00″–75°52′30″ E and 13°30′00″–13°47′30″ N and it has a geographical area of 1,222 sq. kms. It comprises 45 Grama panchayats and 245 villages with a total population of 2,54,093 as per 2001 census. As a whole, the region is a flat plain land with an average elevation of about 590.58m msl (Gazetteer of India, 1981).

Meteorology

The climate of Tarikere taluk is semi arid and enjoys all the three seasons *viz.*, pre-monsoon (February to May), monsoon (June to September) and post-monsoon (October to January). The monthly mean temperature ranges from 8.9 to 38°C. The relative humidity varies between 50 to 84 per cent and it is highest during the month of August and September and lowest humidity observed during the months of January and February.

Precipitation

The precipitation and distribution of rainfall in Tarikere taluk is highly erratic. The annual average rainfall is 980 mm received over 55 rainy days. It varies from as low as 621 mm in the east and as high as 920 mm in west. About 2/3rd of the geological area of Tarikere taluk receives less than 750 mm of annual rainfall. The study area had been receiving rainfall mainly from southwest monsoon and slightly from northeast monsoon.

Geology

Stratigraphically the study area comes under Bababudan belt of Dharwar super group. Bababudan belt is well known for its iron ore (horse shoe shape). Peninsular gneiss, which forms the basement to the younger green stone belts like Shimoga, Chitradurga and Tarikere (Swaminath and Ramakrishnan, 1981). An important gneissic exposure is near Tarikere that transect the structural trend lines of the schist belts while the gneissic foliation is conformable to the schist belt boundary. The tonalitic gneisses occur in Tarikere taluk (Naqvi, 1983).

Pillow breccias and conglomerates are also seen. Metagabbro with titaniferous magnatite is intersheeted with in the metabasites at north of the Tarikere dome of basement gneisses. Serpentinites and talcose schist form a minor component of the Meta volcanic association in the Bababudan group (Fareeddudin *et al.*, 1988). Geological status of Tarikere taluk were represented in Figure 17.1.

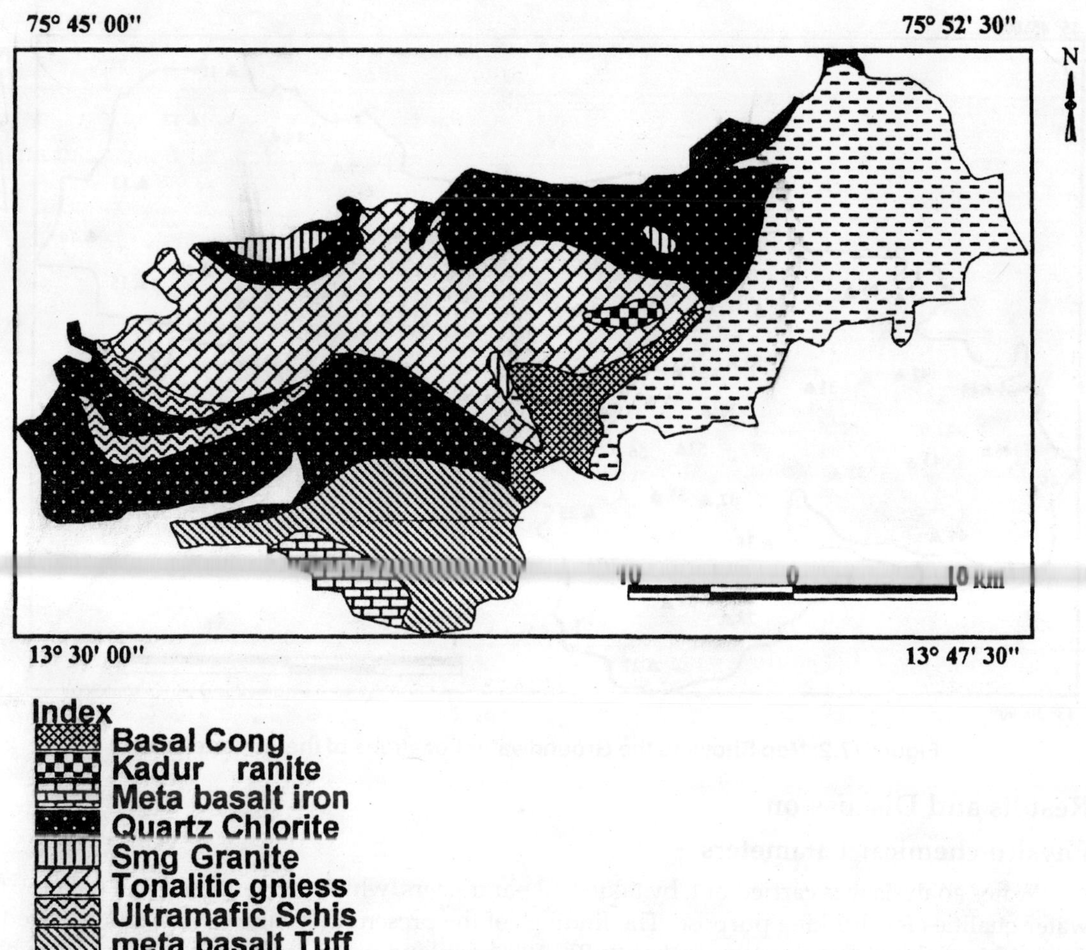

Figure 17.1: The Geological Status of Tarikere Taluk

Analysis of the Samples

The groundwater samples from the 56 sampling sites were collected and analyzed during September 2004 and Sampling locations have been shown in Figure 17.2.

Water samples from the sampling sites were collected from the bore wells. Initially the water was allowed to run for 15 minutes in order to flush out stationary water. Further, the sample bottles were also flushed with water before the samples were collected. The parameters of water such as dissolved oxygen; total dissolved solids, electrical conductivity and pH were analyzed on the spot with the help of water analysis kit (Elico). The remaining parameters were analyzed in the laboratory. Hence, the water was carried to the laboratory in suitable inert bottles. The samples were analyzed using analytical method of APHA (1995) and all analysis was done in triplicate.

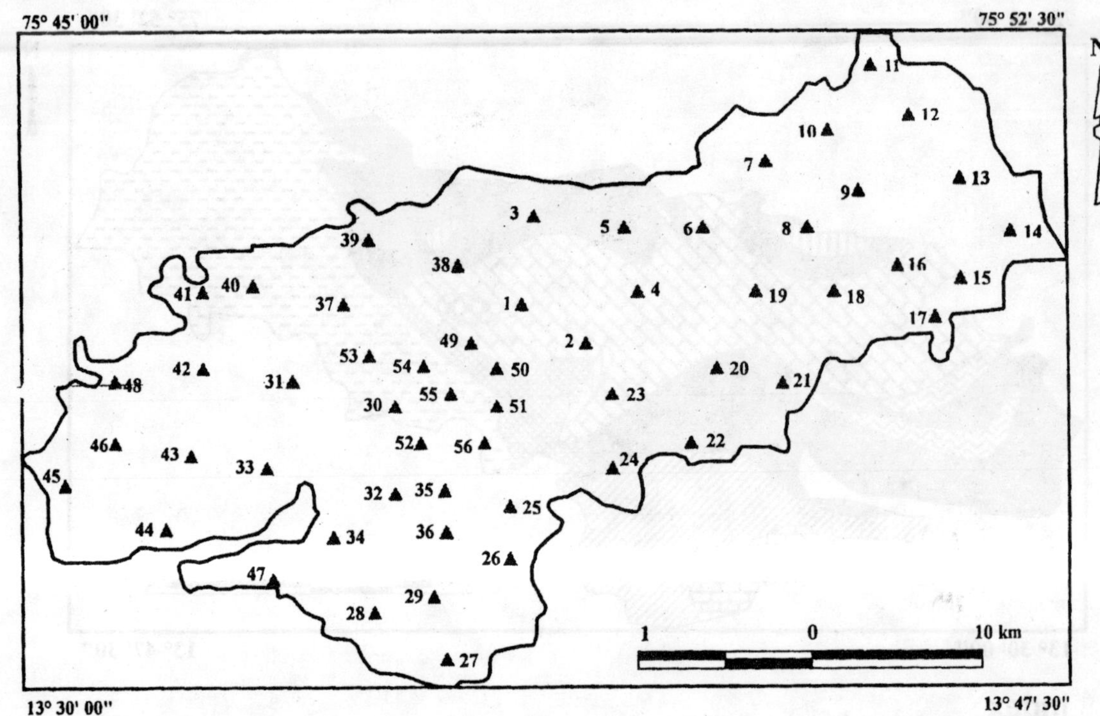

Figure 17.2: Map Showing the Groundwater Locations of the Tarikere Taluk

Results and Discussion

Physico-chemical Parameters

Water analysis was carried out, by taking 12 parameters, which are very essential to know the water qualities for drinking purpose. The findings of the present investigation are summarized in Table 17.1, and it has been compared with BIS 1998 drinking water standards (Table 17.2) which provides the comprehensive picture of the physico-chemical characteristics of groundwater in the study area.

Turbidity (TUR)

In the present study the turbidity values ranged between 0.15 and 86.1 NTU with a mean value of 9.03. The BIS (1998) acceptable limit for turbidity is 25 NTU. It is proved from the present study, 12.5 per cent of total number of samples cross their permissible limit with reference to the BIS (1998) standards. It is observed from the results that these parameters which crossed their permissible limit are unfit for drinking purposes. It causes health problems like gastro-intestinal disorders, headache and also associated with respiratory diseases (Maiti, 1982; Peter, 1974).

pH

In the present investigation, pH values found to be 6.7 to 8.7 with a mean value of 7.54. The recommended value of pH for drinking purposes is between 6.5 to 8.5 (BIS, 1998). In the present study all the water samples analyzed are all well with in the safer limits except in few sampling stations. However, higher values of pH hasten the scale formation in water heaters and reduce the germicidal potential of chlorine (Mohapathra and Purohit, 2000).

Table 17.1: Water Quality Data of Physico-chemical Parameters of the Study Sites During September 2004

Sam. No.	*TUR*	*pH*	*EC*	*TDS*	*DO*	*TH*	*Ca^{2+}*	*Mg^{2+}*	*Cl^-*	*Alk*	*SO_4^{2-}*	*F^-*
S1	11.5	8.35	1558	975	3.8	1325	327.5	218	678	815.5	170.5	1.8
S2	3.3	8.2	2338	1440	5.7	1775	364	228.5	700	295	135	1.55
S3	6.75	7.95	897.5	560	5.15	560	177.5	34.5	265	162	27.1	1.5
S4	0.15	7.85	1150	750	4.4	970.5	332.5	171	409	190	90.5	1.1
S5	7.6	7.2	1751	1051	5.4	1225	395	205.5	670.5	188.5	120.5	1.4
S6	1.2	6.85	1289	782	4.1	845	371	186.5	350	14.7	101.3	1.6
S7	0.15	7.15	449	276	4.5	350.5	119.5	102.1	75.35	146.5	104	1.25
S8	2.7	7.15	1125	718	5.3	790	130.5	108	281.5	231	125.5	1.4
S9	76.15	7.9	1600	970	5	947.5	164.5	173	515	156.5	170.6	1.55
S10	1.25	8.45	715	458	4	470	130	23.6	650	188.5	78.75	1.15
S11	1.7	7.4	544.5	352	4.7	508	186	27.6	616	120	21.8	1
S12	1.15	8.15	1136	771	4.15	970	191	170.5	451	537.5	126	1
S13	0.75	7.9	1699	1056	6.05	972.5	210.5	104.6	750	532.5	133	1
S14	0.2	7.3	1100	670	4.15	759.5	165	125.5	445	172.5	171	1.1
S15	36.5	8.15	1315	815	4.05	735	234.9	160	425	260	118.5	1.15
S16	41	8.3	633.5	421	4.55	395	165.5	110	330	184.5	143	0.9
S17	69.5	6.7	595	370	4.15	505	68	39.95	565	194	22.7	0.9
S18	2.15	8.7	1125	705	5	800	275	9.45	426.5	615	215.2	0.6
S19	1.3	8.4	955	606	4	695	206	21	163.5	119.5	176	0.7
S20	1.6	7.65	1469	950	3.8	1175	417.5	25.35	650	913	117	0.7
S21	1.4	6.7	101.8	68	4.7	85	26.5	3.5	25.65	116	174.5	0.3
S22	1.85	7.25	210	121	6.04	177	24.5	16.3	33.1	130	115.5	0.55
S23	0.3	7.6	1047	636	4.4	594	124	63	241.5	110	98.5	0.45
S24	0.2	6.8	227	138	5.6	270	45	37.5	110.1	46	98.95	0.6
S25	1.3	7.3	1278	758	4.6	970	220.5	108	553	130	165.1	0.725
S26	10.9	8.05	1373	830	4.8	986	180.5	115.5	367	96	136.3	0.625
S27	0.6	7.3	242	147	5.15	385	103	57.05	346	87	128	0.7
S28	0.25	7.15	805	482	5.35	625	135.5	102	467	110	171.5	0.85
S29	0.65	8.1	1608	995	3.65	960	137	131	547	132	100	0.975
S30	0.15	7.1	603	362	5.2	456	122.5	66.2	183.1	144	140	1
S31	0.9	7.25	466.5	280	3.9	400.5	105.5	32.85	285.5	110	246.5	1.05
S32	2.4	6.75	486.5	286	4.8	460	132	92.35	69.35	168	149.1	0.95
S33	6	6.7	778	467	4.12	623	305.5	128.7	70.2	86	104.1	1.05
S34	1.15	7.75	923	554	4.3	644	47.72	110.5	251	68	231	0.85
S35	3.15	7.25	1037	622	4.4	610	127.5	100.5	206	91.8	130.1	0.95
S36	0.2	7.45	987.5	593	4.63	570	177	91.55	266.5	180	145.1	1.05

Contd...

Table 17.1–Contd...

Sam. No.	TUR	pH	EC	TDS	DO	TH	Ca^{2+}	Mg^{2+}	Cl^-	Alk	SO_4^{2-}	F^-
S37	0.2	8.4	840	520	3.75	695	139	84.45	177	82.5	249	1.1
S38	3.45	7.85	545	315	4.05	1700	395.5	229.5	670	145.5	231.5	0.8
S39	1.6	7.95	360	260	3.95	595	83.9	63.35	200.5	79	132	1.55
S40	0.25	8.25	252.5	150	5.1	1160	189	170	735	138.5	215	1.075
S41	0.15	7.7	324	195	4.5	370	76.55	23.05	124.5	44.95	112.1	0.975
S42	10.55	8.35	322.5	178	3.85	1065	181.5	142	535	765.5	171	0.85
S43	1.55	7	452	263	4.6	1125	299.5	119.5	406	74	151.5	0.6
S44	0.5	8.15	784.5	478	4.15	580	279	11.1	285	66.5	141.5	0.95
S45	0.5	7.1	475.5	294	5.1	283	54.45	35.5	156.5	172	161.2	0.8
S46	11.65	7.05	1193	705	4.9	672	234.5	101.5	93	81	139.5	0.9
S47	0.35	6.85	693	425	4	397	108.1	56.7	118	92.1	265.5	1
S48	10.7	7.5	1550	925	3.55	882	212.5	186.2	86.95	75	262	1.05
S49	0.2	7	957	577	5.6	625	78.85	92.5	142	149	152.8	1
S50	1.65	7.05	789	477	4.4	440	91.35	43	175	110	162	1.1
S51	0.6	6.95	900	553	4.4	800	237	45.1	453	126	210.5	1
S52	0.15	7	520.5	315	5.15	640	156.5	56.05	538.5	101	222.5	1.15
S53	86.1	7.05	784.5	478	3.8	1505	520	46.1	652.5	86.7	172	1.1
S54	69.2	7.55	705	416	4.35	219	36.5	25.5	120.5	89	156.4	1.05
S55	1.25	7.9	815	585	4.1	695	215	9.7	685	255	170.3	1.4
S56	7.05	7.8	1200	749	4	695	125	116.5	470	126	57.3	1.2
Mean	9.029	7.542	894.3	551.7	4.552	727.4	185	92.12	361.8	191.1	150.7	1.012
Min	0.15	6.7	101.8	68	3.55	85	24.5	3.5	25.65	14.7	21.8	0.3
Max	86.1	8.7	2338	1440	6.05	1775	520	229.5	750	913	249	1.8

Electrical Conductivity (EC)

The values of electrical conductivity ranged between 101.8 μmhos/cm to 2338 μmhos/cm. Mean values of EC showed 885.31 μmhos/cm. Higher the concentration of acid, base and salts in water, higher will be the EC. The variability of EC could be explained to the natural concentration of ionised substances present in water (Kataria and Jain, 1995). However the higher values of electrical conductivity (>2000 μmhos/cm), may be due to long residence time and lithology. Ballukraya and Ravi (1999) had proved the variation of the conductivity of the water due to the residential times and the geographical features of the sites.

Total Dissolved Solids (TDS)

It is justified from the analytical report TDS values ranges from 68 mg/L to 1440 mg/L with a mean value of 551.53 mg/L. However groundwater chemistry changes as the water flows through the subsurface and the increase in geological environment and dissolved solids and major ions. Chebotarev (1985), Ramababu and Somashekara Rao, (1986) and Joseph (2001) expressed the dissolution of soil particles containing minerals under slightly alkaline condition; favour the TDS concentration in groundwater. However TDS concentration above the permissible limit (1500-ppm) causes gastrointestinal irritation (Shankar and Muttukrishnan, 1994).

Table 17.2: Comparison of Groundwater Quality Data with Drinking Water Standard (BIS, 1998)

Sl.No.	*Parameters*	*BIS (1998)*		*Observed Values*	
		P	*E*	*Minimum*	*Maximum*
1.	Turbidity	5	25	0.15	86.1
2.	pH	6.5	9.2	6.7	8.7
3.	EC	–	–	101.8	2338
4.	TDS	500	1000	68	1440
5	Dissolved oxygen	–	–	3.55	6.05
6.	Total hardness	300	600	85	775
7.	Calcium	75	200	24.5	520
8.	Magnesium	30	100	3.5	229.5
9.	Chloride	250	1000	25.65	750
10.	Alkalinity	200	600	14.7	913
11.	Sulfate	200	400	21.8	249
12.	Fluoride	0.6–1.2	1.5	0.3	1.8

Note: P: Permissible limit; E: Excessive limit.

All paramotors are expressed in mg/L except pH, Turbidity (NTU) and conductivity (μmohs/cm).

Dissolved Oxygen (DO)

The present research findings revealed for the values of DO varies from 3.55 mg/L to 6.05 mg/L, and observed mean value is 4.55 mg/L respectively.

In the present study, all the samples showed for DO values were well with in the permissible limits for drinking and domestic purposes. However, the presence of oxygen demanding pollutants (like organic wastes) causes rapid depletion of dissolved oxygen from water. Oxidizable inorganic substances like hydrogen, sulphide, ammonia, nitrites, iron etc., decrease the dissolved oxygen concentration (Sawant *et al.*, 2000). And also due to physical, chemical and biological activities in water the level of DO may vary (Jameel, 2002)

Total Hardness (TH)

Total hardness levels varied from 85 mg/L to 775 mg/L, associated with a mean value of 727.36 mg/L. The BIS (1998) acceptable limit for total hardness is 600 mg/L. In the present study, revealed that 60.7 per cent of total number of water samples crosses the permissible limits of BIS (1998) drinking water standards. Owing to fact that higher amount of hardness in the study area comes mainly from the leaching of igneous rock and carbonate rocks (dolomite, calcite and limestone). Water containing the soluble salts of calcium and magnesium such as chlorides, sulphates and bicarbonates are also governs the quality of water (Ramaswamy and Rajaguru, 1991). The adverse effects of total hardness are formation of kidney stone and the heart diseases (Sastry and Rathee, 1998). Nevertheless, groundwater chemistry is controlled by the composition of its recharge components as well as by geological and hydrological variations (Narayana and Suresh, 1989).

Calcium (Ca^{2+})

Present investigation reports stated for calcium values ranged from 24.5 mg/L to 520 mg/L, with a mean value of 184.99 mg/L. BIS (1998) acceptable limit for calcium is 200 mg/L. However, in the present study 30.2 per cent of water samples crosses the permissible limit. Presence of higher amount of calcium in the study area may be due to groundwater receives the calcium minerals leached from the rocks and other deposits like limestone, dolomites, calcite, gypsum, amphiboles, feldspar, and clay minerals leaching or weathering of igneous rocks. Sewage and domestic waste are also important sources of calcium (Mishra and Saxena, 1989).

Magnesium (Mg^{2+})

Investigation report reveals for the magnesium values ranged from 3.5mg/L to a 229.5 mg/L with a mean value of 92.11 mg/L.

BIS (1998) acceptable limit for magnesium is 100 mg/L and in the present study 48.2 per cent of the water samples crossed the permissible range. Magnesium arises principally from the weathering of rocks contain ferro-magnesium minerals and some carbonate rocks. High concentration of magnesium proves to be diuretic and laxative (Schroeder 1960).

Chloride (Cl^-)

Chloride is also one of the important parameter to know the quality of water. Anthropogenic sources of chlorides include fertilizer, road salt, human and animal waste. Concentration of chlorides is considered to be an indicator of organic pollution of animal origin (Kumara, 2002). Here Chloride values ranged from 25.65 mg/L to 750 mg/L, however the mean value observed in the present study is 361.82 mg/L.

The BIS (1998) acceptable limit for chloride is 1000 mg/L. In the present investigation, the values of chloride for all the sampling sites are with in the permissible range as prescribed by BIS (1998) drinking water standards. However, dissolving of the soil constituents had contributed the chloride into the groundwater and also the soil characteristics play an important role in contributing the chloride content in the groundwater (Shivasankaran, 1997).

Total Alkalinity

In the present study total alkalinity values ranged from a 14.7 mg/L to 913 mg/L and a mean value of 189.31 mg/L. The BIS (1998) acceptable limit for total alkalinity is 600mg/L. In the present study, the data revealed that 10.7 per cent of water samples in the study area crossed the permissible limit. When alkalinity of water exceeds the permissible limits, it is likely to produce incrustation sediment deposits, difficulties in chlorination, certain physiological effects on human systems etc. (Raviprakash and Rao, 1989). The constituents of alkalinity result from dissolution of mineral substances in the soil and atmosphere contributes to alkalinity in groundwater (Mittal and Verma, 1997).

Sulphate (SO_4^-)

It is very interesting to note that, sulphate values ranged from 21.8 mg/L to 249 mg/L with a mean value of 150.2 mg/L. The BIS (1998) pemlissible range for sulphate is 400 mg/L. In the present investigation, the sulphate values for all the samples are within the prescribed limit of BIS (1998) drinking water standards. Sulphate in groundwater takes place the break down of organic substances in the soil. However, geological, hydrological and geomorphologic characteristics showed a remarkable variation in sulphur content (Alexander, 1961).

Fluoride (F^-)

In the present investigation, fluoride values varied from 0.3 mg/L to 1.8 mg/L with a mean value of 1.01 mg/L, The BIS (1998) acceptable limit for fluoride is 1.5 mg/L. In present study 7.14 per cent of total number of water samples in present investigations have crossed the permissible range as prescribed by BIS (1998) drinking water standards.

Degree of weathering and leachable fluoride in terrain is of great significance for the fluoride present in groundwater than the mere presence of fluoride bearing minerals in rocks (Kumar *et al.*, 2000).

Intake of excess fluoride causes dental, skeletal and non-skeletal fluorosis. The non-skeletal fluorosis can be observed such as gastrointestinal complaints, intermittent diarrhoea and flatulence in expectant and lactating mothers hardworking young adults and children. Therefore, fluorosis has been considered as one of the incurable diseases. Hence, prevention is the only solution for the disease (Hem, 1985).

Conclusion

Currently carried research investigation should give more precise answer on influence of Geomorphological condition than anthropogenic activities in the examined groundwater samples of the study area. Local geological settings may supports the increasing concentration of physico-chemical characteristics in groundwater. The factors like slow circulation, longer period of contact between aquifer and water, dissolving of minerals at the time of weathering, residential time, drainage pattern and surface water link. Porosity of the soil and rock also alters the characteristics of the groundwater.

The high level contents of the parameters observed may be minimized if the groundwater is recharged with the available water in the rainy season. This not only dilutes the constituents of the groundwater but also raises the groundwater level that depletes due to large-scale exploitation.

Groundwater is extremely important to the future economy and growth of rural India. If the resource is to remain available as high quality water for future generation it is important to protect from possible contamination. Hence it is recommended that suitable water quality management is essential to avoid any further contamination.

References

Alexander. 1961. *Introduction to Soil Microbiology*. Wiley, New York, London, p. 171–172.

APHA. 1995. *Standard Methods for the Examination of Water and Wastewater*; 18th edition, AWWA, WPCF, New York, pp. 1120.

Ballukraya, P.N. and R. Ravi. 1999. Characterization of Groundwater the unconfined aquifers of Chennai city. *Indian Journal of Geological Society of India*, 54: 1–11.

BIS. 1998. *Specifications for Drinking Water*, New Delhi, p. 171–178.

Chebotarev. 1985. Metamorphism of natural waters in the crust of weathering. *Geochem. Cosmochim. Acta*, 8: 22–28.

Fareeddudin, A.S. Janardhan and B. Basavalingu. 1988. Sedimentology, minerology and geochemistry of the Kalasapura conglomerate. *Geol. Soc. India. Mem.*, 9: 65–82.

Gazetteer of India, Karnataka State, Chikmagalure district. 1981. p. 8–15 and 630–633.

Hem, J.D. 1985. *Study and Interpretation of the Chemical Characteristics of Natural Water*, 3rd ed. U.S. Geological Survey, Water Supply Paper, 225: 263.

Jameel, A. 2002. Physico-chemical studies in Vyakondan channel water of river Cauvery. *Pollution Research*, 17(2): 111–114.

Joseph, W. 2001. Groundwater chemistry in the valley De Yabucoa alluvial aquifer, South-eastern Pnerto Rico. *AWRA 3rd International Symposium on Tropical Hydrology*, San Juan U.S.A.

Kataria, H.C. and O.P. Jain. 1995. Physico-chemical analysis of river Ajhar. *Indian Journal of Environmental Protection*, 5: 569–571.

Kumar, P. Kumari and L.K.R. Singh. 2000. *Environmental Biology*. P.G. Dept. of Zoology, S.K. University, Dumka, Jharkand, p. 180–182.

Kumar. 2002. *Ecology of Polluted Waters*. 1: 144–180.

Maiti, T.C. 1982. *Science Reporter*, 19: 360–361.

Miller, D.R. 1984. Chemicals in the environment. In: *Effects of Pollutant at the Ecosystem Level*. John Wiley and Sons, Chinchester, pp. 7.

Mishra, S.R. and Saxena. 1989. Industrial effluent pollution at Birla Nagar, Gwalior. *Pollution Research*, 8(2): 77–86.

Mittal, S.K and N. Verma. 1997. Critical analysis of groundwater quality parameters. *Indian Journal of Environmental Protection*, 17: 426–429.

Mohapatra, T.K. and K.M. Purohit. 2000. Qualitative aspects of surface and groundwater for drinking purpose in Paradeep area. *Ecology of Polluted Waters*, 1: 144.

Naqvi, S.M. 1983. Geochemistry of gneisses from Hassan district and adjoining areas, Karnataka, India. In: *Precambrian of South India*, (Eds.) Naqvi, S.M. and J.J.W. Rogers. Geological Society of India Mem., 4: 401–416.

Narayana, A.C. and G.C. Suresh. 1989. Chemical quality of groundwater of Mangalore city, Karnataka. *Indian Journal of Environmental Health*, 31: 228–236.

Ramaswamy, V. and P. Rajaguru. 1991. *Indian Journal of Environmental Health*, 33(2): 187–191.

Rambabu, C. and K. Somashekara Rao. 1986. Studies on the quality of bore well water by Nuzuid. *Indian Journal of Environmental Protection*, 16(7).

Raviprakash, S. and K.G. Rao. 1989. The chemistry of groundwater, Parvada area with regard to their suitability for domestic and irrigation purposes. *Indian Journal of Geochem.*, 4: 39–54.

Sastry, K.V. and P. Rathee. 1988. Physico-chemical and microbiological characteristics of water of village Kanneli, (Distt. Rohtak) Haryana. *Proc. Academic, Environmental Biology*, 7(1): 103–108.

Sawant, C.P., Saxena, G.C. and Shrivastava, V.S. 2000. Trace Metals in and around the Industrial belt. *Ecology Environment and Conservation*, 6(6): 135–137.

Schroeder, H.A. 1960. Relation between hardness of water and death rates from certain chronic and degeneration diseases in the US. *J. Chron. Dis.*, 12: 568–573.

Shankar and Muttukrishan. 1994. *In situ* bioremediation of contaminated groundwater. *Proceedings of National Seminar on EPCR–04*, UBDT Engineering College, Davangere, pp. 32.

Shivashankaran, M.A. 1997. Hydrogeochemical assessment and current status of pollutants in groundwater of Pondichery region, South India. *Ph.D. Thesis*, Anna University, Chennai, p. 80–87.

Swaminath, J. and M. Ramkrishnan. 1981. In: *Early Precambrian Supracrustals of South India*, (Eds.) Swaminath, J. and M. Ramakrishnan. Geol. Surv. Ind. Mem., 3: 23–38.

Chapter 18

Analysis of Fluoride in the Groundwater of Akola District: A Case Study

S.B. Thakare*, A.V. Parwate**, M. Rao***

**Lecturer, College of Engineering and Technology, Babhulgaon (Akola) M.S.*

***Principal, A.V.C. College of Engineering, Mayiladuthurai (Nagai), T.N.*

****Principal, P.R. Engineering College, Vallam (Thanjavur), T.N.*

ABSTRACT

The menace of high fluoride concentration in groundwater resources have now become one of the major health related geo-environmental issue in our country. The problem of fluoride has socio-economic implications. Therefore immense and immediate efforts are required for isolated, secluded of unapproachable population affected with fluorisis without any delay. In this article 100 groundwater samples from open well and bore well were collected from 50 villages in different Talukka of Akola district and analysed for fluoride and concentration. The fluoride and concentration is found to be above the permissible limit (1.00 mg/L) for 29 samples.

Introduction

Fluoride has posed a great threat to human health in our country. First-ever case of fluorosis was detected by the farmers of Andhra Pradesh State, among cattle during early 1930s. Shorett *et al.* published the first report on the endemic fluorosis in India since then endemic fluorosis has been identified in the 17 states of Indian union and is known to prevalent fort the last six decades without receiving much attention to tackle it.

A national survey of drinking water in India indicated that over 43.5 million people living in 1.42 lakh habitations spread over 200 district are at health risk due to water quality problems and fluoride is identified as one of the major contaminants. Fluoride is not found in the nature in free state. It occurs primarily in form of minerals like:

1. Cryolite (Na_2AlF_6)
2. Fluorite (Fluorspar) CaF_2
3. Fluorapatite $Ca_5F(PO_4)_3$
4. Topaz $Al_2SiO_4(OH, F)_2$
5. Sellaite (MgF_2)
6. Villiamite (NaF).

It is also a minor constituents many clay minerals and mica. Fluoride is released to the groundwater through weathering of primary silicates and associated accessory minerals. It is also partly re-deposited in soil profile as carbonate concretions (Kankars) in soluble form of CaF_2 under certain climatic and hydrogeological conditions. Other sources of fluoride in groundwater are the insecticides, herbicides and phosphate fertilizers used in agricultural activities, Fluoride salts commonly in steel, aluminium, brick and title industries may also contribute to groundwater due to industrial spillages. Fluorine in the combined state constitutes 0.078 per cent of the earth' s crust.

Climate and Hydrogeology

Akola district is geographically situated at about 20°44′ North latitude to 21° North latitude and 77°44′ East longitude to 78° East longitude. The district is about 281.90 meters above mean sea level and area of the district is 10.575 sq. km with 11.60 lakhs in habitants. Entire district comprises of rocks like Basalt, Granite, Jambha rock and Black stone. The rivers flowing through the district are Katepurna, Uma, Purna, Shahanur, Vidrupa, Pathar, Nirguna and Man. Hot and dry is the normal climate. Rainfall is during June to September and being received by South-west and North-east monsoon. The average rainfall is about 847 mm. Soil colour ranges from black to red. Depth of water table ranges from 4 mt. to 100 mt. below ground level.

Material and Methods

Considering the present stock of the situation in groundwater quality and possible degradation of the same by environmental pollution, 100 samples of water from open wells and bore wells dispersed over the entire district have been analyzed for quality surveillance, The water table was at its lowest ebb due to extraction from the aquifer to fulfill various daily demands during non-rainy season. So the samples are collected during the month of February 2005 to April 2005. Fluoride was analyzed by Spands method as per the standard methods by uv spectrophotometer in the laboratory.

The study area selected for Fluoride analysis in the groundwater includes 51 villages from all the 7 Talukka of Akola district. It is observed from the analysis given in Table 18.1 that 29 villages have Fluoride ion concentration above the permissible limit of 1 mg/lt. The maximum average Fluoride ion concentration was found in 18 villages above 1.5 mg/lt. and the Fluoride ion concentration below 0.5 mg/lt. is found in about 2 villages *i.e.*, at Panaj in Akot talukka and at Khanapur in Patur talukka. It was found that there is a direct correlation between alkalinity and Fluoride ion concentration.

Conclusion

These findings suggest that the distribution of Fluoride ions in the groundwater of Akola district is growing up. The concentration of Fluoride in groundwater has direct relation to the contact time of water with the Fluoride minerals in rocks.

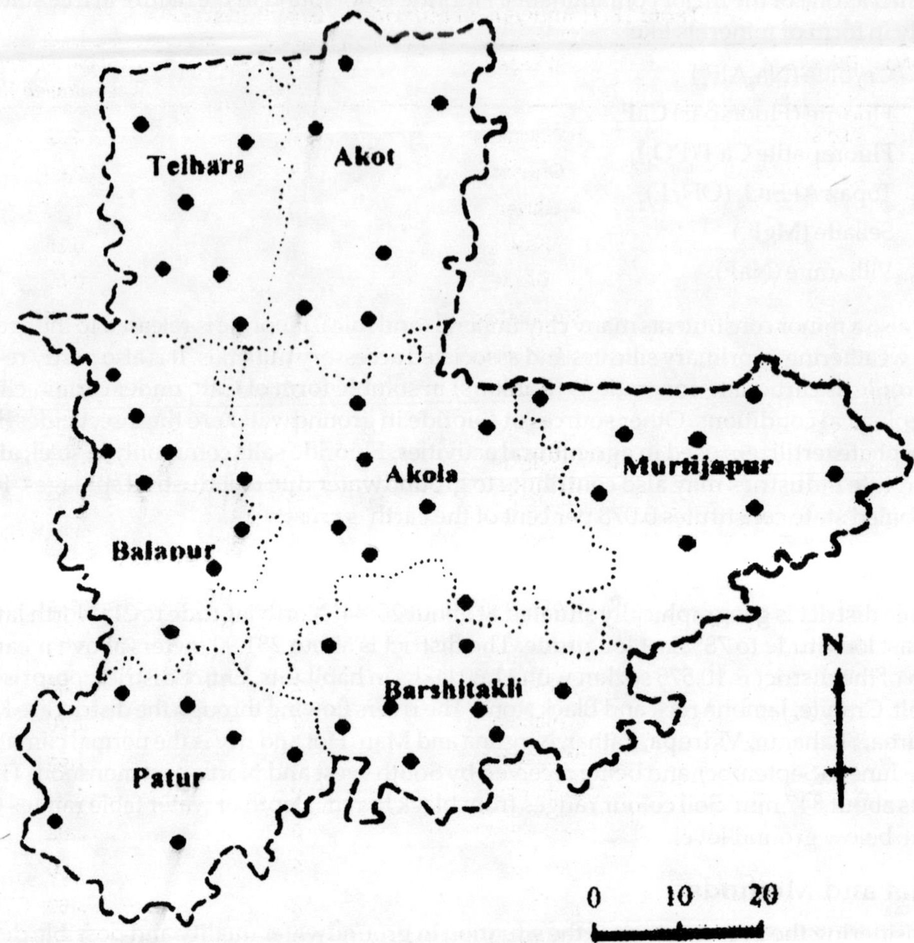

Figure 18.1: Map of Akola District

An important measure to control the concentration of fluoride in groundwater is to adopt techniques to augment the groundwater storage through rainwater harvesting. Artificial recharge is the process by which the groundwater reservoir is augmented at a rate exceeding that under natural conditions. Hydrogeological surveys can help in identifying appropriate sites for construction of artificial recharge structures like percolation tanks, nala bunds, recharge shafts which can go along way in diverting the surplus surface monsoon run-off to groundwater storage. Water conservation techniques like sub-surface dams and roof top rainwater harvesting have special significance in augmenting the groundwater storage which help in improving the quality of groundwater.

In areas where the hydrogeological situation does not favour adoption of any technique for artificial recharge of groundwater and high fluoride occurs in both shallow and deeper aquifers, the only alternative is to either supply safe drinking water from surface/groundwater source at some distance from the village or adopt technique of treatment of water.

Table 18.1: Analysis of Fluoride in Groundwater

Sl.No.	Talukka	Village	No. of Sources Tested	Average Fluoride Concentration mg/l
1.	Akola	Babhulgaon	2	2.5
		Ghusar	2	2.80
		Shivar	2	1.80
		Kasli	2	0.95
		Yeota	2	0.60
		Bakharabad	2	2.80
		Dapura	2	1.30
		Yelwan	2	1.00
2.	Akot	Fattepur	2	2.65
		Dharel	2	0.75
		Akot	2	1.90
		Panaj	2	0.36
		[illegible]	[illegible]	[illegible]
		Salkhed	2	1.60
3.	Balapur	Nimkarda	2	1.40
		Nimbha	2	1.85
		Ridhora	2	1.70
		Wadegaon	2	1.65
		Manarkhed	2	0.52
		Jogalkhed	2	0.83
		Balapur	2	1.60
		Paras	2	0.80
		Barlenga	2	1.50
4.	Barshitakali	Pinjar	2	2.10
		Chelka	2	1.70
		Walpi	2	1.65
		Titwa	2	1.50
		Zodga	2	0.90
		Shindked	2	1.30
		Dhaba	2	0.70
		Kinkhed	2	0.80
5.	Murtizapur	Jamathi	2	0.90
		Mungshi	2	0.85
		Hirpur	2	1.70
		Hatgaon	2	1.58
		Anabhora	2	1.10

Contd...

Table 18.1–Contd...

Sl.No.	Talukka	Village	No. of Sources Tested	Average Fluoride Concentration mg/l
		Kalsara	2	0.70
		Jambha	2	0.50
		Kurum	2	1.30
6.	Patur	Sawargaon	2	0.50
		Khanapur	2	0.40
		Shirla	2	0.80
		Patur	2	1.40
		Chondi	2	0.65
		Hingana	2	3.00
7.	Telhara	Bamberi	2	0.60
		Belched	2	1.25
		Narsipur	2	1.50
		Saundala	2	1.75
		Telhara	2	0.80
		Adgaon (kh)	2	1.50

To Tackle the Fluoride Problem, the Following Approach may be Adopted

Identification and Delineation of Areas Affected

Reconnaissance survey of habitations in fluorosis suspected areas and collection of site-specific data on health and water resources and quality aspects through standardised procedure using appropriately designed questionnaire and format may be completed to identify and delineate the area affected.

In depth analysis of water sources, health status of delineated areas and evaluation of technology:

1. Completion of chemical analysis of water sources along with community water supply sources.
2. Engineering analysis of water quality data so collected.
3. Pre-intervention health assessment.
4. Interaction with community and state level relevant agencies for planning of site-specific fluorosis control measures.
5. Critical evaluation of identified defluoridation technologies through a scientific ranking methodology and economical analysis of defluoridation options vis-à-vis supply from alternate sources for delineated habitations to identify specific best option.

Preparation of Action Plan and Implementation of Solutions

1. Organising community awareness programmes on the health impact of fluorosis and the need for supplying fluoride free water.

2. Demonstration and training programmes for Public Health Engineers, District Health Agencies and Local Operators.
3. Administering the medical and nutritional measures simultaneously making the provision of safe water supply through domestic defluoridation as interim measure.
4. Design of water supply system utilising alternate source or preparation of process package including designs, drawings and specifications for defluoridation plant as best option.

Post Intervention Evaluation

1. Post intervention health assessment on the similar lines to the pre-intervention health assessment.
2. Periodic monitoring of measures taken and improvement in health.
3. Post installation evaluation of the system adopted and recommendations for modification, if necessary, and site-specific O and M provisions for sustained performance.
4. Treatability studies for optimising the process control.

References

APHA. 1995. *Standard Methods for the Examination of Water and Wastewater*, 19th edn. American Public Health Association.

Azbar, N. and Turkman, A. 2000. Defluoridation in drinking waters. *Water Sci. Tech.*, 42: 848–858.

Bhargava, D.S. and Killedar, D.J. 1991. Batch studies of water defluoridation using fishbone charcoal. *J. WPCF*, 63: 848–858.

CPCB. 1994–95. *Laboratory Analytical Techniques*. Series: Lats/9/1994–95. Central Pollution Control Board, New Delhi.

CPHEEO. 1991. *Manual on Water Supply Treatment*. Ministry of Urban Development, New Delhi.

Khabade, S.A. and Mule, M.B. 2003. Studies on physico-chemical parameters of Pundi water reservoir from Tasgaon Tahsil. *Indian J. of Environmental Protection*, 23: 1003–1007.

Susheela, A.K. 1999. Fluorosis management programme in India. *Current Science*, 77: 1250–1256.

Chapter 19

Parasitic Infection and Drinking Water Quality in Lashkar Township (Gwalior) M.P.

Naseem Khan, Asha Mathur** and R. Mathur**

**School of Studies in Zoology, Jiwaji University, Gwalior*

***Department of Zoology, K.R.G. College, Gwalior*

ABSTRACT

Gwalior is an old city, which has grown into populous town having moderate facilities. The old water supply system has become insufficient and therefore most of the new colonies are not getting sufficient tap water supply. With the increase in population, it calls for a fresh study to evaluate the various water sources of the town. In the present study wards no. 31 to 60 of Lashkar town are studied. Various water sources have been recorded and their water tested for potability. Examination of the stool forms an important part in the diagnosis of parasitic infection. 450 stool samples were collected from different wards to find out the protozoal (*E. histolytica,* and *Giardia intestinalis*) and helminthic parasites (*Ascaris lumbricoides, Taenia* sp., *Ancylostoma duodenale, Trichuris trichiura* and *Enterobius vermicularis*). Measures for proper water supply sanitation and hygiene have been suggested.

Keywords: Chemical characteristics, Drinking water, Total hardness, Total solids, Parasitic infection.

Introduction

The urban water supply of Lashkar comes mainly from Moti Jheel which is fed by Tighra Dam, but this dam is insufficient to serve the needs of the people of Lashkar. It is for this reason that number of hand pumps and borewells have been installed throughout the city. To cope with the needs of newly developed colonies, there has been increasing installation of hand pumps by individuals in their premises. The Swarn Rekha Nala which flows through the city carries the sewage but its course through many of the heavily populated colonies leaves impact on the water quality of hand pumps and borewells in the nearby region. Looking to the needs of the day wardwise study of Lashkar township has been attempted.

Materials and Methods

The present study covers the entire urban area of Lashkar township shown in map. All the 30 wards of Lashkar township were studied. Two samples were collected from each ward and wherever necessary, more samples were taken. Total 82 samples were collected from taps, borewells and hand pumps. Water samples from different sources were collected in clean polythene cans in winter, summer and rainy seasons. Stool samples from each ward were collected between 7.00 to 8.00 A.M. for assessing the parasitic infection. 15 stool samples from every infection prone ward were collected in polythene packets and were brought to laboratory and a small portion of feaces was diluted with fresh normal saline. Cysts etc. were identified using concentration method and helminthic parasites were identified using floatation technique (Chatterjee, 1981). Chemical characteristics were determined by the standard methods described by APHA (1989).

Following chemical parameters and their methods were used:

Table 19.1: Chemical Parameters and their Methods

Parameters	*Method*
pH	H^+ ion sensitive electrode method
Conductivity	Conductivity bridge
Carbonates	Phenophthalein and methyl orange indicator
Bicarbonates	Phenophthalein and methyl orange indicator
Total alkalinity	Phenophthalein and methyl orange indicator
Calcium hardness	E.D.T.A. Titrimetric method
Total hardness	E.D.T.A. Titrimetric method
Magnesium hardness	Calculation method
Chlorides	Argentometric method
Sulphates	Turbidimetric method
Phosphates	Stannous chloride method
Sulphides	Titrimetric (Iodine) method
Nitrates	Brucin method
Nitrites	N-(1-Nephthyl) ethylenediamine dihydrochloride method
Silicates	Molybdosilicates method
Sodium	Flame photometric method
Potassium	Flame photometric method
Total Solids	Evaporation method
Total dissolved solids	Evaporation method
Total suspended solids	Calculation method

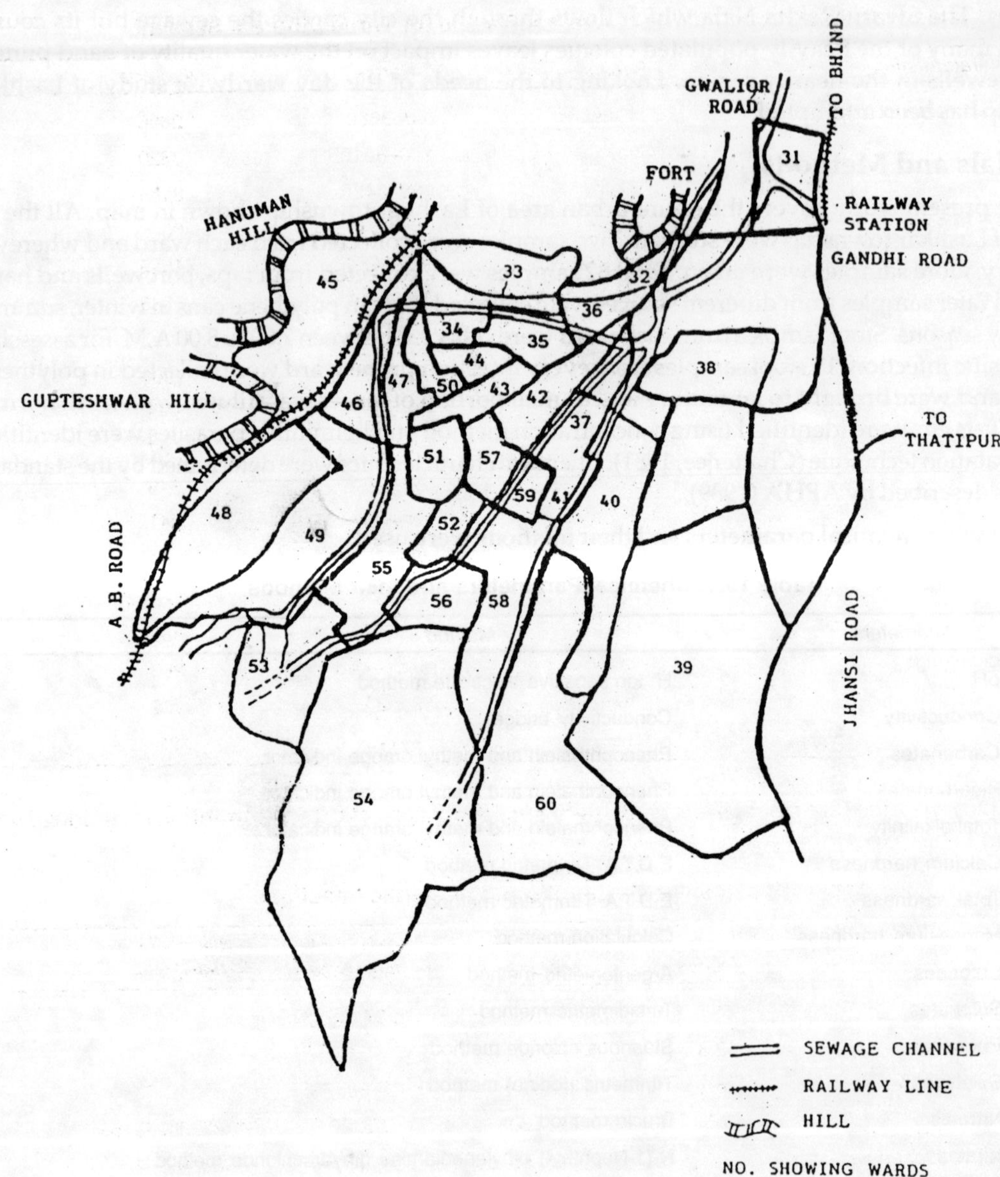

Figure 19.1: Map Showing Study Area of Lashkar Town

Results

Total 82 drinking water sources were selected for the study including taps, hand pumps and borewells. The maintenance of many hand pumps and their neighbourhood was not satisfactory and many hand pumps and borewells were found to be installed at risky sites.

Total solids, TDS, TSS, total alkalinity, total hardness and sodium showed high values. Analytical results of average values for three different seasons for all these parameters are presented in Table 19.2.

Table 19.2: Analytical Results of Average Values for Three Different Seasons

Characteristics	Season	Range	Average
Total solids ($mg.l^{-1}$)	Winter	40–3660	2686
	Summer	120–8640	3320
	Rain	160–8420	3155
TDS ($mg.l^{-1}$)	Winter	20–3640	2046
	Summer	40–2470	1546
	Rain	80–7440	2565
TSS ($mg.l^{-1}$)	Winter	20–2020	750
	Summer	40–6170	2860
	Rain	40–2180	943
Total alkalinity ($mg.l^{-1}$)	Winter	20–494	327
	Summer	40–548	319
	Rain	30–484	341
Total hardness ($mg.l^{-1}$)	Winter	40–1260	361
	Summer	40–930	407
	Rain	68–678	336
Sodium ($mg.l^{-1}$)	Winter	Traces–460	263
	Summer	2–420	242
	Rain	3–145	95

Other parameters were not significantly high. Results of parasitic and helminthic infections are resented in Table 19.3.

Table 19.3: Results of Parasitic and Helmintic Infections

Parasites	% Infection	Wards of Max.	Wards of Min. Incidence	Wards of No Incidence
E. histolytica	21.7	49, 52, 55, 60	31, 37, 38, 39	–
Giardia intestinalis	12.8	46, 49, 52, 60	31	–
Ascaris lumbricoides	15.7	49, 52, 60	31, 38, 59	–
Trichuris trichiura	4.5	52, 60	37, 43	31, 33, 38, 45, 59
Ancylostoma duodenale	7.1	33, 46, 55, 60	34, 43, 45	38, 43, 47
Enterobius vermicularis	3.7	32, 46, 60	37, 40, 44	31, 34, 38, 39, 54, 58

Discussion

Most of the area of Gwalior township is plain except some wards which are contoured in the ases of hillocks. A sewage channel known as Swaran Rekha Nala Passes through the town has three ibutaries. They all merge at different places and proceed towards Gwalior city where it pours in to a ewage pond from where the water is used for irrigation in nearby villages. It may be a source of ontamination for many groundwater sources. There are wide variations in the chemical characteristics of groundwater in Lashkar township. Total solids in the water samples were found high in ward nos. 31, 33, 34, 35, 40, 43, 46, 47, 49, 52, 53, 55, 56, 58, 59 and 60. Out of which 5 wards showed values beyond 2000 mg.l^{-1}. It is interesting to note that most of the wards showing high level of total solids are present near the Swarn Rekha sewage channel. High total dissolved solids content ranging above 2000 mg.l^{-1} was observed in ward no.35 while ward nos. 37, 40, 46, 47, 52, 53 and 55 had values more than 1000 mg/l. Total dissolved solids showed good correlation with conductivity. Suspended solids were found in more than 1000 mg. l^{-1} in ward nos. 31, 33, 35, 53 and 55. Total alkalinity varied from 20 mg. l^{-1} to 494 mg. l^{-1}. Ward nos. 31, 38, 40, 52, 53 and 60 showed high values of alkalinity. Alkalinity seems to be due to bicarbonates present in water. Ward nos. 31, 33, 35, 37, 39, 58 and 60 showed high values of hardness. Which correlated largely with the sources of high solid contents. Sodium also showed great variation ranging from traces to 460 mg.l^{-1}.

In Lashkar township 98 out of 450 (21.7) stool samples were found infected with *E. histolytica*. Maximum incidence was found in ward nos. 49, 52, 55 and 60. Maximum number of *Giardia* was found in ward nos. 46, 49, 52 and 60 and minimum in ward no. 31. Total 12.8 per cent samples were observed infected. Ward nos. 49, 52 and 60 showed high percentage of *Ascaris lumbricoides* and total 15.7 per cent samples were found infected. 4.5 per cent infection of *Trichuris trichiura* was found infected. In ward no. 52 and 60 found maximum incidence of this parasite. 7.1 per cent samples were observed infected with *Ancylostoma duodenale*. Maximum number was found in ward nos. 33, 46, 55 and 60 while ward nos. 31, 38, 43 and 47 showed no evidence of this parasite. 3.7 per cent samples showed infection with *Enterobius vermicularis*. Ward nos. 32, 46 and 60 showed high percentage of infection.

Though there are wide variations in the chemical constituents of different samples, there does not seems to be any serious threat to the water quality except that in some samples it shows more hardness which crosses the maximum permissible limit (600 mg. l^{-1} ICMR 1975, BIS 1992). The high contents of hardness may render it less useful for industrial purpose but for human use it has been used for long time without any health problems.

In this study maximum. percentage of amoebiasis was found, which occurs by contaminated water and food. Maximum incidence of parasitic and helminthic infection was in low income groups due to lack of proper sanitation and hygiene.

Conclusion

In this study we found that there are wide variations in chemical contents of different samples. Some samples had more hardness. Most of the contamination is suspected to be due to Swarn Rekha Nala. The incidence of intestinal parasitic infection can be curtailed with the following precautions:

1. Drinking water should be boiled before consumption.
2. Drinking water source should be installed at least 15 meters away from septic tank and sewer line.
3. Attention should be paid to personal cleanliness and hygienic condition.
4. Avoidance of use of raw vegetables.
5. Soap should be used for hand washing after defecation.
6. Filth should not be allowed to accumulate around the hand pumps and borewells.
7. Water should be treated before supply from borewells.

References

A.P.H.A. 1989. *Standard Methods for the Examination of Water and Wastewater*, 17th edn. American Public Health Association, New York.

B.I.S. 1992. *Bureau of Indian Standards: Drinking Water Specification*, (First Revision) First Reprint Act, 1992. IS 10500 : 1991.

Chatterjee, K.D. 1981. *Parasitology in Relation to Clinical Medicine*, 12th Edn. Chatterjee Medical Publishers, Kolkata, p. 208–212.

I.C.M.R. (Indian Council of Medical Research) 1989. *Manual of Standards of Quality of Drinking Water Supplies*, 2nd Edition. Special Report Series no. 44, New Delhi.

Chapter 20

Energy Dispersive X-ray Spectrometer (EDS) Analysis of Cesspool Environment Soil Samples

J. Subashini, N. Ramamurthy* and G. Jagadeesan***

**Department of Physics, Annamalai University, Annamalai Nagar, Tamil Nadu, India*

***Reader, Toxicological Division, Department of Zoology, Annamalai University – 608 002*

ABSTRACT

An attempt had been made to estimate the trace elements like Na, Mg, Al, Si, P, S, Cl, K, Ca, Mn, Fe, Cu, Zn, Cd and Pb by Energy Dispersive X-ray Spectrometer (EDS) method. With EDS microanalysis device, energy dispersive spectra was recorded and the semi quantitative estimation of major and minor elements present in the soil samples, collected from four different location at three different depths during rainy season in the environment of Cesspools.

Keywords: EDS, Trace elements, Cesspool environment, Soil.

Introduction

Trace elements and heavy metals are inherent components of soils, but today's concern is related to their accumulation from man-made applications (Jean-Marc Boliag and Wieslaw Barabaz, 1979).

With continued application of the cesspool waste in the environment, the ability of the soil to treat it may decline. Even with soil absorption system operating well, groundwater contamination is inevitable because the soil is not equally effective at removing all constituents of the percolate (Viraraghavan, 1976).

Conventional methods of liquid waste disposal for homes in unsewered areas consist of Cesspool treatment followed by subsurface soil disposal (Bouma *et al.*, 1975; Simons and Magdoff, 1979). Large populations in both rural and urban areas rely on cesspool-soil absorption treatment systems as the principal disposer of domestic waste materials (Hagedorn *et al.*, 1981).

Trace elemental analysis, SEM-EDS technique has been used extensively with soils and sediments (Bashar Alsmadi and Peter Fox, 2001). Energy dispersive X-ray fluorescence spectrometry is a popular method for the determination of trace elements in geological and environmental samples (Terrance Hettipathirana, 2004).

The objective of this investigation is to estimate the weight percentage of the elements like Sodium (Na), Magnesium (Mg), Aluminum (Al), Silicon (Si), Phosphate (P), Sulfate (S), Chloride (Cl), Potassium (K), Calcium (Ca), Manganese (Mn), Iron (Fe), Copper (Cu), Zinc (Zn), Cadmium (Cd) and Lead (Pb) accumulating in the Cesspool environment during rainy seasons.

Materials and Methods

Soil samples were collected in the rainy season, at locations near the Cesspools from Chidambaram, Cuddalore District, Tamil Nadu, India. The samples were collected at four different locations near a septic tank leachate at various depths of 0.5 m, 1.0 m and 1.5 m respectively. The soil samples were oven dried at 60°C, ground to a fine powder. The surface of the powder soil sample has been stubbed using the double-sided adhesive carbon tape. Samples are coated with the help of Platinum coater (JEOL Auto Fine coater model JFC 1600, Coating time is [illegible]) and deposited with a thin layer of Platinum on the sample.

Elements present in the samples were estimated using the OXFORD INCA Energy Dispersive X-ray Fluorescence Spectrometer (EDS) attached to the (Scanning Electron Microscope) SEM JEOL model, JSE-5610 LV available at Centralised Instrumentation and Services Laboratory (CISL), Annamalai University.

Results and Discussion

Figure 20.1 and 20.2 show EDS spectrum of the soil samples of location 1 to location 4, in three different (a. 0.1 m, b. 1.0 m and c.1.5 m) depths respectively.

The results obtained from EDS, *i.e.*, weight percentage of trace elements distribution in samples are furnished in (Table 20.1). The energy value of each peak is matched with X-ray emission wavelength for non-diffractive analysis and the elements present in the soils are identified. As all elements have unique configuration of electron energy levels, the energy spectrum obtained from any particular element is characteristic of that particular element.

It is observed from the analysed data, the concentration of the trace elements Al, Si, P, S, Cl, K, Ca, Fe and Cu are present in all sampling points compared to Na, Mg, Mn, Zn, Cd and Pb. Further, it shows the high concentrations of Silicon (Si) *i.e.*, 36.61 to 61.95 per cent. In all the sampling points Si is more than 35 per cent. Aluminum (Al) concentration is 5.49 to 16.39 per cent. Silicon and Aluminum are elements, which are naturally present in the soil.

Sodium values ranged from 0.24 to (1.0 m depth of location 4) to 3.85 per cent (1.5 m depth of location 2) and it is absent in the 1.0 m depth of location 2. Totally the value of Na is less than 5 per cent. The value of Mg varies between 0.05 (1.5 m depth of location 1) to 3.19 per cent (1.5 m depth of location 2). In the 0.5 m depth of location 2, Mg is absent. The values of Phosphate ranged from 1.01 to 4.93 per cent. The highest value of P is at 0.1 m depth of location 4 (4.93 per cent). The lowest value of P is 1.01 per cent (1.5 m of location 4).

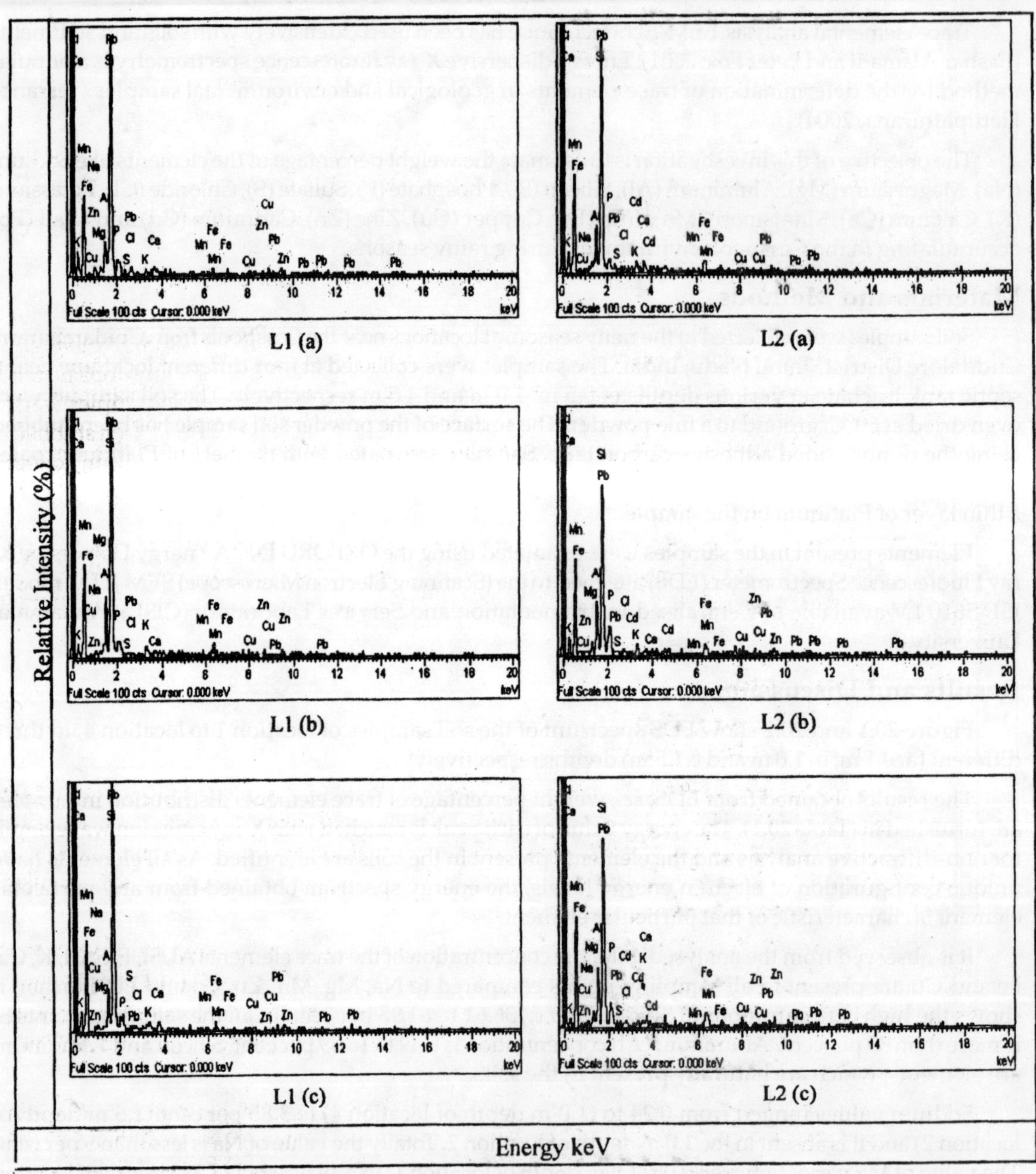

Figure 20.1: Energy Dispersive Spectrum of the Location 1 (L1) and Location 2 (L2) at Depths (a) 0.5 m, (b) 1.0 m and (c) 1.5 m. in Rainy Season

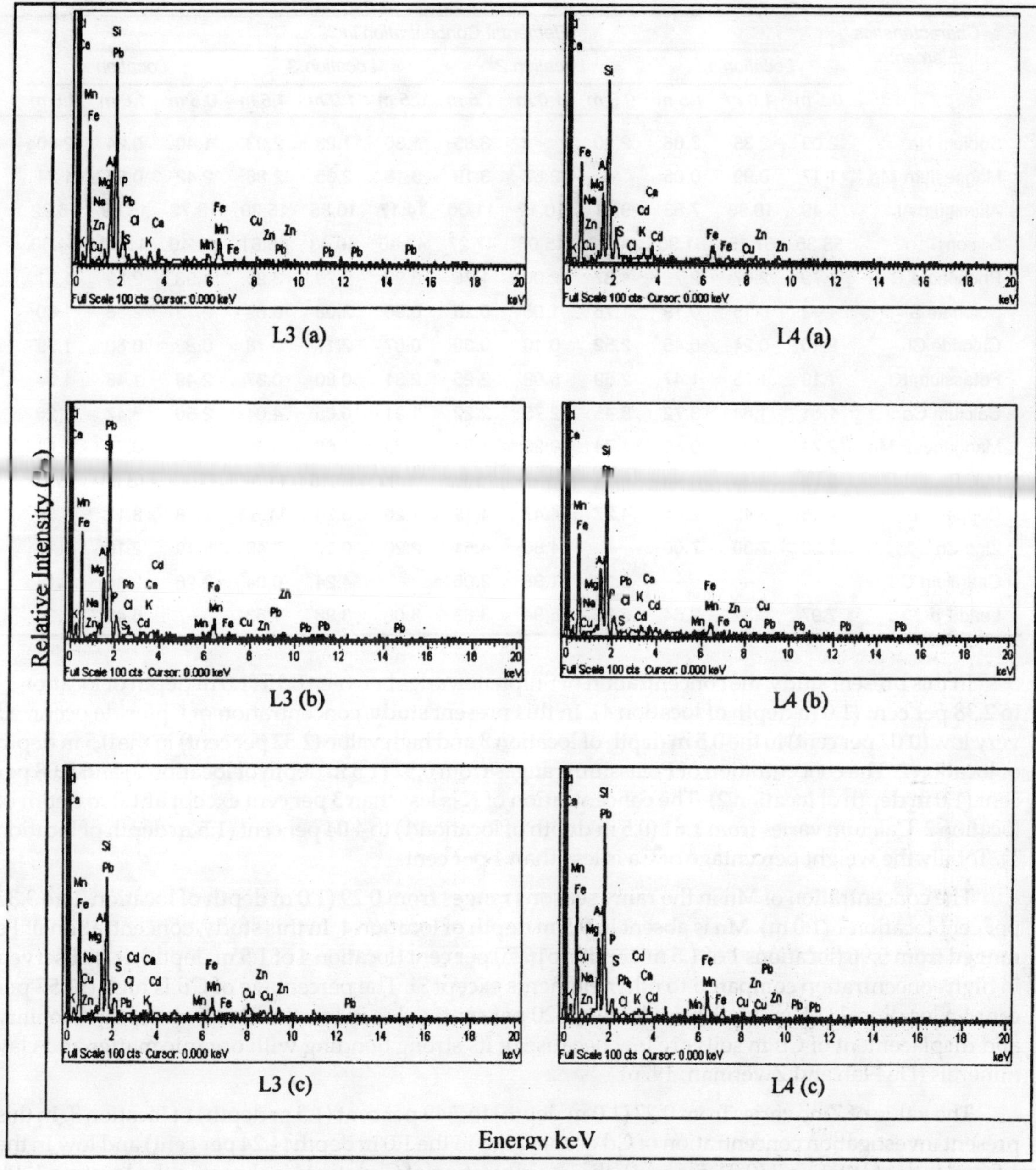

Figure 20.2: Energy Dispersive Spectrum of the Location 3 (L3) and Location 4 (L4) at Depths (a) 0.5 m, (b) 1.0 m and (c) 1.5 m. in Rainy Season

Table 20.1: The Percentage of Trace Elements Present in the Soil Samples at Various Locations and Depths in the Cesspool Environment in Rainy Season

Characteristics Element	*Elemental Concentration in %*											
	Location 1			*Location 2*			*Location 3*			*Location 4*		
	0.5 m	*1.0 m*	*1.5 m*	*0.5 m*	*1.0 m*	*1.5 m*	*0.5 m*	*1.0 m*	*1.5 m*	*0.5 m*	*1.0 m*	*1.5 m*
Sodium Na	2.09	2.35	3.06	2.20	–	3.85	1.30	1.28	2.93	1.40	0.24	2.90
Magnesium Mg	1.17	0.99	0.05	–	2.67	3.19	3.18	2.35	2.88	2.42	0.87	1.24
Aluminum Al	5.49	10.99	7.83	9.81	10.12	11.00	14.17	10.85	15.29	13.72	16.39	16.22
Silicon Si	58.35	57.15	61.95	50.78	48.08	47.27	43.30	50.43	36.61	42.40	41.82	44.60
Phosphate P	1.79	2.17	2.76	4.37	2.00	1.76	3.92	2.79	3.25	4.93	2.19	1.01
Sulphate S	1.12	0.15	0.19	1.76	1.05	0.25	0.50	2.08	0.69	2.13	2.38	1.80
Chloride Cl	2.10	0.24	0.45	2.52	0.10	0.30	0.07	2.17	0.78	0.82	0.60	1.79
Potassium K	1.19	1.75	1.47	2.59	5.08	2.25	2.31	0.80	0.37	2.49	1.48	1.34
Calcium Ca	1.61	1.68	3.72	3.45	2.75	2.22	1.81	0.05	4.04	2.50	3.47	2.26
Manganese Mn	2.74	3.35	0.60	1.74	0.29	1.52	2.39	1.66	0.48	–	0.77	1.11
Iron Fe	5.73	8.33	5.96	8.11	11.65	11.80	15.82	12.15	11.97	12.90	15.58	16.70
Copper Cu	7.15	3.42	2.61	4.27	4.47	4.19	1.20	3.74	11.57	7.78	3.11	6.38
Zinc Zn	1.50	7.30	7.00	–	4.80	4.51	2.20	0.27	7.49	5.19	2.16	1.19
Cadmium Cd	–	–	–	3.10	1.98	2.06	–	4.24	0.04	3.26	2.48	1.15
Lead Pb	7.97	0.32	2.54	5.36	5.94	3.83	8.06	3.98	1.63	–	6.49	0.34

In this present study, the concentration of Sulphate varies between 0.15 (1.0 m depth of location 1) to 2.38 per cent (1.0 m depth of location 4). In this present study, concentration of Chloride occurred very low (0.07 per cent) in the 0.5 m depth of location 3 and high value (2.52 per cent) in the 0.5 m depth of location 2. The concentration of Potassium ranges from 0.37 (1.5 m depth of location 3) and 5.08 per cent (1.0 m depth of location 2). The concentration of K is less than 3 per cent except at 1.0 m depth of location 2. Calcium varies from 1.61 (0.5 m depth of location 1) to 4.04 per cent (1.5 m depth of location 1). Totally the weight percentage of Ca is less than 4 per cent.

The concentration of Mn in the rainy seasons ranges from 0.29 (1.0 m depth of location 2) to 3.35 per cent location 1 (1.0 m). Mn is absent in 0.5 m depth of location 4. In this study, concentration of Fe ranged from 5.96 (locations 1 of 1.5 m depth) to16.70 per cent (location 4 of 1.5 m depth), it is observed in high concentration compared to other elements except Si. The percentage of Cu is high (11.57 per cent) in location 3 (1.5 m depth) and lower (1.20 per cent) value in location 3 (0.5 m depth). Mobility and displacement of Cu in soils are low because of its strong bonding with organic matter and clay minerals (De Han and Zwerman, 1976).

The value of Zinc varies from 0.27 (1.0 m depth) to 7.49 per cent (1.5 m depth) of location 3. In this present investigation concentration of Cd occurs high in the 1.0 m depth (4.24 per cent) and low in the 1.5 m depth of location 3 (0.04 per cent), the concentration of Cadmium is absent in the location 1. Pb varies from 0.32 (1.0 m depth of location 1) to 8.06 per cent (0.5 m depth of location 3) in our samples. Manganese (Mn) is an essential element and functions as a critical component of several enzyme systems.

Sakal *et al.* (1992) and Adhikari *et al.* (1993) have reported that the increased level of heavy metals like Cd, Pb, Zn, Cu, Mn and Fe in surface soil mainly due to irrigated with untreated sewage and industrial effluents.

The highly contaminated groundwater is not suitable for human consumption and other domestic purposes. Som *et al.* (1994) have also similar type of observation when they are analysing in sewage effluents and concluded that although trace metal constitute only a small fraction of sewage effluents, their contents in soils may be significantly raised through long term application of sewage effluents. So proper treatment must be taken prior to the introduction of Cesspool wastes in the environment.

References

Adhikari, S., Gupta, S.K. and Banerjee, S.K. 1993. Trace metal content of city sewage and sludge. *J. Indian. Soc. Soil Sci.*, 41: 170–172.

Bashar, M. Alsmadi and Peter Fox. 2001. Semi-quantitative analysis of changes in soil coatings by scanning electron microscope and energy dispersive X-ray mapping. *Colloids and Surfaces A: Physicochemical and Engineering Aspects*, 194: 249– 261.

Bouma, J., Converse, J.C., Otis, R.J., Walker, W.G. and Ziebell, W.A., 1975. Amount system for on–site disposal of septic tank effluent in slowly permeable soils with seasonally perched water tables. *J. Environ. Qual.*, 4: 382–388.

De Han FA M. and Zwerman, P.J. 1976. Pollution of soil. In: *Soil Chemistry*, (Eds.) G.H. Bolt and M.G.M. Bruggenwert. Elsevier, Amsterdam.

Hagedorn, C., McCoy, E.L. and Rahe, T.M. 1981. The potential for groundwater contamination from septic effluents. *J. Environ. Qual.*, 10: 1–8.

Jean-Marc Bollag and Barabasz, W. 1979. Effects of heavy metals on the denitrification process in soil. *J. Environ. Qual.*, 8: 196–201.

Sakal, R., Prasad, A.K. and Singh, A.P. 1992. Depth wise distribution of trace metals in soils receiving sewage and sludge. *J. Indian Soc. Soil Sci.*, 40: 732–737.

Simons, A.P. and Magdoff, F.R. 1979. Disposal of septic tank effluent in mound and sand filter trench systems on clay soil. *J. Environ. Qual.*, 8: 469–473.

Som, S., Gupta, S.K. and Banerjee, S.K. 1994. Assessment of quality sewage effluents from Howrah sewage treatment plant. *J. Indian Soc. Soil Sci.*, 42: 571–575.

Terrance, D. Hettipathirana, 2004. Simultaneous determination of parts per million level Cr, As, Cd and Pb and major elements in low level contaminated soil using borate fusion and energy dispersive X-ray fluorescence spectrometer with polarized excitation. *Spectrochimica Acta*, Part B, 59: 223–229.

Viraraghavan, T. and R.G. Warnock. 1976. Groundwater quality adjacent to aseptic tank system. *J. Am. Water Works Assoc.*, 68: 611–614.

Chapter 21

Effect of Stocking Density on the Blood Parameters of Goldfish *Carassius auratus*

A. Elezabeth Mary and M. Sakthivel***

**St. Mary's College, Tuticorin – 628 001, Tamil Nadu, India*

***Kamaraj College, Tuticorin – 628 003, Tamil Nadu, India*

ABSTRACT

The impact of stocking density on the blood parameters was studied in *Carassius auratus.* The group reared in the density of 10 individuals/aquarium showed maximum values of all parameters studied. The other groups 1, 5, 20 and 40 individuals/aquarium displayed lower blood values in relation to the density and exposure period. The change in the total RBC of the group with 20 individuals/aquarium was not significant. The RBC reduction was significant in the group with 40 individuals/aquarium. Haemoglobin content, Haemotocrit Thrombocytes and Coagulation time also showed similar trend. But MCV, MCH, MCHC values showed increase. Total WBC count of the higher density group was reduced. Relatively the lymphocytes showed a significant increase in higher density groups.

Keywords: Haematological parameters, Stocking density, Carassius auratus.

Introduction

In fish farming practice, high density culture is often opted to maximize the available water resources to enhance the yield. The interaction of stocking density and water turnover is studied by many authors (Andrews *et al.*, 1971; Sampath, 1985; Jena *et al.*, 2001). Reports on the impact of water quality parameters and other toxic products of metabolism on growth of fish in high density culture are also available (Armstrong and Boyd, 1982; Soderberg *et al.*, 1983; James and Sampath, 2003). Sakthivel (1993) reported the density effect on the blood parameters of *Cyprinus carpio.* Similar studies in ornamental fish is almost lacking. Moreover, the effect of environmental factors on blood will be expressed in advance of reduced growth (Wedemeyer, 1996a). Considering these, the present study was undertaken to ascertain the effect of stocking density on the blood parameters of gold fish, *Carassius auratus*.

Materials and Methods

Carassius auratus (10±2g live weight) were procured from an ornamental fish farm at Sawyerpuram, Tuticorin District, Tamil Nadu and acclimated for 60 days on a pelleted diet with 35 per cent crude protein prepared according to the square method (Hardy, 1980). Test fish were divided into 5 density groups with 1, 5, 10, 20 and 40 individuals/aquarium of 120 L capacity. Triplicates were maintained for each group. Experiments were conducted in glass aquaria. The test fish were fed twice daily at 08.00 and 17.00 hours at 5 per cent body weight every time. The following experimental parameters were maintained throughout the experiment; Temperature 26±1°C; pH 7.1; D.O > 8 mg/l and photoperiod (12LL : 12DD).

Haematological parameters were estimated at a 30 days interval for a period of 120 days. Blood samples were collected in capillary tubes by severing the caudal peduncle. Wintrobe's method (1978) was adopted to estimate the chosen blood parameters. Red cell indices were calculated according to Johansson-Sjöbeck and Larson (1979). Differential count of WBC was made by classical dry smear method. The relative percentage of differential Leucocyte count was made based on the count of 1000 cells from two imprints from each fish.

Results and Discussion

The changes observed in RBC and related parameters of *C. auratus* are given in Table 21.1. The group reared in the density of 10 individual/aquarium showed maximum values of all blood parameters tested. But the other groups displayed gradual reduction in relation to the density and exposure period. In groups with the density of 20 or 40 individuals/aquarium the RBC count decreased from 1.64 to 0.56 × 10^6/mm^3 or 0.34 × 10^6/mm^3 during 120 days exposure. The change in RBC of the group with 20 individuals/aquarium was not statistically significant ($t = 2.9$; $P > 0.05$). But the RBC reduction was statistically significant in the groups with 40 individuals/aquarium ($t = 3.68$; $P < 0.05$). Similar trend was observed in Hb content and haematocrit values. On the other hand, MCV, MCH and MCHC values showed significant increase in higher density groups.

Total WBC decreased from 52.6 to 32.6 × 10^3/mm^3 ($t = 7.21$; $P < 0.01$) at the end of 90 days in the highest density of group of 40 individuals/aquarium. Similar decrease was observed in thrombocytes and coagulation time. Such significant variations occurred only after 120 days in the other groups with 5 or 20 individuals per aquarium whereas, the lymphocytes displayed an opposite trend, for example, in higher density groups (20 or 40 individuals/aquarium) lymphocyte percentage was increased from 37.7 to 67.6 per cent and 37.9 to 71.6 per cent respectively during 120 days exposure (Table 21.2).

The present observations clearly indicate that the increase in density of *C. auratus* has a significant impact on blood parameters. The blood parameters in the smaller groups (1 or 5 individuals) are not much affected. The intermediate group (10 individuals/aquarium) shows fairly good results when compared with other groups of 10 individuals/aquarium. But when the stocking density is further increased from 10 to 40 the RBC is decreased from 1.64 to 0.34 × 10^6/mm^3 and the total WBC is reduced from 52.6 to 32.6 × 10^3/mm^3 during 90 days exposure. This shows that at higher densities the crowding stress over rides the benefits of optimum density.

Table 21.1: Effects of Various Stocking Density on RBC and Related Values in *Carassius auratus*. Each value is the average ($\overline{X}$ ± SD) of 6 individual observations.

Blood Parameters	*Exposure Period (days)*	*Density (No. of individuals/aquarium)*				
		1	*5*	*10*	*20*	*40*
RBC count	0	1.63±0.11	162±0.18	1.64±0.15	1.64±0.19	1.64±0.25
(X 10^6/mm^3)	30	1.58±0.18	1.50±0.26	1.62±0.22	1.31±0.21	1.14±0.32
	60	1.51±0.31	1.32±0.36	1.63±0.18	0.99±0.25	0.84±0.28
	90	1.29±0.25	1.18±0.28	1.64±0.26	0.66±0.31	0.45±0.21
	120	1.18±0.32	1.06±0.32	1.62±0.32	0.56±0.34	0.34±0.18
HB content	0	6.81±0.21	6.78±0.42	6.84±0.32	6.83±0.12	6.82±0.31
(g %)	30	6.67±0.32	6.48±0.21	6.83±0.42	6.10±0.36	6.06±0.18
	60	6.52±0.28	5.81±0.18	6.81±0.15	5.26±0.19	4.81±0.21
	90	5.58±0.18	5.33±0.26	6.82±0.28	3.63±0.16	2.78±0.28
	120	5.26±0.16	4.62±0.12	6.79±0.19	3.17±0.20	2.30±0.33
Haematocrit	0	26.23±1.23	26.42±1.24	27.36±1.18	27.64±1.29	27.68±1.18
(%)	30	25.48±1.34	24.86±1.28	27.62±1.15	22.73±1.26	21.84±1.21
	60	24.52±1.39	22.41±1.38	27.51±1.27	18.85±1.23	16.31±1.24
	90	21.18±1.14	19.18±1.35	27.49±1.33	11.80±1.22	9.05±1.32
	120	19.86±1.15	16.25±1.36	27.52±1.35	10.22±1.11	6.94±1.36
MCV (fl)	0	160.92±1.31	163.09±1.14	166.83±1.11	168.54±1.36	168.78±1.61
	30	161.27±1.62	165.73±1.31	170.49±1.21	173.51±1.21	191.58±1.82
	60	162.38±1.91	169.77±1.76	168.77±1.36	190.40±1.19	194.17±1.73
	90	165.47±1.21	171.02±1.69	167.62±1.55	178.79±1.53	201.11±1.32
	120	168.31±1.43	177.36±1.48	169.88±1.90	182.5±1.28	204.12±1.18
MCH (pg)	0	41.78±0.84	41.85±0.72	41.71±0.63	41.65±0.92	41.59±0.68
	30	42.22±0.64	43.20±0.68	42.16±0.72	46.56±0.61	53.16±0.71
	60	43.18±0.38	44.02±0.32	41.78±0.81	53.13±0.72	57.26±0.36
	90	43.59±0.09	45.17±0.52	41.59±0.58	55.00±0.31	61.78±0.63
	120	44.58±0.92	48.87±0.19	41.91±0.29	56.60±0.42	67.65±0.32
MCHC (g/l)	0	25.96±0.28	25.66±0.78	25.01±0.72	24.71±0.82	24.64±0.94
	30	26.18±0.72	26.07±0.62	24.73±0.66	26.84±0.64	24.75±0.62
	60	25.59±0.66	27.26±0.58	24.76±0.51	27.90±0.58	29.49±0.78
	90	27.46±0.65	28.35±0.79	24.81±0.65	30.76±0.38	30.72±0.42
	120	28.49±0.73	29.46±0.82	24.67±0.74	31.02±0.73	33.14±0.38

The decline in leucocrit value obtained in the present study corroborates with the findings of Sinha, Ranjana *et al.*, (1999) in *Labeo rohita* under crowded conditions. Leucocrit is regarded as a sensitive indicator of physiological stress. The stress response of salmonids and fishes normally include a moderate to severe leucopenia. (Srivastava and Agarwal, 1977; McLeay and Gordon, 1977).

Table 21.2: Effects of Various Stocking Density on Total WBC, DLC and Coagulation Time in *Carassius auratus*. Each value is the average ($\bar{X}$ ± SD) of 6 individual observations.

Parameters	*Exposure Period (days)*	*Density (No. of individuals/aquarium)*				
		1	*5*	*10*	*20*	*40*
Total WBC	0	52.8±3.32	52.4±2.19	52.8±3.14	52.6±2.06	52.6±3.055
(X 10^3/mm^3)	30	51.3±1.80	50.1±2.23	52.8±2.05	45.6±3.29	42.5±2.10
	60	50.7±2.32	49.4±3.19	53.0±1.18	41.3±4.34	39.4±1.82
	90	49.1±3.46	47.9±2.08	52.7±3.19	38.6±3.14	35.2±1.19
	120	48.2±1.28	46.5±2.36	52.5±2.19	36.3±4.80	32.6±1.28
Lymphocyte	0	37.6±1.21	36.9±1.32	37.8±1.24	37.7±1.05	37.9±1.02
(%)	30	39.0±1.05	39.8±1.22	37.6±1.16	41.3±1.34	43.1±1.32
	60	41.3±1.14	46.8±1.08	37.9±1.08	49.4±1.14	48.2±1.21
	90	46.0±1.62	53.9±1.36	36.9±1.34	57.2±1.46	66.5±1.43
	120	53.9±1.44	57.1±1.44	37.6±1.42	67.6±1.06	71.6±1.36
Neutrophil	0	22.5±1.12	22.6±1.34	22.6±1.32	22.8±1.19	22.6±1.26
(%)	30	21.7±1.18	21.3±1.05	22.0±1.30	19.7±1.14	18.1±1.10
	60	20.5±1.02	19.6±1.15	22.5±1.24	16.5±1.05	16.3±1.22
	90	20.3±1.36	17.7±1.28	22.3±1.08	13.7±1.32	13.3±1.19
	120	17.2±1.08	15.2±1.19	22.5±1.06	10.2±1.28	10.7±1.08
Basophil	0	24.9±1.02	24.7±1.05	24.8±1.08	24.8±1.24	24.9±1.25
(%)	30	22.2±1.21	20.6±1.15	25.0±1.32	18.3±1.28	18.1±1.31
	60	19.3±1.36	16.7±1.23	24.7±1.43	13.5±1.32	13.2±1.18
	90	16.0±1.26	12.8±1.41	24.8±1.03	10.2±1.36	8.6±1.08
	120	12.8±1.44	8.6±1.29	25.2±1.15	6.8±1.31	6.1±1.19
Thrombocyte	0	18.9±1.15	18.9±1.08	19.2±1.05	18.8±1.36	19.0±1.11
(× 10^3/mm^3)	30	16.5±1.09	15.3±1.24	18.8±1.41	13.7±1.05	16.3±1.44
	60	14.8±1.12	12.4±1.44	18.8±1.36	10.9±1.11	9.2±1.08
	90	12.5±1.42	11.5±1.32	18.9±1.26	8.2±1.24	7.6±1.25
	120	10.6±1.06	8.6±1.08	19.1±1.28	6.4±1.32	5.1±1.34
Coagulation	0	22.61±1.05	22.63±1.08	22.58±1.28	22.68±1.06	22.72±1.07
time (sec)	30	22.38±1.09	20.41±1.22	22.62±1.09	19.74±1.03	29.01±1.04
	60	20.51±1.22	19.91±1.06	22.47±0.09	17.33±1.08	16.44±1.05
	90	20.18±1.08	17.34±1.08	21.92±1.11	16.15±1.26	15.95±1.21
	120	19.62±1.26	15.24±1.12	22.48±0.05	15.63±1.18	15.04±1.40

In the higher density groups while the lymphocytes a significant reduction in total RBC or WBC is observed in higher density groups (20 or 40 individuals/aquarium). This adverse effect on blood parameters of higher density groups may be due to the physiological stress caused by overcrowding.

It is well established that the intensity of competition for food is directly related to the population density (Weatherley, 1966). Though overcrowding may affect the health and growth of fish, an optimum degree of grouping exists among fish populations (Seng-Keh Teng and Thia-Engchua, 1978). Such a grouping effect is reflected in the present study also. The blood parameters of the fish at the lower stocking densities of 1 or 5 individuals/aquarium do not show significantly better values than those of the fish at a comparatively higher density and basophils are increased, the thrombocytes and neutrophils are significantly reduced. This finding is supported by similar haematological effects due to physiological stress caused by dietary nutrient deficiency (Mahajan and Agarwal, 1979) or by industrial effluents (McLeay, 1973) or crowding (Sakthivel, 1993) reported earlier.

The increase in lymphocytes observed in higher density group may be due to the stimulated lymphopoiesis or due to an increased release of lymphocytes from lymphomycloid tissues (Ellis, 1976) under physiological stress caused by overcrowding.

References

Armstrong, M.S. and Boyd, C.E. 1982. Oxygen transfer calculation for a tractor powered wheel aerator. *Trans. Amer. Fish. Soc.*, 111: 363–366.

Andrews, J.W., Knight, L.H., Page, J.W., Matsuda, Y. and Brown, E. 1971. Interactions of stocking density and water turnover on growth and food conversion of channel catfish reared in intensively stocked tanks. *Prog. Fish Cult.*, 33: 197–203.

Ellis, A.E. 1976. Leucocytes and related cells in the plaice *Pleuronectus platessa*. *J. Fish. Biol.*, 8: 143–156.

Hardy, R. 1980. Fish feed formulation. In: *Fish Feed Technology, Agriculture Development and Coordination Programme*, FAO/ADCP/Rep/80/11: 233–239.

James, R. and Sampath, K. 2003. Effect of stocking density on growth and fertility in the ornamental fish, *Xiphophorus helleri*. *Indian J. Fish*, 50(3): 339–347.

Jena, J.K., Ayyappan, S. and Aravinda Krishnan, P.K. and Hick Mudul. 2001. Comparative evaluation of growth, survival and production of carp species at different stocking densities under polyculture. *Indian J. Fish*, 48(1): 17–25.

Johansson-Sjobeck and Larson, A. 1979. Effect of inorganic lead on d-aminolevulinic acid dehydratase activity and haematological variables in the rainbow trout, *Salmo gairdneri*. *Arch. Environ. Contam. Toxicol.*, 8: 419–431.

McLeay, D.J. and Gordon, M.R. 1977. Leucocrit: A simple haematological technique for measuring acute stress in salmonid fish, including stressful concentration of pulp mill effluent. *J. Fish. Res. Bd. Can.*, 34: 2164–2175.

McLeay, D.J. 1973. Effects of 12h and 25 day exposure to kraft pulp mill effluent on the blood and tissues of juvenile Coho salmon. *J. Fish. Res. Bd. Can.*, 30: 395–400.

Sakthivel. 1994. Effect of stocking density on the blood parameters of the fish *Cyprinus crapio* L. *Environment and Ecology*, 12(2): 424–428.

Sampath, K. 1985. Food intake, conversion and surfacing activity as a function of density and water change frequency in an air breathing fish, *Channa striatus*. *Aquaculture*, 46: 201–213.

Seng-Keh Teng and Thia-Eng Chua. 1978. Length-weight relationship and growth of young estuary grouper, (*Epinephelus salmonides maxwell*) caught from the middle Bank, Penang straits, Sch. Biol. Sci. Univ. Sains Malasia Penang, Tech. Rep. USM/IFM/CTE No. 26: 15.

Sinha, Ranjana, Dutta, Tansuree and Das, M.K. 1999. Impact of stress on blood parameters and interrenals of *Labeo rohita*. In: Eco-friendly management of resources for doubling fish production: Strategies for 21st century. *Proceedings of the National Seminar*, December 22–23. Inland Fisheries Society of India, Barrackpore.

Soderberg, R.W., Flynn, J.B., Schmittou, H.R. 1983. Effects of ammonia on growth and survival of rainbow trout in intensive static water culture. *Trans. Amer. Fish. Soc.*, 112: 448– 451.

Srivastava, A.K. and U. Agarwal. 1997. Involvement of pituitary internal axis and cholenergic mechanism during cold shock leucocyte sequence in a freshwater tropical teleost, *Colisa fasciatus*. *Archs. Anat. Microsc.*, 66: 97–108.

Weatherley, A.K. 1966. Ecology of fish growth. *Nature*, 212: 1321–1324.

Wedemeyer, G. 1996a. Rearing conditions effects on fish in intensive culture. In: *The Physiology of Fish* (Edn.) Chapman Hall, New York.

Wintrobe, M.M. 1978. *Clinical Haematology*, 7th edn. Philadelphis Lea and Feluger Publ.

Chapter 22

Food and Feeding Habits of the Gobiid Fish *Pseudapocryptes lanceolatus* (Bloch and Schneider, 1801) of the Vasista Godavari Estuary, East Coast of India

K.V.C.S. Appa Rao and K. Sreeramulu*

Department of Zoology, Andhra University, Visakhapatnam

ABSTRACT

Investigations on the food and feeding habits of *Pseudapocryptes lanceolatus* (Bloch. and Schneider, 1801) show that the first mainly feed on planktonic organism. The diet of males and females did not differ. Qualitative analysis of the diet composition of *P. lanceolatus* revealed that the fish is phytoplanktonic bottom feeder.

Keywords: *Pseudapocryptes lanceolatus, Vasista Godavari Estuary, Food and Feeding habits.*

Introduction

Pseudapocryptes lanceolatus (Bloch and Schneider) is an important food fish of the east coast estuaries of India. Food and feeding habits of gobiid fishes were studied by Blaber (1977), Sarker and Al-daham (1980). In Indian waters, similar works were carried out among others by Venkateswarlu (1966), Geevarghese (1983 and 1984), Mehta *et al.*, (1990) and Sankar Rao (2001). In gobioids, herbivores have a long and coiled gut thrice the length of their body, whereas carnivores possess short gut

* Lecturer in Zoology, Sri Y.N. College, Narsapur, West Godavari District – 534275.

(Venkateswarlu, 1966). No attempt has been made on the food and feeding habits of the *P. lanceolatus*. The present study deals with the food and feeding habits of the lanceolate goby *P. lanceolatus* of the Vasista Godavari estuarine region.

Material and Methods

Generally the fish is captured by employing stake nets, laid across the Vasista Godavari estuary and also by using traps in canals and creeks around. Samples (30–50 numbers) for the present investigations were procured from such fish stocks once in a month from Narsapur during December 2000 to November 2001. On bringing the fish to the laboratory belly of each fish was cut open, alimentary canal removed and stomach dissected and its contents were emptied into a petri dish. The food items were examined under a binocular microscope.

Various methods are employed in the food analysis of fishes. Pearse (1915) first used volumetric method. Point method was adopted by Swynnerton and Worthington (1940). Frequency of occurrence method (Hynes, 1950) was adopted to decipher the relative importance of various items in the food number of stomachs (containing food) during each month. A number of modifications and combinations of different methods are used to express feeding intensity, employing different indices (Natarajan and Jhingran, 1962). One of such important method is index of preponderance (IP). This method combines both volumetric and frequency of occurrence methods giving rise to accuracy in grading various food elements. Details of methods are: (1) per cent volume and per cent occurrence of each food item were calculated and respective indices (V_1–Volumetric index and O_1–Occurrence index) for all the food items are assessed at (2) Subsequently, a combined index for each food item is calculated using the relation.

$$I_1 = \frac{V_1 O_1}{\Sigma V_1 O_1} \times 100$$

where,

I_1 is a composite index, *i.e.*, Index of preponderance, V_1 is volumetric index of food item and 1 and O_1 is the occurrence index of food item 1.

(3) Another method of indices is the Gastrosomatic index (GSI) which is expressed as:

$$GSI = \frac{\text{Weight of Stomach Contents}}{\text{Weight of Fish}} \times 100$$

Results and Discussion

A total of 493 stomachs of *P. lanceolatus* (Bloch and Schneider) were examined during the entire course of study. Out of them, 368 stomachs contained food though in varying quantities and stages of digestion where as the remaining 125 guts were empty (Tables 22.2 and 22.3).

Feeding Intensity

Seasonal variation in feeding intensities was ascertained based on monthly analysis of the data. Per cent occurrence of empty stomachs was found to be quite high during almost all the months (Tables 22.2 and 22.3).

Table 22.1: Total Points and Gastro Somatic Index (GSI) in *P. lanceolatus*

Month	*Specimens Examined*	*Specimens with food*	*Total Points*	*Average Points*	*Average Feeding*	*GSI*
December 2000	45	42	42	93.33	0.466	3.747
January 2001	50	33	33	64.00	0.250	1.984
February	45	42	42	93.33	0.375	3.760
March	50	27	27	54.00	0.287	0.888
April	50	28	28	56.00	0.268	1.118
May	30	16	16	53.33	0.281	1.502
June	40	34	34	85.00	0.449	5.910
July	45	28	28	62.22	0.348	1.551
August	40	31	31	77.50	0.379	2.567
September	40	35	35	87.50	0.329	2.358
October	40	27	27	67.50	0.315	2.919
November	30	25	25	83.33	0 330	2.934

Table 22.2: Occurrence of Fullness in the Stomachs of *P. lanceolatus*

Month	*Specimens Examined*	*Condition of Stomachs*						
		Gorged	*Full*	*¾ Full*	*½ Full*	*¼ Full*	*Trace*	*Empty*
December 2000	45	6 (13.33)	–	–	9 (20)	27 (60)	–	3 (6.67)
January 2001	50	–	2 (4.44)	–	–	33 (66)	–	17 (34)
February	45	1 (2.22)	–	–	11 (24.44)	28 (62.2)	–	3 (6.67)
March	50	–	–	–	4 (8)	23 (46)	–	23 (46)
April	50	–	–	–	6 (12)	22 (44)	–	22 (44)
May	30	–	–	–	2 (6.66)	14 (46.67)	–	14 (46.67)
June	40	2 (5.00)	--	–	19 (47.5)	13 (32.5)	–	6 (15)
July	45	–	–	1 (2.22)	8 (17.78)	19 (42.22)	–	17 (37.78)
August	40	1 (2.50)	–	3 (7.5)	6 (15)	21 (52.5)	–	9 (22.5)
September	40	–	–	2 (5)	7 (7.75)	56 (65)	–	5 (12.5)
October	40	–	–	–	7 (7.75)	20 (50)	–	13 (32.5)
November	30	–	–	–	8 (26.67)	17 (56.67)	–	5 (16.66)

Table 22.3: Feeding Condition in *P. lanceolatus*

Month	*Well Fed (Active)*	*Moderately Fed*	*Poorly Fed*	*Empty*
December 2000	13.33	20.00	60.00	6.67
January 2001	–	–	66.00	34.00
February	6.67	24.44	62.22	6.67
March	–	8.00	46.00	46.00
April	–	12.00	44.00	44.00
.May	–	6.66	46.67	46.67
June	5.00	47.50	32.50	15.00
July	2.22	17.78	42.22	37.78
August	10.00	15.00	52.50	22.50
September	5.00	17.50	65.00	12.50
October	–	17.50	50.00	32.50
November	–	26.67	56.67	16.66
Total	**42.22**	**213.05**	**623.78**	**320.95**
Average	[illegible]	**17.75**	**51.98**	**26.75**

Food Composition

Analysis of the gut contents revealed that *P. lanceolatus* consumes only planktonic organisms and is thus a planktovorous fish. Diet of the fish generally comprised of phytoplankton belonging to algae and diatoms besides sand grains. No difference was observed between the food of male and female fishes.

Qualitative Variation

Qualitatively no variation was noticed between the male and female fishes. The gut contents could be identified into five categories, *i.e., Pleurosigma* sps., *Navicula* sps., *Nitzchia* sps., *Oscillatoria* sps., sand grains and digested material *Plemo sigma* sps. and *Nitzchia* sps. constituted main food (Table 22.4).

Analysis of Points

Month wise pooled data of the gut contents indicated that a maximum of 93.33 points per stomach occurred in the month of December, 2000 and February, 2001, followed by 87.50 points in September while minimum points of 53.00 were obtained during May, followed by 54.00 points in March (Table 22.1). The study also showed that on average only 17.75 per cent of the fishes existed in moderately fed condition. Based on the fullness of stomach, it was found that fishes with well fed stomachs were relatively few than remaining categories. A maximum of 13.33 per cent well fed stomachs were encountered during December 2000 followed by 10 per cent in August, 2001 (Table 22.3).

Analysis of Volume

Volume of each item in the food material was calculated and they were 43.32 per cent for *Pleurosigma* sps., 30.30 per cent *Nitzchia* sps., 7.67 per cent for *Navicula* sps., 0.53 per cent for *Oscillatoria*, 4.15 per cent for sand grains and 14.03 per cent for digested matter (Table 22.4).

Table 22.4: Food Composition of *P. lanceolatus* in Terms of Volume (V) Frequency of Occurrence (O) and Index of Preponderance (IP)

Food Items	Volume		Frequency of Occurrence		Index of Preponderance	
	Per cent	Rank	Per cent	Rank	Per cent	Rank
Pleurosigma sp.	43.32	1	27.67	2	49.66	1
Navicula sp.	7.67	4	18.62	3	5.90	4
Nitzchia sp.	30.30	2	28.45	1	35.58	2
Oscillatoria sp.	0.53	6	2.41	6	0.05	6
Sand grains	4.15	5	10.09	5	1.68	5
Digested matter	14.03	3	12.76	4	7.13	3
Total	**100**		**100**		**100**	

Table 22.5: Index of Preponderance of Food Items of *P. lanceolatus*

Food Item	V_1	O_1	V_1O_1	IP	Rank
Pleurosigma sp.	43.32	27.67	1208.35	49.66	1
Navicula sp.	7.67	18.62	143.56	5.90	4
Nitzchia sp.	30.30	28.45	865.73	35.58	2
Oscillatoria sp.	0.53	2.41	1.28	0.05	6
Sand grains	4.15	10.09	40.97	1.68.	5
Digested matter	14.03	12.76	173.54	7.13	3
	100	**100**	**2433.43**	**1.000**	

Analysis by Frequency of Occurrence

Different food items occurred in different frequencies in the food material and they are *Nitzchia* sps. 28.45 per cent, *Pleurosigma* sps. 27.67 per cent, *Navicula* sps. 18.62 per cent, digested matter 12.75 per cent, sand grains 10.09 per cent and *Oscillatoria* sps. 2.41 per cent (Table 22.4).

Analysis of Index of Preponderance

By Index of Preponderance, *Pleurosigma* sps. ranked first with 49.66 per cent, *Nitzichia* sps. second with 35.58 per cent, digested matter with 7.13, *Navicula* sps. fourth with 5.90 per cent, sand grains, fifth with 1.68 per cent and *Oscillatoria* sps., least with 0.05 per cent (Table 22.4).

Gastrosomatic Index

In gastrosomatic index is higher in the months of December, 2000, February, June, October and November months of 2001. It is lower in the month of January, March, April, May, July, August, September months of 2001 (Table 22.1).

Generally, fishes feed intensively during post monsoon and winter periods. This habit is attributed to the abundance of food supply (Hardy, 1936) and to the combined effect of food supply and temperature (Ricker, 1937).

Gastro somatic indices (GSI) of the fish for different months revealed that the feeding intensity in both the sexes attained peaks twice during a year. The first peak was observed (Figure 22.1) in the month of February while the second peak in the month of June. Stomachs of a majority of fishes were with ¾ and ½ full during these two months. The first peak corresponded to the post spawning period and the second peak to the abundant production of phytoplankton.

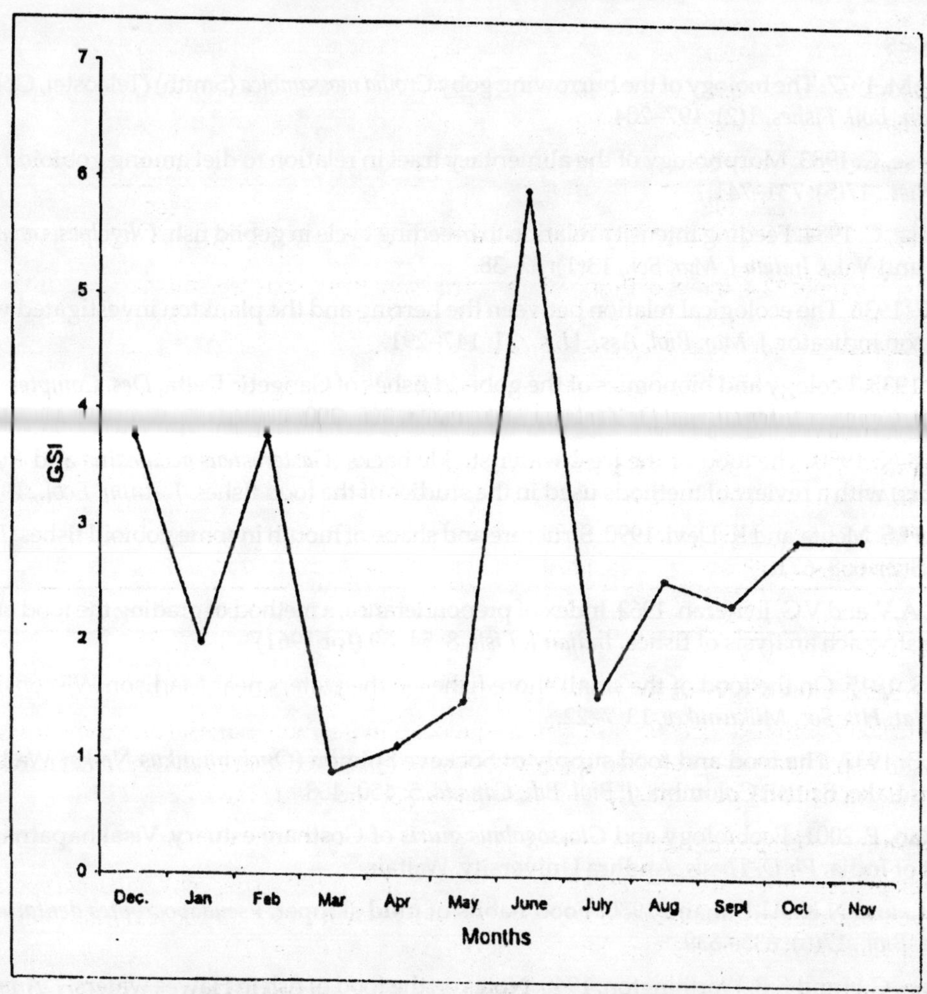

Figure 22.1: Gastro Somatic Index (GSI) in *P. lanceolatus*

An analysis of the food and feeding habits of *P. lanceolatus* based on various methods clearly ndicated that this fish is a pure herbivore chiefly feeding on Phytoplankton both during juvenile and dult stages of its life history. *Pleurosigma* sps. and *Nitzchia* sps., were the two diatoms that constituted ts main food. All the three methods of food analysis, namely volumetric, frequency of occurrence and ndex of preponderance methods confirmed this finding. However, according to Hora (1935) *P. nceolatus* inhabiting the Gangetic estuaries is a planktonic feeder during juvenile stages but a mud eeder during adult stages. Such variation in the food and food feeding habits of a single species from two distinct geographical regions is rather uncommon and may be attributed to competition for food coupled with isolation.

Acknowledgements

The authors are grateful to Andhra University for providing facilities. One of the authors is grateful to UGC for the award of a Teacher Fellowship during the tenure.

References

Blabber, S.J.M. 1977. The biology of the burrowing goby *Croilia mossambica* (Smith) (Teleostei, Gobiidae). *Environ. Boil. Fishes*, 1(2): 197–204.

Geevarghese, C. 1983. Morphology of the alimentary tract in relation to diet among gobioid fishes. *J. Nat. Hist.*, 17(5): 731–741.

Geevarghese, C. 1984. Feeding intensity relation to breeding cycle in gobiid fish, *Oligolpesis acutipennis* (Cuv. and Val.). *Indian J. Mar. Sci.*, 13(1): 37–38.

Hardy, A.E. 1936. The ecological relation between the herring and the plankton investigated with the plankton indicator. *J. Mar. Biol. Ass., U.K.*, 21: 147–291.

Hora, S.L. 1935. Ecology and bionomics of the gobioid fishes of Gangetic Delta. *Des. Comptes. Rendus Du XII. Congress International De Zoologie-Lisbonne*, p. 265–290.

Hynes, H.B.N. 1950. The food of the freshwater stickle backs (*Gasterosteus acculeatus* and *Pygosteus pungitis*) with a review of methods used in the studies of the food fishes. *J. Anim. Ecol.*, 19: 36–58.

Mehta, R., H.S. Mehta and K. Devi. 1990. Structure and shape of mouth in some gobioid fishes. *Environ. Ecol.*, 8(2): 668–671.

Natarajan, A.V. and V.G. Jhingran. 1962. Index of preponderance, a method of grading the food elements in the stomach analysis of fishes. *Indian J. Fish*, 8: 54–59 (for 1961).

Pearse, A.S. 1915. On the food of the small shore fishes in the waters near Madison Wisconsin. *Bull. Wis. Nat. Hit. Soc. Milkwaukee*, 13: 7–22.

Ricker, W.E. 1937. The food and food supply of Sockeye Solmon (*Onchorhynkus Nerkar* Walbum) in Cultus Lake, British Colombia. *J. Biol. Bd., Canada*, 5: 450–468.

Sankara Rao, P. 2001. Ecobiology and *Glossogobius giuris* of Gosthani estuary, Visakhapatnam, East Coast of India. *Ph.D. Thesis*, Andhra University, Waltair.

Sarker, A.L. and N.K. Al-Daham. 1980. Food habits of mud skipper, *Pseudapocryptes dentatus* (Val.). *J. Fish. Biol.*, 17(6): 635– 639.

Swynnerton, G.H and E.B. Worthington. 1940. Notes on the food of fish in Hawes waters. *J. Anim. Ecol.*, 9: 283–287.

Venkateswarlu, T. 1966. A note on the oxygen carrying capacity of the blood in three gobioid fishes. *Naturwiss*, 12: 313.

Chapter 23

Physico-chemical Studies on Pollution in River Sengar at District Etawah (U.P.)

K.K. Saxena*, Raj Narayan and Yogesh Babu Dixit

P.G. Department of Zoology, Janta Mahavidyalaya, Ajitmal, Auraiya (U.P.)

ABSTRACT

The river Sengar at District (Etawah) is an important source of water supply in the last few years. This study of physico-chemical characteristics of water of the river Sengar revealed that the water temperature always followed the atmospheric temperature with a minimum of 18.5ºC during winter and the maximum being 33.0ºC in summer. The transparency fluctuated between 19.0 to 50.9 cm. The colour of the water was green in summer and winter but muddy in rainy season. The minimum value of pH was 7.0 during rainy season and elevated upto 8.7 during winter season. The minimum DO recorded was 4.0 mg/l while the maximum value of 8.5 mg/l was recorded during winter season. The CO_2 was recorded between 8.3.to 14.0 mg/l. The BOD were recorded minimum value was 4.3 mg/l in winter season and the maximum value was recorded during summer while the COD were found between 26.0 to 44.1 mg/l.

Keywords: Physico-chemical, Pollution, River Sengar, Etawah (U.P.).

Introduction

The water quality in terms of physico-chemical characteristics was studied to assess the river water suitability for drinking purpose. Most of the Indian rivers and their tributaries are being used as sites for disposal of domestic sewage and industrial waste leading to the grate problem of water pollution.

---

* Postal Address: Param Hans Kuti (Bari), Street No. 8, Ashok Nagar, Etawah – 206001, (U.P.), India.

The river Sengar originates from and flows through agricultural lands known as gangatic plains and confluences with the river Yamuna. The river has its stretch of about 300 km. As per geographical preview, Shikohabad and several villages discharge their domestic and industrial waste and thus make this river highly polluted. The river reduces to a slow spill channel during summers. Moreover, pollution load is increasing at a fast rate due rapid industrialization and urbanisation.

This investigations on the river Sengar were conducted fortnightly at five sampling stations (A to E) it showed the effect of pollution on the physico-chemical aspects of the river water at district Etawah (U.P.) in summer, rainy and winter seasons (2003). A few physico-chemical parameters were studied in the field and rest were done in the laboratory.

Materials and Methods

The river Sengar passes through varied topographical and meteorological conditions, hence it provides heterogeneous ecosystem, there are definite interrelationships between plankton and the physico-chemical factors of the river Sengar and an analysis of data reveals the fluctuations there of and their bearing on the plankton population.

Five sampling stations A to E were selected on the basis of varied ecological condition of the exploration zone due to domestic and industrial wastes. Station A was setup near the entry point in district Etawah at Dhanua village. Station 'B' was setup near the confluence of its tributary 'Sirsa' river at Chakwa which is about 30 km. away from the first sampling station, while station 'C' near the village Morhi which is about 20 km. ahead from sampling station 'B' and the station 'D' is located near village Jua. The distance covered upto this station is about 15 km. from station 'C'. The sampling site 'E' was set up near Dibiyapur. It is about 15 km. away from the sampling station 'D'.

The water samples were collected fortnightly by immersing a wide mouthed bottle at the sub surface level. Samples were collected during morning hours in between 8.30 to 10.30 A.M. Brown sample bottles having 250 ml. capacity were used for collecting samples for dissolved oxygen. The sample of the water for physico-chemical characteristics were analysed following Standard Methods (APHA *et al.*, 1995) using HACH model DR-EL/4 Kit with a digital pH meter.

Results and Discussion

The growth reproduction and the development of biota is influenced by physico-chemical factors, *i.e.*, temperature. Transparency, pH, dissolved oxygen, free carbon dioxide.

During the present investigation period, the water temperature of the river stretch explored ranged between 18.5°C to 33.0°C. The minimum was recorded during winter and the maximum during summer season. The minimum and maximum temperature of the river shows a clear correlation with atmospheric temperature. Chacko and Ganapati (1949) reported about the water temperature which followed the atmospheric temperature being hottest during May and June and lowest during December and January. Transparency is an important limiting factor in the population and distribution of freshwater plankton. Transparency in the river Sengar was low in rainy season due to heavy rains while it increased in winter and summer at station 'B' and 'D' respectively. Joshi *et al.* (1973b) found that the transparency of the water of the river Bhagirathi at Garwal Himalaya did not show wider annual variations. Colour of the Sengar river water was muddy during rainy season while green colour was in winter and summer seasons. Green colour indicated fish transparency and correlated with increased plankton production.

The pH plays an important role on the biotic life of a river system. It has been suggested by Verma *et al.* (1977); Rai and Munshi (1979) while Chakraborty *et al.* (1959) reported that pH value of the river Yamuna at Allahabad during 1954–55. The value of Hydrogen ion concentration was low (7.0) in monsoon season and highest value was (8.7) in winter season.

Table 23.1: Physico-chemical Characteristics of Water of River Sengar at Various Sampling Stations in Different Seasons (2003)

Paramotors	*Seasons*	*Sampling stations*				
		A	*B*	*C*	*D*	*E*
Temperature (ºC)	S	29.8	31.0	30.0	29.5	33.0
	R	22.0	23.0	22.8	22.4	24.0
	W	18.5	19.1	18.8	19.0	20.0
Transparency (cm)	S	37.0	35.5	36.0	38.0	35.5
	R	21.0	20.5	20.5	19.0	17.5
	W	50.0	50.5	49.0	49.5	47.5
Colour	S	Green	Green	Green	Green	Green
	R	Muddy	Muddy	Muddy	Muddy	Muddy
	W	[illegible]	[illegible]	Green	Green	Green
pH	S	7.6	7.3	7.5	7.2	7.6
	R	7.0	7.5	7.8	7.2	7.1
	W	8.0	8.7	8.3	8.1	8.2
Dissolved oxygen (mg/l)	S	4.9	5.0	4.8	4.2	4.0.
	R	7.6	5.0	5.3	6.1	5.3
	W	8.5	6.6	6.0	6.5	7.4
Carbon dioxide (mg/l)	S	I4.0	9.8	13.6	10.9	14.1
	R	9.8	13.2	11.9	10.3	12.6
	W	8.3	13.0	10.8	10.5	10.2
BOD (mg/l)	S	10.6	10.2	10.8	16.5	14.5
	R	7.7	9.9	7.6	8.5	8.2
	W	5.5	5.2	4.4	4.3	4.3
COD (mg/l)	S	44.1	41.5	40.8	42.1	39.8
	R	26.0	38.5	35.2	37.5	38.8
	W	32.0	30.0	33.2	33.0	31.0

S: Summer; R: Rainy; W: Winter.

Dissolved oxygen plays a vital role in supporting aquatic life in the running water, which is susceptible to the slight environmental changes in D.O. concentration in the river water may be due to polluted affluent, which lowers the dissolved oxygen. In the present investigation the D.O. concentration varied from minimum of (4.0 mg/l) to a maximum of (8.5. mg/l). The high value of D.O. was recorded during the winter season where temperature was low, and low value of D.O. was observed during the summer season when the temperature was high and river flow was also low. This finding about D.O.

are in conformity with George *et al.* (1966) and Saxena *et al.* (1966). Carbon dioxide is one of the essential constituents of an aquatic ecosystem and the abundance of it exerts certain specific effects upon aquatic bioata. The river Sengar exhibited maximum carbon dioxide as in summer but low concentration were recorded during winter season. In the present investigation carbon dioxide was found to be inversely proportional to the dissolved oxygen. The same correlation between carbon dioxide and dissolved oxygen was reported earlier by Ray *et al.* (1966) and Prakash *et al.* (1978)

The BOD and COD are the significant parameters of pollution (Odum, 1971). The higher load of them produces obnoxious conditions (Verma *et al.*, 1977) and in the open drains they are directly concerning the productivity (Bewtra and Radhacharan, 1968). The value of BOD was minimum in winter and maximum in summer at all the five stations.

The chemical oxygen demand in the river Sengar was higher than that of BOD. The maximum values were recorded in summer and minimum were in rainy season. In river Yamuna COD was also found higher than BOD by Joshi and Bisht (1993) from which it may be concluded that same degree of biodegradable oxygen demanding pollutants are present in the water.

Acknowledgements

Authors are greatly indebted to Dr. R.R.S. Chauhan, Reader, Department of Zoology, Janta Mahavidyalaya, Ajitmal (Auraiya) for his constant guidance, suggestions and necessary help to this work. Our sincere appreciation goes to Arvind Kumar Dubey, Head of the Department for providing necessary laboratory facilities.

References

APHA, AWWA, WPCF. 1995. *Standard Methods for the Examination of Water and Wastewater*, 19th edn. American Public Health Society Pub. Washington, D.C. pp. 874.

Bewtra, J.K. and Radhacharan. 1968. Further observations on progression of biochemical demand. *Environ. Hlth.*, 10: 51–60.

Chacko, P.I. and Ganapati, S.V. 1949. Some observations on the Adyar river with special reference to its hydrobiological conditions. *Indian Geogr. J.*, 24(3): 25–49.

Chakrabarty, R.D., Roy, P. and Singh, S.B. 1959. A quantitative study of the physico-chemical conditions of the river Yamuna at Allahabad in 1954–55. *Indian J. Fish*, 6(1): 186–203.

George, M.G., Qasim, S.Z. and Siddiqui, A.Q. 1966. A limnological survey of the river Kali with special reference on fish mortality. *Environ. Hlth.*, 8(4): 262–269.

Joshi, B.D., Pathak, J.K., Singh Y.N., Bight, R.C.S. and Joshi, N. 1993b. Phytoplankton production in the snow bed river Bhagirathi in the Garhwal Himalaya. *Him. J. Env., Zoo.*, 7: 60–63.

Joshi, B.D. and Bisht, R.C.S. 1963. Some aspects of physico-chemical characteristic of western Ganga canal near Jawalpur at Hardwar. *Him. J. Env. Zool.*, 7: 76–82

Odum, E.P. 1971. *Fundamentals of Ecology*, 3rd edn. W.B. Saunders Co., Philadelphia. pp. 574.

Prakash, C., Rawat, D.C. and Grover, P.P. 1978. Ecological study of river Yamuna (Poiya Ghat to Jawahar Bridge) IAWPC. *TECH. Ann.*, 5: 32–45.

Rai, D.N. and Munshi, J.S.D. 1979. Observation on diurnal changes of some physico-chemical factors of three tropical swamps of Darbhanga (North Bihar) India. *Comp. Physiol. Ecol.*, 4(2): 52–55.

Ray, P., Singh, S.B. and Sehgal, K.L. 1966. A study of some aspects of ecology of the river Ganga and Yamuna at Allahabad (U.P.) in 1958–59. *Proc. Nat. Acad. Soc. India*, (B), 36(3): 235–272.

Saxena, K.L., Chakrabarty, R.N., Khan, A.Q., Chattopadhya, S.N. and Harish Chandra. 1966. Pollution studies of the river Ganga, near Kanpur. *Environmental Health*, 8: 270–285.

Verma, S.R., Tayagi, A.K. and Dalela, R.C. 1977. Studies on the characteristic and disposal problem of industrial effluents with reference to ISI Standards (Part II). *Ind. J. Environ. Hlth.*, 19 (3): 165–175.

Chapter 24

Distribution of Nutrients at Different Seasons in Tharangambadi–Vanjur Coasts, South East Coast of India

P. Martin Deva Prasath* and T. Hidayathullakhan**

**Post Graduate Department of Chemistry, TBML College, Porayar – 609 307, Nagapattinam District, Tamil Nadu, South India*

***Post Graduate Department of Chemistry, Khadir Mohideen College, Adiramapattinam, Tamil Nadu, South India*

ABSTRACT

Investigations on the physico-chemical factors at two different stations namely, Tharangambadi (Station 1) and Vanjur (Station 2), lying along the southeast coast of India showed spatial and temporal differences for all the parameters observed. Of the two stations studied, Station 1 recorded higher values of surface water temperature, pH and salinity than Station 2. Station 2 recorded higher concentrations in nutrients of NO_2, SiO_3 whereas Station 1 recorded higher concentrations in nutrients of PO_4, NO_3.

Keywords: Nutrients, Tharangambadi, Vanjur, India, Marine water quality.

Introduction

Information on the coastal hydrography of Nagapattinam coast is essential to know the pollution status and consequent productivity of the ecosystem. Little information is available on the physico-chemical aspects of the Nagapattinam coast in southeast coast of India. There is no detail and long-term (continuous monitoring) study in this area. The present communication reports on the seasonal distribution of some hydrographical parameters and micronutrients in selected two stations like Tranquebar and Vanjur coast in Nagapattinam district, southeast coast of India. The studies were carried out for the period of two years from January 2000 to December 2001. Since Nagapattinam is one of the main fishing centers of (South) Tamil Nadu, it is essential to know the hydrographical characters of the coast. For this would provide valuable information on the nature of the coastal waters.

Materials and Methods

Two stations were selected for the present study. Station 1 is Tranquebar coast (Lat. 11°2′N; Long. 79°49′ E) and station 2 is Vanjur coast (Lat. 10°2′ N; Long. 79°59′ E) in Nagapattinam district of Tamil Nadu state. The distance between the stations is 20 km. Both the stations are occupied with sandy shore beach and receive freshwater from the river Uppanar and Bravadayan river. Station 1 is clean control site were there is no industrial pollution. Station 2 is crowded with 36 small chemical industries, which came up in a short span of 10 years.

Surface water samples were collected at monthly intervals from the study areas for the estimation of various physico-chemical parameters for the period of two years from January 2000 to December 2001. Temperature (air and surface water) was measured using a standard centigrade thermometer. Salinity was estimated with the help of a Salinometer model E-2 and pH was measured by using a Elico pH meter. The nutrients like dissolved inorganic phosphate, reactive silicate, nitrate and nitrite was measured by adopting the method described by Strickland and Parsons (1972). Rainfall data were collected from statistical department of Nagapattinam.

For the sake of convenience and easy interpretation a calendar year (of study) was divided into four seasons *viz.*, Postmonsoon (January–March), Summer (April–June), Premonsoon (July–September), Monsoon (October–December) seasons, which centers round the north east monsoon. During this monsoon only, Tamil Nadu part of India (where the present study areas are situated) gets the bulk rainfall.

Results

Monthly variation in the physico-chemcial parameters like rainfall, air and surface water temperature, salinity, pH and nutrients in station 1 and station 2 are shown in Figures 24.1 and 24.2.

Rainfall

Annual total rainfall during first and second year was 1043.3 mm and 1397 mm. There was no rainfall during February and March 2002 of the second year. Maximum rainfall was recorded in November 2000 (234.8 mm) and December 2001 (389.6 mm) for the first and second year respectively. The monthly mean rainfall for the first and second years was 86.94 mm and 116.4 mm respectively.

Air Temperature

Air temperature varied from 28.2°C–37.1°C and 29.3°C–37.9°C during first and second year respectively at station 1. At station 2, it varied from 28.2°C–37.9°C and 29.2°C–37.8°C during first and second year respectively. The minimum temperature was recorded in December in both the years and stations.

Surface Water Temperature

Surface water temperature varied from 22°C–35.4°C and 25.6–33.6° C during first and second year respectively at station 1. At station 2, it varied from 23.6°C–35.6°C and 24.2°C–33.1°C during first and second year respectively.

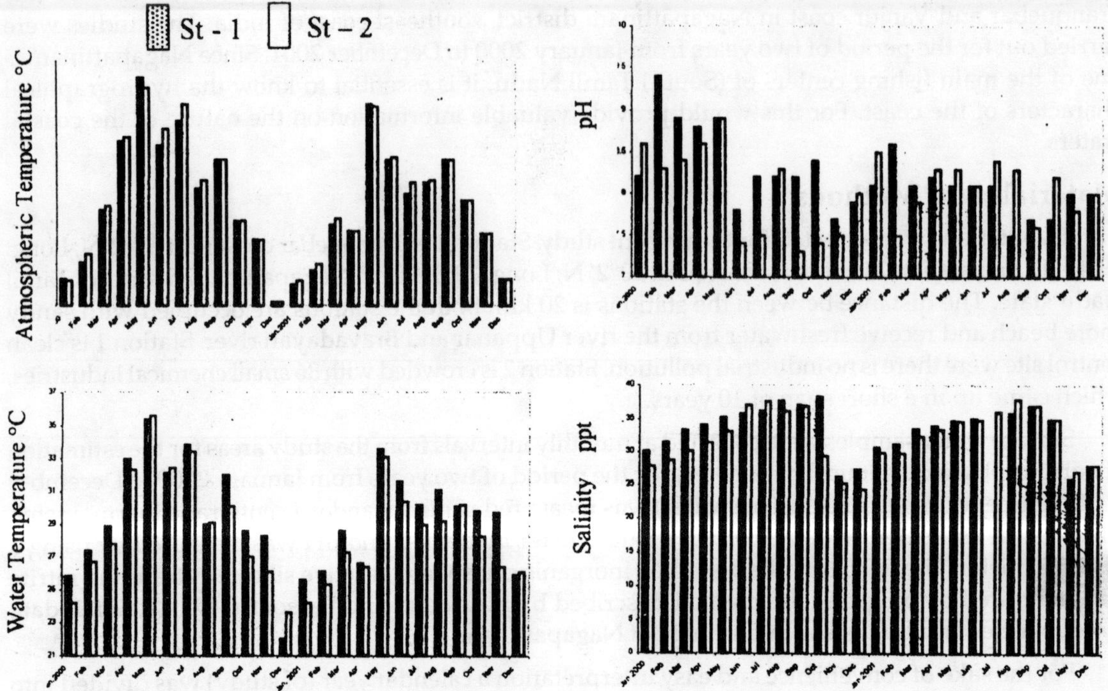

Figure 24.1: Monthly Variations in Temperature (Air and Surface Water), pH and Salinity

pH

pH ranges from 7.7–9.4 and 7.7–8.6 at station 1 and at station 2, it varied from 7.3–8.9 and 7.5–8.5 during first and second year respectively.

Salinity

Salinity varied from 26.2–38.2 per cent and 26–36 per cent at station 1 during first and second year respectively. At station 2, it varied from 24.4–36.9 per cent and 27–36.1 per cent during first and second year of the study.

Table 24.1: Simple Correlation Co-Efficient (R) for Physico-chemical Parameters at Tharangambadi (Station 1) and Vanjur (Station 2)

Parameters	*'r' Value (Station 1)*	*'r' Value (Station 2)*
Surface water temperature × Atmospheric Temperature	0.9575***	0.9836***
Surface water temperature × pH	0.0757	– 0.3385
Atmospheric temperature × pH	0.0346	0.3314
Salinity × Atmospheric temperature	0.6868***	0.4480**
Salinity × Surface water temperature	0.5910***	0.4651**
Salinity × pH	0.0436	– 0.1095
Salinity × Rainfall	– 0.7557	– 0.8562

***: 0.001 level; **: 0.01 level; *: 0.05 level.

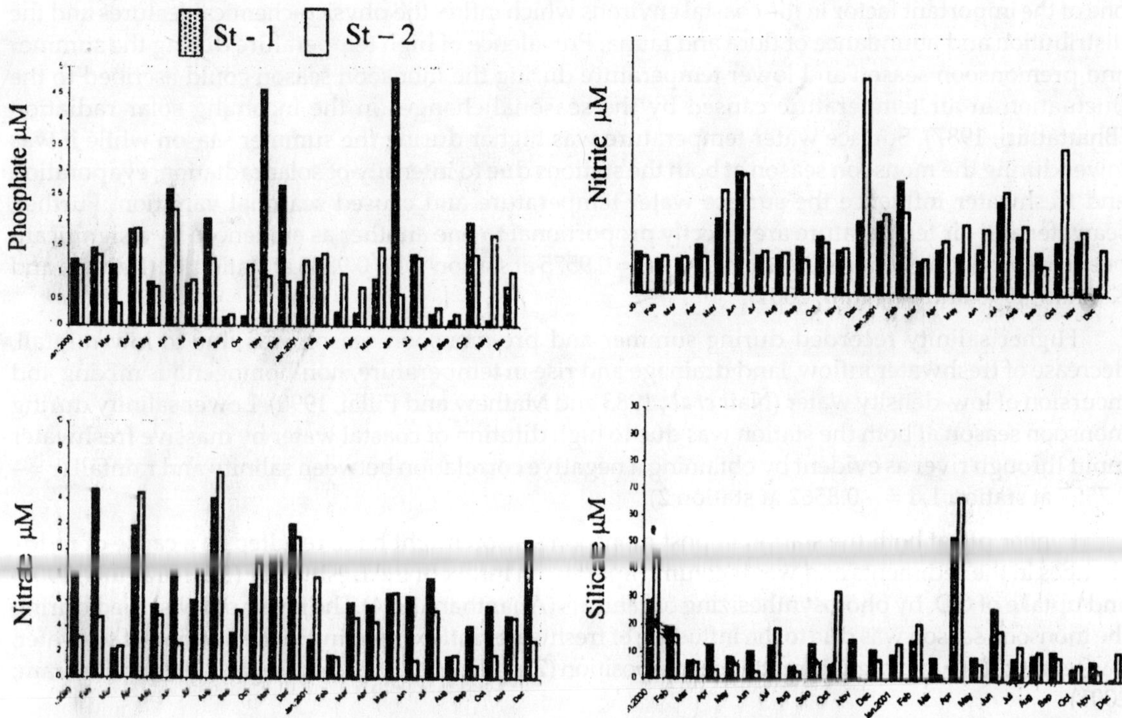

Figure 24.2: Monthly Variations in Nutrients Concentration

Phosphate

Concentration of inorganic phosphate varied from 0.16–4.58 µM and 0.11–4.82 µM at station 1 during first and second year respectively. At station 2, it varied from 0.21–2.22 µM and 0.23–1.37 µM during first and second year respectively.

Nitrate

Nitrate concentration varied from 2.19–14.72 µM and 1.69–10.83 µM at station 1, during first and second year respectively. At station 2, it varied from 2.67–10.04 µM and 1.01–6.67 µM during first and second year respectively.

Nitrite

Concentration of nitrite varied from 0.64–2.37 µM and 0.14–2.21 µM at station 1 and 0.65–4.21 µM and 0.69–2.86 µM at station 2 during first and second year respectively at both the stations.

Silicate

Concentration of silicate varied from 4.54–28.52 µM and 0.64–52.57 µM at station 1 and 0.13–44.24 µM and 4.01–20.69 µM at station 2 during first and second year respectively at both the stations.

Discussion

Monsoon season recorded more rainfall *i.e.*, 39.75 per cent during first year and 72.3 per cent during second year study period, owing to influence of the northeast monsoon which can bring about profound changes in the hydrography due to floods and runway water (Qasim, 1990). Temperature is one of the important factor in the coastal environs which influx the physico-chemical features and the distribution and abundance of flora and fauna. Prevalence of high temperature during the summer and premonsoon season and lower temperature during the monsoon season could ascribed to the fluctuation in air temperature caused by the seasonal changes in the incoming solar radiation (Bhattathiri, 1987). Surface water temperature was higher during the summer season while it was lower during the monsoon season at both the stations due to intensity of solar radiation; evaporation and freshwater influence the surface water temperature and caused seasonal variation. Further, seawater and air temperature are directly proportional to one another as evidenced by a significant positive correlation obtained from this two ($r = 0.9575$ at station 1, $r = 0.9836$ at station 1), (Kannan and Kannan, 1996 and Rajaram, 2001).

Higher salinity recorded during summer and pre-monsoon season was due to low rainfall, decrease of freshwater inflow, land drainage and rise in temperature, non-homogenous mixing and incursion of low-density water (Nair *et al.*, 1983 and Mathew and Pillai, 1990). Lower salinity during monsoon season at both the station was due to high dilution of coastal water by massive freshwater input through river as evident by obtaining a negative correlation between salinity and rainfall ($r = -0.7557$ at station 1, $r = -0.8562$ at station 2).

Higher pH at both the station in post monsoon season might have resulted as a cause of redox changes in the sediments and water column apart from influx of the freshwater (Kumaraguru, 1980) and uptake of CO_2 by photosynthesizing organisms (Ananthan, 1994).The lower pH observed during the monsoon season was due to the influence of freshwater influx resulting in the dilution of seawater, low temperature and organic matter decomposition (Zingde, *et al.*, 1987; Rajasegar, 1999 and Rajaram, 2001).

Higher concentration of nutrients like phosphate, nitrate, nitrite and silicate was observed during premonsoon and monsoon season at both the stations due to the land runoff, following rains and flushing of domestic effluents and influx of river water, land runoff from irrigation channels and release of minerals from sediments (Chandran and Ramamoorthy, 1984). Higher phosphate concentrations are found in both the coasts because of the strong wind stirring action, which blows from the phosphate rich sediments. Seasonal variation and higher concentration of nitrate are due to phytoplankton excretion, oxidation of ammonia, reduction of nitrite (Kannan and Kannan, 1996) and air-sea interaction for exchange of chemical elements (Mathew and Pillai, 1990). Higher level of silicate during monsoon season at both the coasts is due to heavy influx of freshwater derived from land drainage carrying silicate leached out from rock, shrimp farm discharges in station 1 and industrial effluent discharges in station 2 (Ramaraju *et al.*, 1987; Zingde *et al.*, 1987 and Sarma *et al.*, 1988).

The lower level of nutrients noted during postmonsoon and summer season at both the coasts are due to decreased land runoff, higher salinity, high pH, less freshwater inputs, utilization of phytoplankton (high density), bacterial decomposition and nitrifying bacterial activity (Mathew and Pillai, 1990).

References

Ananthan, G. 1994. Plankton ecology and heavy metal studies in the marine environs of Pondicherry, India. *Ph.D. Thesis*, A.U., India, pp. 125.

Bhattathiri, P.M.A. 1987. Laccadive Sea: Its environmental characteristics. In: *Contribution in Marine Sciences*, Dr. S.Z. Qasim, sixtieth birthday felicitation volume, p. 105–115.

Chandran, R. and Ramamoorthy, K. 1984. Hydrobiological studies in the gradient zone of the Vellar estuary: II. Nutrients. *Mahasagar-Bull. Natl. Inst. Oceanoger.*, 17(3): 133–140.

Kannan, R. and L. Kanna. 1996. Physico-chemical characteristics of seaweed beds of Palk Bay, Southeast coast of India. *Indian J. Mar. Sci.*, 25: 358–362.

Kumaraguru, A.K. 1980. Studies on the chemical and biological transport of heavy metal pollutants copper, zinc and mercury in Vellar estuary and the toxicity of three pollutants to some estuarine fish and shellfish. *Ph.D. Thesis*, Annamalai University, Parangipettai.

Mathew, L. and Pillai, V.N. 1990. Chemical characteristics of the waters around Andamans during late winter. *Proc. First Workshop Scient. Result*. FORV Sagar Sampada, 5–8 June, 1989, p. 15–18.

Nair, V.R., S. N. Gajbhiye and Sayed, F.Y. 1983. Organic carbon content of zooplankton from the near shore water of Bombay, India. *J. Mar. Sci.*, 12: 183–185.

Qasim, S.Z. 1990. Adaptation in phytoplankton to changing condition in tropical estuaries. *Mahasagar-Bull. Natl. Inst. Oceangogr.*, 13(2): 117–124.

Rajaram, R. 2001. Studies on hydrobiology and mercury content in Uppanar estuary, Cuddalore, Southeast coast of India. *Ph.D. Thesis*, A.U, India, pp. 128.

Rajasegar, M. 1998. Environmental inventory of Vellar estuary (Southeast coast of India) in relation to shrimp farming. *Ph.D. Thesis*, A.U., India, pp. 95.

Ramaraju, V.S., Sarma V.V., Narasimha Rao, T.V and Vijayakumar, R. 1987. Variation of physico-chemical characteristics with tide in Visakhapatnam harbour water, east coast of India. *Indian J. Mar. Sci.*, 16: 218–222.

Sarma, V.V., Narasimha Rao T.V., Ramaraju V.S., Vijayakumar R. and Suguna, C. 1988. Interrelationship and distribution of hydro-chemical parameters in coastal waters off Vishkhapatnam, east coast of India. *Mahasagar-Bull. Inst. Oceanogr.*, 21(4): 197–207.

Strickland, J.D.H. and T.R. Parson, 1972. A practical handbook of seawater analysis. *Bull. Fish. Res. Bd., Canada*, 167: 310.

Zingde, M.D., S.A.H. Abidi, P. Sarma and M.A. Rokade, 1987. Base water quality off Tha. In: *Contribution in Marine Sciences*. Dr. S.Z. Qasim, sixtieth birthday felicitation volume, p. 304–318.

Chapter 25

Impact of Garbage Dumping on the Groundwater Quality of Madurai City: A Case Study

S. Sheerin and Mary Esther Rani***

**Anja College, Sivakasi*

***Lady Doak College, Madurai, E-mail: ranimam@yahoo.com*

ABSTRACT

The quality of groundwater near the selected garbage dumping sites at Madurai city, Tamil Nadu was studied. A total of 30 representative undergroundwater samples, were collected from the different localities. Various parameters namely pH, Temperature, Turbidity, Alkalinity, Total dissolved solids, Calcium, Total hardness, COD, BOD, Chloride, Nitrogen, Nitrite, Nitrate, Sulphate, Phosphorous, Sodium, Potassium, Iron, Zinc and Chromium had been determined to evaluate its suitability for domestic application. The microbial flora was also studied in all water samples. The result of the investigation revealed that the quality had been affected.

Introduction

Now in the era of economic growth, groundwater is becoming contaminated with hazardous substances from thousands of legally and illegally constructed and operated land fills, surface impoundments and septic systems. The major problem with groundwater is that once contaminated it is very difficult to restore it. Intensive use of natural resources and various human activities are posing a great threat to groundwater quality. Solid water management is an important factor of environmental hygiene and needs to be integrated with total environmental planning. The major groundwater pollutants are municipal solid waste (MSW) and other liquid waste. Although several steps have been taken by municipal authorities, for management of these wastes, only a small fraction of them is being properly utilized or disposed off in a scientific manner.

Due to the prevailing conditions of water scarcity for corporation supply people of Madurai depend on borewell water sources for their needs. So in the present study more emphasis was given to borewell water analysis.

Methodology

The sample used in the study were collected in Madurai and their details are given in Table 25.1

Table 25.1: Details of Groundwater Collected for the Study

Areas	*Sites*	*Depth of the Borewells in Feet*
Chockalinga nagar	Site 1	350–450
Ellisenagar	Site 2	250–460
Mattuthavani	Site 3	280–300

A total of 30 representative undergroundwater samples were collected near selected sites of Madurai city where they are used for drinking as well as other domestic purposes by the residents of the area. Sampling was done during August 2003. Samples were collected in pre cleaned sterilized and well dried polyethylene bottles. The collected samples were stored in an ice box and brought to laboratory. Analytical grade chemicals were used to prepare reagents. Double distilled water was used to prepare all the reagents, dilution standards and dilution water. The physico-chemical analysis was carried out according to standard methods (APHA 1989). The various analyzed parameters include pH, Electrical conductivity (EC), Total dissolved solids (TDS), Total hardness (TH), Turbidity, Acidity, Alkalinity, Calcium, BOD, COD, Chloride, Iron, Nitrogen, Nitrite, Magnesium, Potassium, Zinc and Chromium and EC of water samples were measured by using a pH meter and a conductivity meter, respectively. Temperature was determined by using the thermometer. Turbidity was determined by Nephelometer. Acidity and Alkalinity was determined by Titration method. Total dissolved solids was determined by using the following formula:

$$TDS (g/l) = A-B/V \times 100$$

BOD and COD was determined by titration method. This was expressed as mg/l. Total hardness was found out by titration method. Estimation of chloride was done titrimetrically. Calcium, Sodium, Potassium, Iron, Zinc and Chromium was determined by Atomic absorption spectrophotometry. Those was expressed as ppm units. Nitrogen was determined spectrophotometrically at 635 nm. Nitrate was estimated by Brucine method. Nitrite was done by EDTA method. Sulphate was estimated by turbidimetric method. Phosphorous was estimated spectrophotometerically at 690 nm.

Biochemical analysis was carried out by Indole production test, Methyl red test, Voges Proskauer test, Citrate Utilization test, Lactose Utilization test, Gelatin Utilization test, Triple sugar iron test, Urease agar test, Catalase test and Oxdiase test.

Microbial Study

Total microbial content of samples were found out. (Nutrient Agar for bacteria, Rose bengal agar for Fungi, Oat meal agar for Actinomycetes).

After two days of incubation, the number of colonies per plate were counted and calculated for gm of ml sample.

For Identification and Characterization of Microbes

Lactophenol cotton blue staining method for fungi and Actinomycetes. Gram's staining method for bacteria was used.

Results and Discussion

The dumping site at Chcokalinganagar is about 20 years old and it is considered to be the oldest dumping site of Madurai city than two other sites.

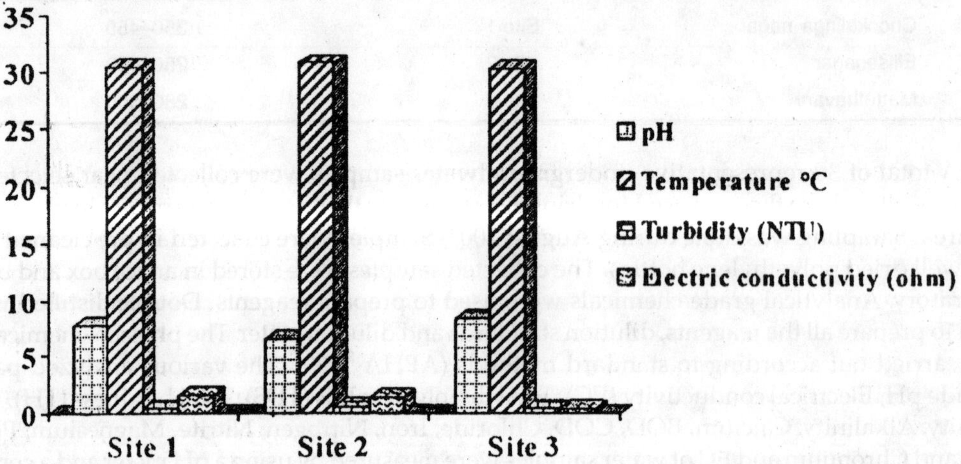

Figure 25.1: Groundwater Quality Nearby Garbage Dumping Site

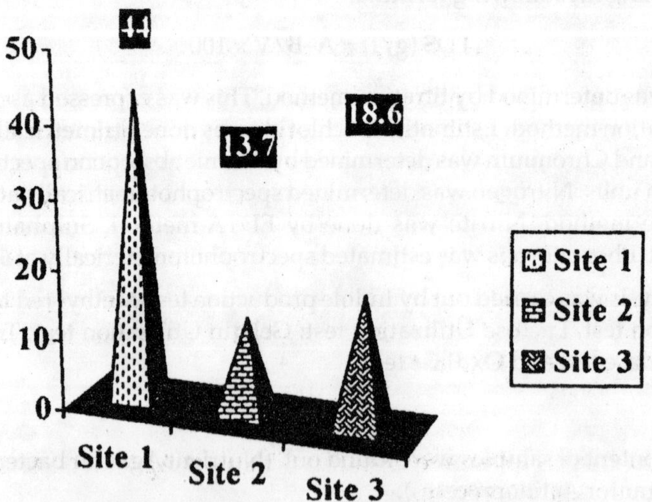

Figure 25.2: BOD Content of Groundwater Nearby Garbage Dumping Site

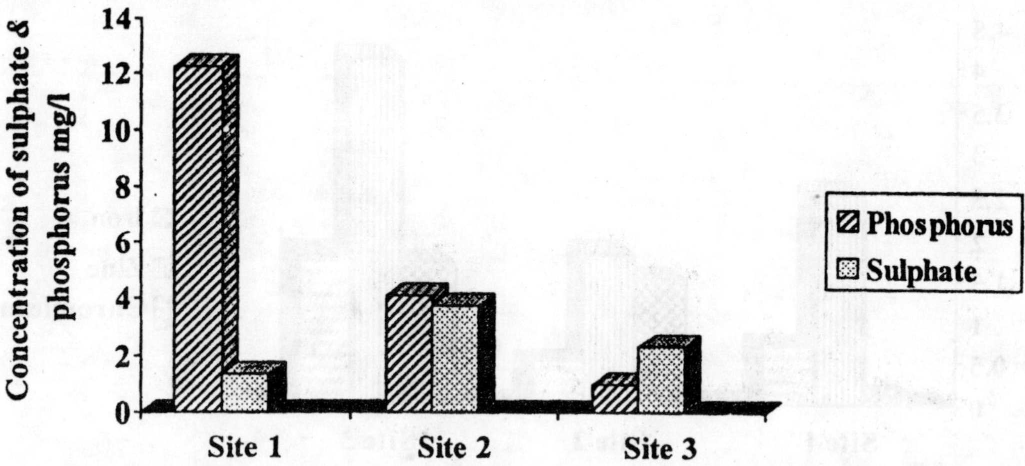

Figure 25.3: Sulphate vs. Phosphorus of Groundwater Nearby Garbage Dumping Site

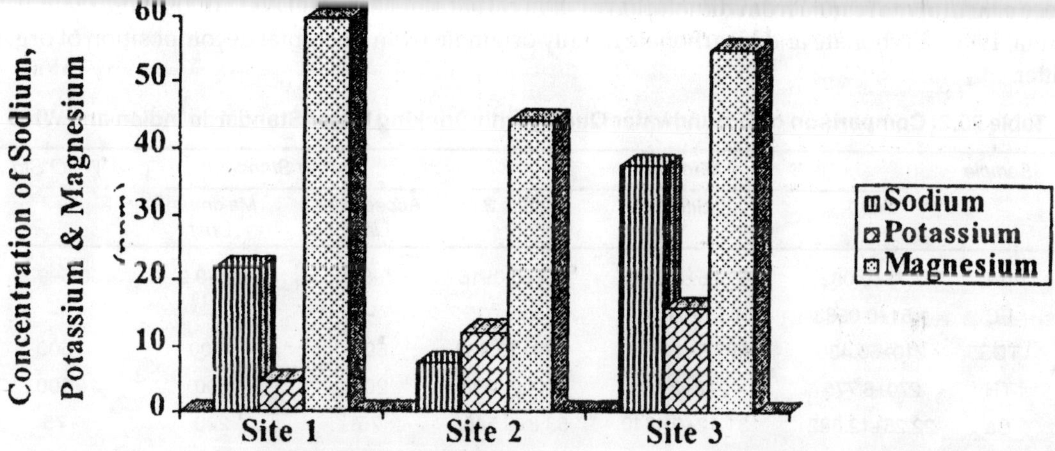

Figure 25.4: Comparison of Sodium, Potassium, Magnesium of Groundwater Samples in ppm

The physico-chemical characteristic of the analyzed water samples had been presented in (Table 25.2). Water samples varied greatly in quality with respect to different chemical constituents. The pH in the studied area lies in the range of 6.3–8.5. All the water samples were slightly alkaline in nature. The EC of the undergroundwater at Madurai city varied between 1.40 1.51 ohm and TDS varied from 210–627 mg/l. EC and TDS signified the inorganic pollution. According to WHO table limit for TDS in under groundwater sample. Except Site 2, apart from that other 2 sites are well within permissible limit. Total hardness ranged from 230–322 mg/l.

In the present study turbidity ranged from 0.20–0.67 NTU. Temperature ranged from 29.7–31.4°C. Calcium content in the studied area is well with in the permissible limit (< 200 pm). The Magnesium content ranged from 44.15–57.72 ppm all the three site sample fall with in WHO standard limit. The constituent of alkalinity in natural system mainly include carbonate, bicarbonate and hydroxide.

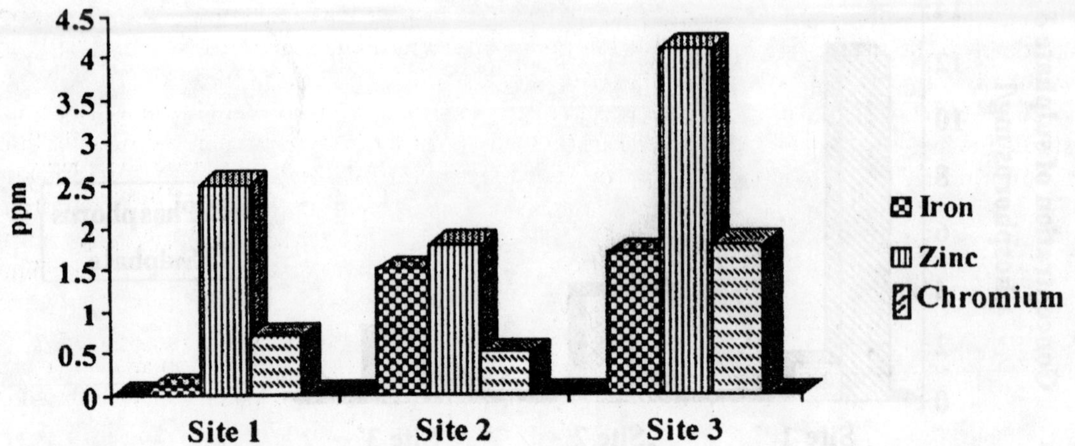

Figure 25.5: Comparison of Iron, Zinc and Chromium of Groundwater Near Dumping Sites

These constituents results from dissolution of mineral substances in soil and atmosphere (Mittal and Verma, 1997). Carbonate and bicarbonate mainly originate from microbial decomposition of organic matter.

Table 25.2: Comparison of Groundwater Quality with Drinking Water Standards Indian and WHO

Sample	Madurai			Indian Standard		WHO Std.
	Site 1	Site 2	Site 3	Acceptable Limit	Maximum Limit	
pH	7.75±0.082	6.625±0.0116	8.2±0.0916	7–8.2	6.5–9.2	6.5–9.2
EC	1.51±0.0583	1.51±0.0182	1.4±0.035	–	–	–
TDS	210±50.436	627.5±11.615	379±13.432	500	1500	500
TH	270±6.775	248±2.15	311±1.398	200	600	500
Ca	22.25±12.582	31.12±0.4840	53.8±0.5463	75	200	75
Mg	57.72±0.940	44.15±0.453	54.31±0.620	30	–	150
K	4.55±0.088	10.45±0.727	15.15±0.841	–	–	200
Na	20.45±0.546	30.4±0.456	41.35±1.310	–	–	200
Cl	62.41±0.277	50.49±0.204	57.11±0.601	200	1000	500
F	0.2±0.012	1.5±0.097	1.7±0.37	1	1.5	–
NO_3	$0.4±6.7 \times 10^{-3}$	$0.86±9.21 \times 10^{-0.3}$	0.16±0.01	–	–	–
TA	18.24±0.176	27.48±0.208	15.36±0.599	–	–	–

In the present study, acidity ranged from 6.53–27.23 ppm in all three area under groundwater which fall with in acceptable limit. In the present study, total alkalinity ranged from 12.06–27.48 ppm. which is in the acceptable limit. Sodium and Potassium are naturally occurring elements of groundwater. Industrial and domestic waste also adds sodium to groundwater. Sodium content in the studied area ranged from 20.45–41.35 ppm. When the Sodium content is more than 50 ppm it makes the water unsuitable for drinking purpose. Sodium content of three site water sample had been found to fall with in the permissible limit. The potassium content in the studied area ranged from 4.5–15.5 ppm. Chloride in excess imparts a salty taste to water and people who are not accustomed to high Chloride in water are subjected to laxative effect (Prakash Rao, 1989). The Chloride in the studied area ranges from 50.49–62.41 mg/ml which is within acceptable limit. The sulphate content in the water sample of studied area ranged from 1.31–3.75 ppm which had been found well with in permissible limit. Nitrate is a poisonous component of undergroundwater, which is of natural as well as of anthropogenic origin. The excess amount of nitrate in water cause a disease in children called Blue Baby syndrome (Methaemoglobinemia). The permissible limit of nitrate in undergroundwater is 10 ppm. The nitrate content in present study ranged from 0.16–0.86 mg/l. The nitrate content of three sample area had been found to be well with in permissible limit. BOD content ranged from 13.7–44.0 mg/ml which exceeded desirable limit of BOD. The COD in the studied area ranges from 0.2–0.3 mg/l, it was present below desirable limit. In the present study Nitrite ranged from 0.43–2.35 mg/l. In human being excess amount of nitrite lead to the oxidation of normal hemoglobin to methaemoglobin which was unable to transport of O_2.

The Nitrogen content of all the three area groundwater had been found to be higher than permissible limit because of high amount of waste from sawmills and animal wastes from slaughter houses and skin processing unit.

The Iron in the studied area ranged from 0.19–1.7 ppm, the chromium content of 3 area groundwater had been found to be within permissible limit. In the present study, Zinc ranged from 1.8–4.1 ppm. Zinc content outside the desirable limit might cause astringent and opalescence in water (Tiwari and Kumar, 2002).

The groundwater samples of all the three sites were highly contaminated by bacterial, than fungi and actinomycetes (Table 25.3). Site 2 groundwater were highly polluted by bacteria because of their ability to grow in the favourable environmental condition and soil condition, etc. Years of dumping in the same area led to the migration organisms into groundwater. Open air defection and use of non-sanitary latrines were the cause for contamination of the groundwater. Actinomycetes were dominant in Site 1 because of the soil texture, and soil drilling activities which enhanced the growth of actinomycetes.

Table 25.3: Population of Micro Organisms Present in the Groundwater

Name of the Organisms	*Number of Organisms per ml of Groundwater Samples*		
	Site 1	*Site 2*	*Site 3*
Fungi	65	29	10
Bacteria	1500	1100	1115
Actinomycetes	8	10	14

Based on the morphological, biochemical tests the bacteria were identified (Table 25.4). Except Site 2 the other two sites were contaminated with faecal matter which was proved by the presence of *Escherichia coli* in the groundwater. *Pseudomonas* sp. and *Proteus* sp. was present only in the Site 2 and *Streptococcus* sp. was observed only in Site 3. Presence of these microorganisms indicate the contamination of groundwater which might be due to the influence of near by garbage dumping sites.

Table 25.4

Characteristic of Bacteria

Name of Site	*Morphological Characters of Bacteria*	*Identification*
Site 1	Single, Motile cremy, Round	Gram Negative Rods
Site 2	Motile, Straight, Rod shaped, White	Gram Negative Rods
	Non motile, Round, Pinhead, Yellow, Entire	Gram Positive Cocci
Site 3	Motile, Singly, White, Round, Entire	Gram Negative Rods

Characteristic of Fungi

Name of Site	*Morphological Characters*	*Identification*
Site 1	White Coloured Mycelium becomes greenish later	*Aspergillus flavus*
	White cottony mycelium and black dots of sporangia	*Rhizopus*
	White Coloured Mycelium becomes greenish later	*Aspergillus flavus*
	White cottony mycelium and black dots of sporangia	*Rhizopus*
	Purple Brown to Black Conidial heads	*Aspergillus niger*
Site 2	White Coloured Mycelium becomes greenish later	*Aspergillus flavus*
	White cottony mycelium and black dots of sporangia	*Rhizopus*
	White Coloured Mycelium becomes greenish later	*Aspergillus flavus*
	White cottony mycelium and black dots of sporangia	*Rhizopus*
	Purple Brown to Black Conidial heads	*Aspergillus niger*
Site 3	White Cottony Mycelium and blackdots of Sporangia	*Rhizopus*
	White Coloured Mycelium becomes greenish later	*Aspergillus flavus*
	White cottony mycelium and black dots of sporangia	*Rhizopus*
	Purple Brown to Block Conidial heads	*Aspergillus niger*

Conclusion

Influence of garbage dumping on the physico-chemical characteristics and microbial flora of the near by groundwater was studied. It was found that the amount of total hardness, Nitrate, Nitrite, Sulphate, Phosphorus, Zinc, Chromium, Chloride and COD were within the permissible limit. Apart from that Nitrogen, BOD and TDS were more than permissible limit. These findings are due to the influence of garbage dumping sites. Microbial flora included such as *Aspergillus flavus, Aspergillus niger, Mucor* and *Rhizopus*, bacteria such as *Escherichia coli, Pseudomonas* sp., *Proteus* sp., *Staphylococcus* sp. and *Streptococcus* sp. The presence of the *Escherichia coli* was the indication of faecal contamination of groundwater. The presence of various microbial population could be influenced by the organic and inorganic content of groundwater. In all the three sites the influence of garbage dumping of physico-chemical and microbial quality was observed in groundwater.

References

Alanmild. 1993. *Soils and the Environment: An Introduction*, Cambridge University Press, London.

APHA. 1989. *Standard Methods for the Examination of Water and Wastewater*, 17th edn. Washington, D.C.

BIS. 1991. *Drinking Water Specification*. Bureau of Indian Standards, New Delhi.

Bishnoi, M.S. and Bishnoi, N.R. 2003, Assessment of groundwater quality of Jind city. *Environmental Protection*, Kalpana Corporation, Hisar.

Hussain, Jakir and Ikbal Hussain. 2002. Evaluation drinking water quality of the villages situated near river Banas, Rajasthan. *Env. J.*, p. 640–645.

Manivasakam, N. 1987. *Industrial Effluents: Origin, Characteristics, Effects Analysis and Treatment*. Sakthi Publications, Coimbatore.

Masood, Alam and Anwar Ahamad. 2001. Water quality in and around industrialised city of Delhi East and Sahibabad. *Env. J.*, p. 900–904.

Mishra, N.K. and Sahoo, H.K. 2003. Evaluation of Groundwater quality in and around Deogarh. *Env. J.*, p. 667–672.

Mittal, S.K. and N. Verma. 1997. Critical analysis of groundwater quality parameters. *Env. J.*, 17(6): 426–429.

Pandey, Vardana. 1992. *Environmental Pollution*. Himalaya Publishing House, India.

Prakash, S.R. and Krishna Rao. 1989. The chemistry of groundwater in Paravada area with regard to their suitability for domestic and irrigational purposes. *Env. J. Geochemist.*, 8(6): 421–424.

Sarma, H.P. and Bhattacharyaa. 2001. Quality of drinking water in the Darrang district with particular reference to Managaldai sub division. *Ph.D. Thesis*, Gauhati University.

Saxena, M.M. 1990. *Environmental Analysis of Water, Soil and Air*. Agro-botanical Publishers, India.

Shrivastava, V.S. 1996. A case study of wastewater collected from GIDC area of Surat city. *Poll. Res.*, 15(3): 315–316.

Trivedy, R.K. and Goel, P.K. 1984. *Chemical and Biological Methods for Water Pollution Studies*. Environmental Publications, Karad.

Usepa. 1985. Elevated blood pressure and high sodium levels in the public drinking water. *Archives of Env. Hlth.*, 35: 200.

Chapter 26

Occurrence of a Cyanophycean Bloom in Mallapura Tank Near Chitradurga, Karnataka

A.B. Banakar, B.R. Kiran,* R. Purushothama,* E.T. Puttaiah* and S. Manjappa***

**Department of Environmental Science, Kuvempu University, Shankaraghatta – 577451, Karnataka*

***University BDT College of Engineering, Davanagere, Karnataka*

ABSTRACT

The present study deals with ecological aspects of cyanophycean bloom in relation to certain physico-chemical factors from Mallapura tank, near Chitraduraga of Karnataka during October 2004 to March 2005. Physico-chemical variables on the growth of this bloom both collectively and individually include alkaline pH, higher phosphate. Chloride, sulfate, E.C. and temperature. The major bulk of the bloom was represented by the genus *Arthospira.*

Introduction

Algae are involved in water pollution in a number of ways. Phytoplankton, such as *Microcystis, Anabena, Pandorina,* and *Euglena* etc. produce thick massive blooms in inland waters of the tropics (Round 1965; Pandey 1973). The algal blooms can create severe problems in the maintenance of water supplies and in meeting the ever-increasing demand for potable water.

Cyanophycean algae are distributed all over the world and found in marine as well as freshwater habitats. They grow luxuriantly in eutrophic lentic water bodies. Blue green algae have reported to accumulate heavy metals and therefore ameliorate the toxicity of flyash with increased fertility (Rai *et al.*, 2000). Some non-heterocystous blue green algae like Spirulina are used as single cell protein and in wastewater treatment considerable work has been done on blue green algae of fresh water habitats of Uttar Pradesh, India (Prasad and Srinivas, 1965; Tiwari, 1979; Ahmed, 1972; Chaturvedi and Pandey, 1976). Our article discusses the ecological aspects of various cyanophycean blooms and their significance in evaluating the trophic status of the habitats.

Study Area

Mallapura tank is situated at 76°22′19″ North latitude and 14°13′30″ East longitude with a water spread area of 91.84 hectares. The water storage capacity is 189 MCFT (Million Cubic feet) surface water samples were collected from Mallapura tank during October 2004 to March 2005. The filtered samples were analyzed for various physico-chemical parameters by standard methods (Gilterman and Clymo, 1969; APHA 1998) simultaneously 100 ml of the sample was collected separately and preserved in 4 per cent formaldehyde solution. Using Sedgewick-rafter counter cell did the quantitative estimation of algae and finally, the number of plankton was calculated per litre. The data was statically analyzed by following the methods of Snedecor and Cochran (1967).

Results and Discussion

The data presented in Table 26.1 shows the physico-chemical properties of water from this tank. Table 26.2 shows the occurrence of blue green blooms during this study. In this investigation the species of *Aphanocapsa*, *Oscillatoria*, *Merismopedia*, *Spirulina* and *Nostoc* was occurring rarely in this waterbody. While, *Arthrospira* bloom was dominant in all the months.

Table 26.1: Physico-chemical Complexes of Mallapura Tank

Parameters	*Oct*	*Nov*	*Dec*	*Jan*	*Feb*	*Mar*
pH	8.2	8.8	8.6	8.5	8.2	9.6
Total Hardness	216	240	230	242	252	248
E.C	1120	1160	1190	1300	1880	2200
Water temperature	25	26.5	28.5	29	29.5	31.5
Sulfate	48	54	107.30	198.4	255.2	269.7
Calcium	49.699	56.112	42.48	28.05	34.46	38.47
Magnesium	20.93	22.69	28.96	26.47	39.45	100.8
D.O	9.73	13.186	9.986	13.586	5.186	4.8
Iron	0.027	0.021	0.085	0.085	0.116	0.006
Phosphate	0.24	0.62	0.87	0.70	0.85	1.02
Nitrate	0.005	0.007	0.03	0.035	0.015	0.02
Chloride	107.78	326.2	358.82	468.03	496.39	787.15

Note: All the parameters are in mg/L except pH, temperature (°C) E.C. (μmhos/cm).

Table 26.2: Occurrence of Cyanophycean Blooms in Mallapura Tank (Organisms/Litre)

October	*November*	*December*	*January*	*February*	*March*
2.7×10^5	3.575×10^5	3.634×10^5	6.362×10^5	2.162×10^5	5.434×10^5

The water temperature of the sampling sites ranged between 25–31.5°C, EC 1120–2200 μmhos/cm, pH 8.2–9.6, total hardness 216–252 mg/L, Calcium 28.0–56.11 mg/L, magnesium

20.93–100.8 mg/L, Chloride 107.78–787.15 mg/L, phosphate 0.24–1.02 mg/L, Dissolved oxygen 4.8–13.58 mg/L, in the current study 5 genera of blue greens have been identified of which *Arthrospira* bloom was dominant (90 per cent) as compared to other Cyanophycean bloom.

According Tucker *et al.* (1984) high temperature supports the luxuriant growth of blue greens. In this study, the temperature fluctuated between 25–31.5°C. The dominance of blue green algae with high temperature support the work of Munawar (1974) who stated that the blue green algal forms develop profusely towards the end of monsoon when the temperature rise or during winter (October) when the temperature was fairly high, ranges between 25–32° C.

In the present water body the Cyanophycean blooms as a whole exhibit a direct relationship between dominance of algal species with nitrates. Moreover this tank much harbour having less quantities of nitrate confirming the result of Pearsallis (1932) and Zafar (1967), that cyanophycean are capable of growing rapidly in minimal quantities of nitrate. It is well known that blue green algae occur more abundantly in hard water than in soft water bodies and the major factors that influence their growth are light water temperature, high pH, DO, Chloride, Phosphate and low Nitrate.

In general dense blooms occur in nutrient rich water and many blooms are highly concentrated. Blooms develop frequently during certain times of the year in temperate region or almost any time of the year observed in the tropics (Reynolds and Walsby, 1975) the abundance of any Phytoplankton organism in a water body depends not only on such factors as light, temperature and content of nutrient, but also on the changes which the plankton population by its own growth may bring about in its environment. Certain organic substances liberated by the algae themselves may be important for successful growth (Brook, 1957) on the basis of the experimental work, Lam and Silver (1979) concluded that (i) the effective uptake of nutrients by one species may cause it to dominate over other algae or (ii) the liberation of inhibitory substances by one alga may eliminate the other species.

The temporary water bodies such as rainwater pools, the area is very small and if they have high nutrient loading (hypertrophic condition), the species with fastest rates of growth can develop into a bloom for a short period. After a certain period of growth these blooms disappear suddenly from the habitat perhaps, due to the following reasons:

1. Exhaustion of essential nutrients
2. Concentration of certain elements which may become toxic
3. Extracellular substances released by the organism may be of inhibitory nature (autoinhibition)
4. Ion binding effect of certain extra cellular substances which will alter the ionic composition of the medium.

Uhlmanns (1971) models of growth of Phytoplankton in various habitats indicate that small species of algae are capable of dominating at high rates of nutrient loading and short retention times by virtue of their rapid rates of growth.

Like the preceding species, the blue green alga *Arthrospira* sp. also occurs in polluted environment where the medium is alkaline (pH 9.6) and very rich in chloride, total hardness and electrical conductivity. It is interesting to note that although the habitat shows all the characteristics of eutropic waters, its oxygen content is very high and this could attributed to the vigorous photosynthetic activity of alga at the time of collection which may be further substantiated by the very high pH value during the study.

Conclusion

1. Nutrient loading of water bodies favour bloom formation
2. Algal blooms can develop at any time of the year
3. Species with fastest growth rate in a short span develop into blooms
4. Depending on the type of environment and nature constituents of the medium bloom forming species vary greatly
5. Extracellular substances released by the algae also play a significant role in bloom formation as they have great biological significance in ecology
6. Well oxygenated waters are more favorable to blooms.

References

Ahmed, M.R. 1972. Algae of Allen forest Lake Kanpur India. *Nova Hedwigia*, 23(1): 125–129.

APHA. 1985. *Standard Methods for the Examination of Water and Wastewater*, 16th edition. USA.

Brook, A.J. 1957. *New Biology*, 23: 36–101.

Chaturvedi, V.K. and U.C. Pandey. 1976. A list of blue green and green algae from Rohilkhand division, U.P., India *IV. Phykos*, 15(1–2): 127–137.

Golterman, H.L and R.S Clymo. 1069. *Methods for Chemical Analysis of Freshwater.* I.B.P Hand Book No. 8. Blackwell Scientific Publications, Oxford, London.

Hyderabad, India, IV. The biocenose, periodicity and species composition of unicellular and colonial phytoplankton in polluted and unpolluted environments. *Hydrobiologia*, 45(1).

Lam, W.Y. Catherine and W.B. Silvester. 1979. *Hydrobiologia*, 63: 135–143.

Munawar, M. 1974. *Limnological Studies on Freshwater Ponds.*

Pearsall, W.H. 1932. Phytoplankton in the English lakes II. *J. Ecology*, 20: 241–262.

Prasad, B.N. and P.N. Srivastava. 1965. Some interesting morphological features in *Talypothrix* species. *Hydrobiologia*, 26(3–4): 517–520.

Reynolds, C.S. and A.E. Walsby, 1975. *Biol. Rev.*, 50: 437–481.

Snedecor, G.W. and W.G. Cochran. 1967. *Statistical Methods*, 6th edition. Oxford Publishing Co., New Delhi.

Tiwari, G.L. 1979. A study of the blue green algae from paddy field soils of India IV. Taxonomic consideration of Nostocales and Stigonematotales. *Nova Hedwigia*, 63: 133–159.

Tucker, C.S., Loyed, S.W. and R.L. Bush. 1984. Relationship between Phytoplankton periodicity and concentration of total and unionized ammonia in channel catfish. *Hydrobiologia*, II(1): 75–79.

Uhlmann, D. 1971. Mitt. der Intern. Ver. Fur Theor. And Angew. Limnol., 19: 190–224.

Zafar, A.R. 1967. On ecology of algae in certain fishponds of Hyderabad, India II. *Hydrobiologia*, 30(1): 6–112.

Chapter 27

Physico-chemical Parameters and Elemental Analysis of the Soils of Sugarcane Fields with and without Red Rot Disease Incidence

S. Velmurugan, R. Narayanaswamy** and S. Ravi***

**Department of Physics, Annamalai University, Annamalai Nagar – 608 002, Tamil Nadu, India*

***Engineering Physics Section, Faculty of Engineering and Technology, Annamalai University, Annamalai Nagar, Tamil Nadu – 608 002, India*

ABSTRACT

In this study, an attempt has been made to find any possible variation in physico-chemical parameters of the soils collected from various sugarcane fields with red rot disease prone and disease resistant areas. Also the concentrations of four important trace elements *viz.*, Fe, Mn, Cu and Zn are estimated using atomic absorption spectrometer. The results of the physico-chemical parameters, the concentration of trace elements and their results are discussed.

Keywords: Sugarcane, Soil, Red rot disease, Physico-chemical parameters and AAS.

Introduction

The modern agriculture has considerably raised the output, but created problems like land degradation, atmospheric and water pollutions. Frequent and excess use of chemical fertilizers have created imbalance of nutrients in soil and adverse effect on soil physico-chemical properties and ecology (Ramesh *et al.*, 2002).

The plant growth and yield primarily depends upon the nutrient status of soil (Zende, 1999). Now nutrients level is decreasing continuously in Indian soils due to extensive agricultural activities to meet the demand of increasing population (Aneg Singh *et al.* 2001).

Sugarcane is one of the most important cash crops of India. It is a long duration crop and intercropping is also recommended in the crop. The soil under sugarcane cultivation gets exhausted very soon. The nutrient uptake is also very high in this crop. So this soil, under continuous sugarcane cultivation show deficiencies of different elements such as N, P, K and S (Aneg Singh *et al.*, 2001).

Diseases are one of major constraints in the profitable cultivation of sugarcane. Of these, red rot disease caused by "collectrotrichum falcutam went" is most destructive, because the fungal growth is encouraged by variations in temperature, pH, nutrients and environmental conditions (Raid an Lentini, 2002 and Sheokumar and Singh 2003). Influence of environmental temperature has been studied by various researchers (Anilkumar *et al.*, 1998). But yet, only limited information is available on the role of infected soil in initiating red rot infection (Agnihotri *et al.*, 1979). Hence, the present study has been taken up to determine and to compare various physico-chemical parameters and some major trace elements with respect to the disease prone and resistant soils from sugarcane fields.

Materials and Methods

The soil samples are collected from red rot disease prone and resistant sugarcane growing areas during the year 2003–2004. The sample areas are Bannari Amman Sugar Ltd., Sathiyamangalam, Erode district (Red rot disease resistant area) and EID Parry (I) Ltd., (red rot disease prone area) Nellikuppam, Cuddalore district, Tamil Nadu, India.

Representative composite soil samples are taken from the study area upto 15 cm depth as per standard procedure of IARI (Indian Agricultural Research Institute) New Delhi (Tripathy *et al.*, 1990). The location of sampling sites are randomized to avoid biasing in results. The collected samples after coning, quartering and sieving are used for various spectral investigations. The physical parameters of soil, like particle size, bulk density and pore space are determined by standard methods (Piper, 1966) and the chemical parameters of soil, like pH, EC, available N, available P, available K and organic carbon are determined by standard methods (Page, 1982). The soil samples are digested with diacid mixture using 10:4 nitric acid and perchloric acid. The resultant soil samples are subjected to chemical analysis for the estimating Fe, Mn, Cu and Zn through atomic absorption spectrometer (Linday and Norwell, 1978). This analysis has been carried out in soil testing laboratory, Cuddalore, Tamil Nadu, India.

Results and Discussion

Table 27.1 shows the physical parameters of red rot disease prone sugarcane growing soils. Table 27.2 shows the above parameters of red rot diseases resistant soils.

Soil texture is an important parameter. With the help of it, mineralogical and chemical composition are defined. The soil texture describes the size distribution of particles in soil. It indicates the relative percentages of sandy, silt and clay fraction of soil. In this study, the soil texture of all the samples fall under the type sandy clay loam. The total biomass of plants grown in sandy clay loam are always significantly greater, when compared with plants grown in gravel and clay (Paula Power, 1996).

All bulk density values are nearly equal both for red rot disease prone and resistant sugarcane growing areas soils. When bulk density values are more than 1.75 mg m^{-3} in sandy clay loam soil, the root growth of all the crops would be restricted (Richard, 2003).

Table 27.1: The Physical Parameters of Red Rot Disease Prone Sugarcane Growing Area Soils

*Sample No.**	*Sandy %*	*Clay %*	*Silt %*	*Textural Class*	*Bulk Density mg m⁻³*	*Pore Space %*
Sample 1	51.8	27.1	21.1	Scl	1.51	39.4
Sample 2	55.4	29.0	15.6	Scl	1.42	40.0
Sample 3	47.9	27.6	24.5	Scl	1.54	42.5
Sample 4	54.5	23.0	22.5	Scl	1.42	41.4
Sample 5	57.5	23.5	19.0	Scl	1.53	38.4

Note: * The samples collected from red rot disease prone sites.

Sc'l: Sandy clay loam.

Table 27.2: The Physical Parameters of Red Rot Disease Resistant Sugarcane Growing Area Soils

*Sample No.**	*Sandy %*	*Clay %*	*Silt %*	*Textural Class*	*Bulk Density mg m⁻³*	*Pore Space %*
Sample 6	66.4	29.5	4.1	Scl	1.42	48.5
Sample 7	63.4	28.2	7.9	Scl	1.44	47.8
Sample 8	64.3	21.9	13.8	Scl	1.46	45.5
Sample 9	73.2	22.0	4.8	Scl	1.53	43.0
Sample 10	72.3	29.8	2.9	Scl	1.56	42.1

Note: ** The samples collected from red rot disease resistant sites.

Scl: Sandy clay loam.

The pore space of the soils in disease prone areas are very much lower when compared to red rot disease resistant area soils.

The chemical parameters of both red rot disease prone and resistant sugarcane growing soils are shown in Tables 27.3 and 27.4. In this study regarding the red rot disease resistant soils, it has been found that chemical parameters in electrical conductivity, available P, available K and organic carbon are high compared to red rot disease prone sugarcane growing soils, except available N. These results are in close conformity with the various observations of Jayapaul *et al.*, 2002 and Kumar *et al.*, 1999.

Table 27.3: The Chemical Parameters of Red Rot Disease Prone Sugarcane Growing Area Soils

*Sample No.**	*pH*	*EC (dsm⁻¹)*	*Available N (kg/ha)*	*Available P (kg/ha)*	*Available K (kg/ha)*	*Organic Carbon %*
Sample 1	8.25	0.40	261.85	53.70	376.75	1.34
Sample 2	7.60	0.30	275.50	48.15	444.70	1.80
Sample 3	8.25	0.30	252.65	69.10	338.65	0.99
Sample 4	7.70	0.15	272.95	87.40	290.30	1.51
Sample 5	8.30	0.25	243.35	84.55	297.70	0.65

Note: * The samples collected from red rot disease prone sites.

Table 27.4: The Chemical Parameters of Red Rot Disease Resistant Sugarcane Growing Area Soils

*Sample No.***	*pH*	*EC (dsm^{-1})*	*Available N (kg/ha)*	*Available P (kg/ha)*	*Available K (kg/ha)*	*Organic Carbon %*
Sample 6	7.35	0.51	221.10	123.50	716.5	2.42
Sample 7	7.45	0.79	228.5	123.50	1238.4	2.23
Sample 8	7.50	0.78	161.6	123.50	1050.1	2.54
Sample 9	7.35	0.59	140.75	123.50	488.0	2.35
Sample 10	7.40	0.40	150.65	123.50	420.0	2.01

Note: ** The samples collected from red rot disease resistant sites.

The pH of soils in red rot disease prone and resistant area soils are found to be primarily alkaline with some gradation (alkaline and weakly alkaline). It is recorded as 7.6 to 8.3 and 7.3 to 7.5 for red rot disease prone and resistant area soils respectively. The optimum range of pH values are 6.6 to 7.4, but sugarcanes will be grown in soils, with pH in ranges from 5 to 8.5 (Faostat, 2001).

Higher values of Fe (52.17 to 96.30 ppm) Mn (20.75 to 50.0 ppm), Cu (4.27 to 10.08 ppm) and Zn (2.1 to 5.28 ppm) have been found in red rot disease resistant sugarcane growing soils where as lower values of Fe (12.5 to 29.5 ppm), Mn (1.45 to 22.5 ppm), Cu (2.75 to 3.75 ppm) and Zn (1.5 to 3.2 ppm) are found in the red rot disease prone sugarcane growing soils as seen from Tables 27.5 and Table 27.6. Similar results have been reported by Sheokumar *et al.*, 2003 and Rajakumar, 2002.

Table 27.5: The Trace Elements of Red Rot Disease Prone Sugarcane Growing Area Soils

*Sample No.**	*Iron (ppm)*	*Manganese (ppm)*	*Copper (ppm)*	*Zinc (ppm)*
Sample 1	12.5	1.95	3.75	1.8
Sample 2	29.5	3.25	3.1	2.0
Sample 3	19.6	12.15	3.25	2.5
Sample 4	20.75	1.45	3.0	1.5
Sample 5	26.7	22.35	2.75	3.2

Note: * The samples collected from red rot disease sites.

Table 27.6: The Trace Elements of Red Rot Disease Resistant Sugarcane Growing Area Soils

*Sample No.***	*Iron (ppm)*	*Manganese (ppm)*	*Copper (ppm)*	*Zinc (ppm)*
Sample 6	53.32	22.5	4.27	2.58
Sample 7	52.17	25.19	4.77	2.04
Sample 8	81.11	20.75	9.41	5.09
Sample 9	62.41	50.0	6.34	2.1
Sample 10	96.30	22.0	10.08	5.27

Note: ** The samples collected from red rot disease resistant sugarcane growing soils.

From this study, it is seen that the observed concentrations of micronutrients (Fe, Mn, Cu, Zn) are higher in red rot disease resistant sugarcane growing soils. It leads to the fact that with proper management of micronutrient levels in soils *i.e.*, concentration of Fe, Mn, Cu and Zn are to be increased in order to control the disease. However the nitrogen content is found high in the disease prone areas when compared to diseased areas. So it is inferred that a higher value of nitrogen favours disease proliferation. Hence reduction of nitrogen may be one of the factors responsible in the reduction of the disease incidence.

References

Agnihotri, V.P., Budhayaraja, T.R and Krushana Singh. 1979. Role of diseased setts and soils in the annual recurrence of red rot in sugarcane. *Indian Sugar J.*, p. 263–265.

Aneg Singh, A.K., Gupta, R.N., Srivastava, K., Lal and Singh, S.B. 2001. Nutrient status of sugarcane growing soils in central Uttar Pradesh, India. *Sugar Tech.*, 3: 117–119.

Anil Kumar, Satyavir and Ram Niwas. 1998. Relationship between climatic factors and initiation of red rot infection in sugarcane. *Indian Phytopath*, 51(4): 370–372.

Crop Water Management Sugarcane. 2001. FAOSTAT publication.

Jayapaul, G.S., Duraisingh, R., Senthilvel, T and Joseph, M. 2000. Influence of population an stage of incorporation of intercrossed green manure (Daincha) and nitrogen levels on yield and quality of sugarcane. *Indian Sug.*, 50(1): 989–991.

Kumar, S., Singh, N.N.P., Dwivedi, N.B and Kumar, S. 1999. Changes in chemical composition of cane juice due to pattotype red rot pathagne. *Bharathiya Sugar*, 24: 9–17.

Linday, N.L and Norwell, W.A. 1978. Development of DTPA soil test-for Zn, Fe, Mn and Cu. *Soil Sci. Soc. Am. Proc.*, 33: 62–68.

Page, A.L. 1982. *Methods of Soil Analysis* (Part 2). Soil Science Society, America Madison, Wisconsion.

Paula Power. 1996. Growth of texas wildrice (*Zizania texana*) in three sediments from the san marcos River. *J. Agual. Plant Manage.*, 34: 21–23.

Piper, C.S. 1966. *Soil and Plant Analysis*. Hans Publication, Bombay.

Raid, R.N and Lentini, R.S. 2002. *Sugarcane Red Rot Disease*. UF/IFAS Publication.

Rajakumar, N. 2002. Spectral analysis of red rot disease incidence in sugarcane. *Ph.D. Thesis*. Department of Physics, Annamalai University.

Ramesh, T., Chinnusamy, C. and Sankaran, N. 2002. Influence of integrated Nitrogen and Phosphorus management on cane productivity and soil fertility status in sugarcane. *Cooperative Sugar*, 34: 7–38.

Richard, C. 2003. Cane planter (Soil compaction). *Sugar Journal*, p. 27–28.

Sheokumar, S. and Singh, R.R. 2003. Management of nutritional disorders in sugarcane. *Cooperative Sugar*, 34: 973–975.

Tripathy, D.P., Gurdeep, S. and D.C. 1990. Panigrahi proceeding of the 7th National symposium of environmental, India school of mines, Dhansand, 826 004: 204– 205.

Zende, G.K. 1999. Recent trend in basic and applied research on soils and their utility in soils and crop management practices. *Bharathiya Sugar*, p. 9–29.

Chapter 28

Impact of Fungicide Validacin-3L on Bioenergetics of the Freshwater Fish Silver Carp *Hypophthalmichthys molitrix*

S. Athikesavan, S. Vincent and B. Velmurugan

Unit of Environmental Health and Biotechnology, PG and Research Department of Zoology, LIFE, Loyola College, Chennai – 600 034, India

ABSTRACT

Fungicide Validacin-3L was investigated in the present study. Silver carp (*Hypophthalmichthys molitrix*) were selected for the bioassay experiments. Fishes were exposed to four sublethal concentrations of validacian-3L for 21 days, and bioenergetics aspects such as food consumption, absorption, convertion, metabolized, efficiency and net conversion efficiency were analysed. The bioenergetics values are reduced compared to control fishes in validacian-3L treated fishes.

***Keywords:** Fungicide Validacin-3L, bioenergetics, Hypophthalmichthys molitrix.*

Introduction

Pollution of aquatic system by chemicals used in industries including those of textiles, tanneries, rayon, paper, synthetic drugs, oil refineries, photofilms, insecticides, fungicides, Agricultural drainage water upon entering the water body have a fatal effect on fish (Metelev *et al.*, 1983). Used in fungi control, these fungicides have contaminated the ecosystem and non-target species. Also, the fungicides stay in the environment for long period and remain a permanent threat to human health and safety (Revelle and Revelle, 1974).

Aquatic pollutants such as pesticides and metals reduce the food intake in fishes (Arunachalam *et al.*, 1980). Industrial effluents have shown to affect food intake and growth in many fishes (Stones and Livington, 1978). The food consumption was decreased with increasing concentration of superphosphate and the feeding rate was also lower at the higher concentration (Khillare and Wagh, 1987). Feeding rate in *Macrones keletiusi* when exposed to sublethal concentration was observed to decrease with increasing concentration of Rogor (Hameed and Vadamailai, 1986).

Growth is a broad index of various physiological functions and may be affected by decrease in food consumption or absorption and conservation efficiency: an increase in metabolism may also influence the growth rate. Growth directly depends on food consumption. The inverse relationship between the feeding rate and concentration suggests that the dye stuff effluents bring about growth reduction through lowered food consumption (Sakthivel and Sampath, 1989).

This present study was conducted to determine the sublethal effect of bioenergetics status of Validacin-3L in silver carp *Hypophthalmichthys molitrix*. The observation of food utilization parameter *viz.*, food consumption efficiency, absorption effiency, conversion efficiency, with exposed different concentration of Validacin-3L.

Materials and Methods

The freshwater fish *Hypophthalmichthys molitrix* were obtained from the Tamil Nadu fish seed farm, near Poondy Reservoir. Fish fingerlings (6±1 gm/wt and measuring 5.5±0.5 cm) were selected for the experiment. In the laboratory fishes were acclimated in tap water in aquaria with 100 l capacity for 10 days before the commencement of the experiment. Fish were fed *ad libitum* with fed with commercial pellet. The fungicide Validacin-3L used in static bioassay stock solution was prepared using in double distilled water. For the bioenergetics studies fishes exposed to sublethal concentrations of 0.01, 0.02, 0.03 and 0.04 ppm for 21 days.

The scheme of energy budget followed is IBP formula (Petrusewicz and Macfudyen, 1970), represented as:

Average food consumed = Food given – Unfed

Average food absorption was estimated as the difference between the average food consumed and that of faeces.

Average food absorbed = Average food consumed – Faeces

Growth or food converted was calculated by substracting final weight of the fish and initial weight of the fish.

Average food converted = Final weight of the fish – Initial weight of the fish

Metabolism was calculated as the difference between average food absorbed and average food converted.

Average food metabolized = Average food absorbed – Average food converted

$$\textbf{Absorption efficiency} = \frac{\text{Absorption rate}}{\text{Consumption rate}} \times 100$$

$$\textbf{Net conversion efficiency} = \frac{\text{Absorption rate}}{\text{Consumption rate}} \times 100$$

Results and Discussion

Food utilization parameters and efficiencies of the fish *Hypophthalmichthys molitrix* alteration in food utilization under the stress of sublethal concentration of the fungicide Validacin-3L was recorded in, Table 28.1. Minimum food consumption was observed in *H. molitrix* treated with at 0.04 ppm, 1.246 g/dry weights after 21 days for exposure of Validacin-3L. Food consumption was maximum found in control fish (1.324mg/g/dry weight) after 21 days. Consumption rate was decreased with increasing concentrations of the fungicide. The decrease in food absorption with increase in concentration of the fungicide, the quantity of faeces egested by the fish decreased as 1.057 to 1.051 mg/g/21 days. Food conversion also decreases with increase in the concentration of the fungicide from 0.112 to 0.080 mg/g/21 days. Similarly with increase in concentration of the fungicide the energy spent on metabolism decreased from 0.963 to 0.943 mg/g/21 days for 0.01 to 0.04 ppm fungicide. The absorption efficiency of the fish decreased with the increase in fungicide concentration from 83.62 to 80.07 for fishes treated with 0.01 to 0.04 ppm respectively. The net conversion efficiency shows decreased in fishes treated with Validacin-3L, the values varies from 8.494 to 6.42 for 0.01 to 0.04 ppm respectively.

Table 28.1: Effect of Validacin-3L on Food Utilization in *Hypophthalmichthys molitrix* Food Consumed, Food Absorbed, Food Converted, Food Metabolized, Absorption Efficiency and Net Conversion Efficiency Expressed in mg/g/21 Days

Different Concentrations	*Food Consumed*	*Food Absorbed*	*Food Converted*	*Food Metabolized*	*Absorption Efficiency*	*Net Conversion Efficiency*
Control	1.324±0.003	1.064±0.002	0.123±0.034	0.980±0.01	85.38±0.32	9.07±0.008
0.01	1.296±0.001	1.057±0.094	0.112±0.011	0.963±0.078	83.62±0.310	8.494±0.12
0.02	1.276±0.012	1.054±0.02	0.10±0.01	0.95±0.002	82.28±0.210	7.83±0.521
0.03	1.257±0.01	1.052±0.021	0.090±0.003	0.947±0.02	81.77±0.42	7.15±0.01
0.04	1.246±0.001	1.051±0.02	0.080±0.01	0.943±0.003	80.07±0.062	6.42±0.04

Values are means of three replication ± S.E.

Feeding behavior is often a sensitive indicator of toxicity (Little *et al.*, 1985). Feeding initiation, the major behavioral changes, as observed in the present investigation a reduction in consumption from 1.296 to 1.246 gm/dry weight/21 days of Validacin-3L concentration is comparable to the observations of Bull and McNerney (1974) in the *Coho salmon* exposed to fenitrothion. In the present study lowered food intake which was inversely proportional to toxicants similar to the findings of (Palanisamy and Ravichandran, 1990).

A loss of appetite might have reduced the glucose level and in turn, the feeding (Colgat, 1973). According to Sarkar (1988), fertilizers could cause damage to the buccal epithelium, rendering it unable to detect the palatability of the food material. According to Arunachalam *et al.*, (1980) aquatic pollutants such as fungicides and metals reduce the food intake in fishes, with the increase in the concentration and decreasing body weight food consumption rate for the fish decreased gradually.

The variation in feeding efficiency with difference in concentration of fungicide in accordance with the results of earlier studies (Haniffa and Venkatachalam, 1980), food type is known to significantly affect food utilization. In conclusion the present study reveals that the fungicide Validacin-3L affects the bioenergetics aspects on dose dependent manner in fish silver carp *Hypophthalmichthys molitrix*.

References

Arunachalam, S.K., Jayalakshmi and Abobucker, S. 1980. Toxic and sublethal effects of Carbaryl on fish. *Arch. Environ. Contam. Toxicol.*, 9: 307–316.

Bull, C.J. and McNerney. 1974. Behaviour of Juvenile Coho Salmon *Oncorhynchus mykiss* exposed to sumithion (Fenitrothion) an organophosphate insecticide. *J. Fish. Res. Bd. Can.*, 31: 1872–1876.

Colgat, P. 1973. Motivatiponal analysis of fish feeding behaviour. *J. Fish. Res. Bd. Can.*, 45: 38–66.

Hameed, P.S. and Vadamalai., P. 1986. Effect of sublethal concentration of dimethoate EC 30 on feeding, growth, O_2 consumption and activity in *Macrones kelatins* (Dumeril). *J. Environ. Biol.*, p. 277–284.

Haniffa, M.N. and Venkatachalam. 1980. Effect of food quality on energy budget and chemical composition of the grass carp *Ctenopharyngodon idella. Proceeding of the International Symposium on Conservation: Inputs from Life Sciences*. University Kebangsaan, Malayasia, p. 163–173.

Khillarre, Y.K. and Wagh, S.B. 1987. Effect of Agricultural fertilizer superphosphate on food and growth in fish, *Barbus stigma. Poll. Res.*, 6(1): 25–28.

Little, B.E., Flerou, B.A. and Ruzhinskaya, N.V. 1985. Behavioural approaches in aquatic toxicity investigation, a review. In: *Toxic Substances in the Aquatic Environment: An International Aspect*, (Eds.) Jr. P.M Mehrle, R.H. Gray and R.C. Kendal. Amer. Fish. Soc., Water Quality Section, Bethesda, M.D., p. 72–98.

Metelev, V.V., Kansev, A.K. and Dzasokhova, N.G. 1983. *Water Toxicology*, Amerind, New Delhi, pp. 216.

Palanisamy, R. and Ravichandran, P. 1990. Sublethal effects of Diammonium phosphate on food utilization in *Labeo rohita. Proc. Sump. Trends in Ectoxical.*, p. 279–284.

Petrusewicz, K. and Macfadyen, 1970. *Productivity of Terrestrial Animals: Principles and Methods*. Black Well, Oxford.

Revelle and Revelle. 1974. *Source book on the Environment*. Houghton Miffin Company, London, p. 227–228.

Sakthivel, M. and Sampath, K.1989. Effect of textile dye stuff effluent on food utilization of *Cyprinus carpio* (Linn). *Ind. J. Exp. Biol.*, 27: 1032–1034.

Sarkar, S.K. 1988. Influence of Ammonium sulphate on the feeding rate of fish under multivariate temperature. *Comp. Physiol. Ecol.*, 13(1): 30–33.

Stoner Allam and. Livingston, V. 1978. Respiration, growth and food conversion efficiency of Pin Fish (*Lageodon rhomboids*) exposed to bleaked kraft mill effluent. Environ. Poll., p. 207–207.

Chapter 29

BGA Diversity in Paddy Fields and Wetlands of Satna (M.P.)

Rashmi Singh and Priti Samdariya

P.G. Department of Botany, Government P.G. (Autonomous) College, Satna – 485 001, M.P., India

ABSTRACT

The present study deals with the distribution of Blue Green Algae (BGA) in paddy fields and wetlands at different localities of Satna (M.P.). The paddy fields are best thriving grounds for growth of BGA because of the prevailing high temperature, high humidity and water logged condition during the rice growing season. Importance of BGA in soil fertility and nitrogen economy of soil has been well documented in tropical countries. During the course of present investigation a total of 112 unicellular, non-heterocystous and heterocystous species of BGA belonging to genera *Microcystis, Chroococcus, Gloeocapsa, Gloeothece, Aphanocapsa, Aphanothece, Merismopedia, Dactylococcopsis, Chlorogloea, Arthrospira, Spirulina, Oscillatoria, Porphyrosiphon Phormidium, Lyngbya, Schizothrix, Microcoleus, Anabaena, Nostoc, Aulosira, Scytonema, Tolypothrix, Microchaete, Calothrix, Plectonema, Gloeotrichia, Haplosiphon, Westiellopsis* etc. were recorded. It is the first step in the utilization of these form in nitrogen economy of our paddy field soils.

Keywords: BGA diversity, Paddy fields, Wetlands.

Introduction

Cyanobacteria are the earliest autotrophic organisms evolved around 3.2 billion years ago in Precambrian geological age. These organisms still common in tropical regions of the world because they prefer high temperature, alkaline pH and enough water. So they are common in India also.

Cyanobacteria are nitrogen fixing alga which contributes a lot, to maintain the soil status of the rice fields (De, 1939; Singh, 1961; Mitra, 1951). So they have also been used as biofertilizers (Venkataraman, 1972). Blue green algal flora of India now consists of 100 genera and 1500 taxa. Of which about 64 genera and 800 taxa of Cyanobacteria are recorded from rice fields of India. The present communication revealed that there are 33 genera and 112 species common in wetlands and rice fields of Satna.

Materials and Methods

The study site Satna district is located between 23°58′ and 25°12′ North latitude and 80°21′ and 80°23′ East longitude in the North-Eastern part of Madhya Pradesh. The district occupies all most the central part of Vindhyan region and is surrounded from district Banda of U.P. (North), Rewa (East), Shahdol (South) and Panna (West).

The algal materials were collected from four different sites of Satna district after an interval of a week. These sites are Satna river (Site 1), Uchehra river (Site 2), Maihar Pond (Site 3), and Site 4 belongs to different rice fields of Satna district. After collecting the algal material were preserved in 5 per cent formaline or in F.A.A. and a portion of the same was brought to the laboratory in living condition for microscopic examination and identification. The systematic identification was done by the help of binocular research microscope at Botany Department of Satna and with the help of standard works by Fritsch (1935), Smith (1933), Desikachary (1959), Geitler (1932) and Prescott (1951).

Table 29.1: Taxonomic Categorization of Cyanobacteria Encountered in the Rice Field Soils and Wetlands of Satna (M.P.)

Sl.No.	*Order*	*Family*	*Genera*	*Species*
1.	Chroococcales	Chroococcaceae	8	25
		Entophysalidaceae	1	1
2.	Nostocales	1. Oscillatoriaceae	8	37
		2. Nostocaceae		
		Sub-family–Anabaenae	5	13
		Sub-family–Scytonemataceae	3	7
		3. Microchaetaceae	2	7
		4. Rivulariceae	4	14
3	Stignematales	Stigonemataceae	2	8
	3 Orders	7 Families	33	112

Results and Discussion

The following Blue green algae collected and identified from four different localities of paddy fields and wetlands of Satna district. Blue green algae in paddy fields and wetlands of Satna included under three broad categories: unicellular forms, filamentous non-heterocystous forms and filamentous heterocystous forms. A considerable degree of modifications and diversification results in variety of thallus organization.

Unicellular Blue Green Algal Forms

***Microcystis aeruginosa* (Blooms in standing water)**

M. robusta (Ponds and moist soil)

M. elongata (Moist soil of rice fields)

M. bengalensis (Plankton in standing water)

M. flosaquae (Plankton in wetland)

M. marginata (Moist soil of rice fields)

***Chroococcus minutus* (Plankton in rice fields)**

C. turgidus (Moist soil of rice fields)

C. pallidus (Plankton in wetland)

***Gloeocapsa decorticans* (Moist soil-of rice fields)**

G. rupestris (Moist soil of rice fields)

G. polydermatica (Plankton in ponds)

G. crepidinum (Plankton in ponds)

G. montana (Moist soil)

***Gloeothece rupestris* (Moist soil of rice fields)**

G. samoensis (Moist soil)

***Aphanocapsa pulchra* (Plankton in wetland)**

A. roseana (Plankton)

A. koordersi (Plankton)

A. biformis (Moist soil of rice fields)

***Apanothece pallida* (Free floating plankton as well as moist soil of rice fields)**

A. saxicola (Plankton in rice fields)

A. stagnina (Free floating in rice fields)

A. microscorpica (Submerged soil of a pond)

***Merismopedia glauca* (Plankton in standing water)**

M. tenuissima (Plankton in standing water)

***Dactylococcopsis fascicularis* (Plankton)**

***Chlorogloea fritschii* (Plankton)**

Filamentous Blue Green Algal Forms (Non-heterocystous)

***Arthrospira massartii* (Plankton)**

***Spirulina princeps* (In paddy fields)**

S. major (Free floating in wetlands).

***Oscillatoria princeps* (Free floating)**

O. obtusa (Plankton)

O. okeni (Moist soil)

O. pseudogeminata V. *unigranulata* (Plankton in moist soil)

O. proteus. (Free floating)

O. subbrevis (Moist soil)

O. limosa (Plankton in pond)

O. proboscidea (Plankton in pond)

O. terebreformis (Plankton in wetland)

O. amphibia (Plankton)

O. acuta (Free floating form)

O. cortiana (Free floating form)

O. annae (Moist soil)

O. boryana (Moist soil of paddy fields)

O. chlorina (Plankton)

O. curviceps (Plankton)

O. laete virens V. *minimus* (Plankton)

O. schultzii (Free floating in paddy fields)

***Porphyrosiphon notarisii* (Moist soil of rice fields)**

***Phormidium ambigum* (Moist soil)**

P. retzii (Moist soil)

P. tenue (Moist soil)

P. corium (Moist soil)

P. uncinatum (Free floating)

P. anamala (Moist soil)

***Lyngbya birgei* (Free floating in paddy fields)**

L. gracilis (Standing water)

L. limnetica (Moist soil)

L. major (Moist soil)

L. allorgei (On stones in river)

***Schizothrix lacustris* (In a ditch)**

S. friesii (Moist soil in ditch)

***Microcoleus chthonoplastes* (Moist soil)**

M. acutissimus (Moist soil)

Heterocystous Blue Green Algal Forms

***Cylindrospermum muscicola* (Moist soil of rice field and in ponds)**

***Nostoc paludosum* (Paddy field soil)**

N. punctiformae (In wetland of Uchehera)

N. commune (Moist soil of rice field and in ponds)

N. hatei (Epiphyte in a ponds)

N. humifuscum (Moist soil of rice field and in ponds)

***Anabaena fertilissima* (Moist soil of rice field and in ponds)**

A. variabilis (In wetlands and paddy fields)

A. oryzae (Moist soil of rice fields)

A. ambigua (In paddy fields)

A. iyengarii (In wetlands)

A. circinalis V. crassa (In wetlands)

***Pseud-anabaena schmidlei* (Free floating in a pond)**

P. catenata (Plankton in a pond)

***Aulosira fertilissima* (Paddy fields)**

A. prolifica (Dense bloom on a pond)

***Plectonema nostocorum* (Paddy fields)**

***Raphidiopsis indica* (Plankton in a pond)**

***Scytonema hofmanni* (On moist soil)**

S. bohneri (On moist soil)

S. guyanense V. minus (In wetlands)

S. javanicum (On moist soil of rice fields)

S. tolypotrichoides (In a pond)

S. coactile (On stones in wetlands)

***Tolypothrix rechingeri* (On moist soil)**

T. byssoidea (On moist soil)

***Microchaete uberrima* (With *Spirogyra* in paddy fields)**

M. loktakensis (On moist soil of rice fields)

M. tenera (On wet soil of rice fields)

***Calothrix marchica* (In a tank)**

C. geitonos (Epiphyte on hydrophytic plants of a pond)

C. castellii (Free floating in a pond)

C. fusca (In ponds)

C. javanica (On the shell of snails in a pond)

C. brevissima (Epiphyte on angiospermic plants in a pond)

C. brevissima V. *monilliforme* (In a ditch)

***Dichothrix orsiniana* (On moist stones of a river)**

D. fusca (On moist stones of a river)

***Rivularia dura* (From a pond)**

***Gloeotrichia ghosei* (Free floating in paddy fields)**

G. raciborskii (In rain water ponds and ditches)

G. intermedia (Epiphyte on hydrophytic plants of a pond)

G. raciborskii V. *kashiense* (In a ditch)

G. longicauda (From a pond)

G. natans (Free floating in a pond)

***Hapalosiphon welwitschii* (In paddy fields)**

***Westiellopsis prolifica* (On moist soil)**

The algal flora of Satna district is fairly rich still unexplored, therefore a systematic work to describe the BGA members of this district were undertaken and seemed that the locality was adjourned with 33 genera and 112 species belonging to unicellular, heterocystous and non heterocystous forms.

The taxonomic investigations on the floristics of blue green algae were also made by Pandey, 1965; Sankaram 1972; Tiwari, 1972, 1975; Laloraya and Mitra 1973 and Singh and Singh, 1989; Anand, 1990; Ahluwalia *et al.*, 1992; Kaushik and Subhashini, 1995. Several Nitrogen fixing blue green algae were isolated from the soils of paddy fields of Satna district. During present investigation it was found that these blue green algae are normally improved the physico-chemical properties of tropical soil. Therefore they may be used as biofertilizers in rice fields of Satna.

References

Ahluwalia, A.S., Dahiya, S. and Reena. 1992. Role of blue green algae in increasing the growth and yield of rice. In: *Proc. Nat. Symp. Cyanobacterial Nitrogen Fixation.* Associated Publ. Co., New Delhi, p. 307–313.

Anand. N. 1990. *A Handbook of Blue Green Algae.* Bishen Singh Mahendra Pal Singh, Dehradun, p. 1–79.

De, P.K. 1939. The role of blue green algae in nitrogen fixation in rice fields. *Proc. R. Soc. Lond. Ser. B*, 127: 121–140.

Desikachary, T.V. 1959. *Cyanophyta: Monograph of Algae.* I.C.A.R., New Delhi, India, pp. 686.

Fristch, F.E. 1935. *The Structure and Reproduction of the Algae* I. Cambridge, pp. 791.

Geitler, L. 1932. Cynophyceae in Rabenhorst's Krypto gamenoflora. *Leipzig*, 14: 1196.

Laloraya, V.K. and Mitra, S.K. 1973. Studies on the blue green algae of the paddy fields of India. Part I: Cultural studies, general considerations and distribution pattern of the blue green algae in paddy fields of India. *Nova. Hedwigia*, Belheft, pp. 47.

Mitra, A.K. 1951. Algal flora of certain Indian soils. *Indian J. Agric. Sci.*, 21(4): 357–373.

Prescott, G.W. 1951. Algae of Western great lakes area. *Gran. Book Inst. of Science.* Bull. 31. Bloom field. Hills Misch, 946 p. 136 pl.

Pandey, D.C. 1965. A study on the algae from paddy field soil of Ballia and Ghazipur districts of U.P. India Part II (A) Taxonomic considerations. Cyanophyceae. *Nova Hedwigia*, 10: 177–209.

Sankaram, V. 1972. Taxonomic studies in the genus *Nostoc* Vaucher. *Ph.D. Thesis*, Univ. Madras.

Singh, A.L and Singh, P.L. 1989. *Nitrogen Fixation in Indian Rice Fields.* Agro-Botanical Publishers, Bikaner, pp. 236.

Singh, R.N. 1961. *Role of BGA in Nitrogen Economy of Indian Agriculture.* I.C.A.R., New Delhi.

Smith, G.M. 1933. *Freshwater Algae of the United States*, 1st ed. New York. London, p. 716, 449f.

Tiwari, G.L. 1972. A study of the blue-green algae from paddy field soils of India. *Hydrobiologia*, 29(3): 335–350.

Tiwari, G.L. 1975. A study of the blue-green from paddy field soils of India. Part II: Taxonomic considerations of non heterocystous BGA. *Nova Hedwigia*, 26: 765–797.

Venkataraman, G.S. 1972. *Algal Fertilizers and Rice Cultivation*. Today and Tomorrow's Printers and Publishers. Faridabad, Haryana, pp. 75.

Chapter 30

Effect of Earthworm Exudate on Growth and Yield of *Tagetes erecta L.* (Family: Compositae)

Shweta, Deepika Sharma, Sonal and Kiran Kumar

Vermiculture Unit, Department of Zoology, D.S. College, Aligarh – 202 001, UP, India

ABSTRACT

The effect of earthworm exudate on plant height, number of branches, number of flowers and total yield of *Tegetes erecta* was recorded. The treatment of control (T_0), 0.5 per cent urea (T_1), 50 per cent vermiwash (T_2) and 100 per cent vermiwash (T_3) was given at regular intervals. The result showed that the application of vermiwash enhanced the plant height, number of flowers and total yield. Among all the four treatments, T_2 was found better for plant height, budding, and total yield and T_3 was better for branching as well as number of flowers. The application of fifty per cent vermiwash was found very superior than hundred per cent.

Keywords: *Eisenia foetida, Vermiwash, Tagetes erecta, Flowering, Earthworm exudate.*

Introduction

Tegetes erecta gained popularity amongst gardeners and flower dealers on account of its easy culture and wide adaptability. With their habit of free flowering, short duration to produce marketable flowers, long shelf life and attractive colours. *Tagetes* sps. have become recent sensation in Indian floriculture.

Dhingra and Dhingra (1956) reported that all parts of *Tagetes* sps. contain essential oil which can readily be extracted by steam distillation. Sharma *et al.* (1961) reported that its oil has a pronounced odour and acts us a repellent to flies.

The growth and development of *Tagetes* sps. can be stimulated by using chemical growth regulators. However, the application of these chemicals are not ecofriendly. An alternative, more economical and natural growth stimulator is sought after to improve the growth and development in *Tagetes* sp.

Culturing of earthworms on organic waste to produce the organic manure (vermicompost) is an upcoming venture. This is not only for curbing the organic pollution caused by accumulation of decomposable organic waste but also to promote use of organic manure in cultivable lands for soil sustainability.

The importance of earthworms in soil fertility have been made since Darwin era. Earthworms are physically an aerator, crusher, mixer, chemically a degrader and biologically a stimulator in the decomposition system. They along with soil microorganisms play a major role in degrading organic wastes materials and thus maintain the nutrient flux in the system.

Research programmes initiated in the early 1980's aimed at using earthworm to breakdown organic wastes. Plant residues and dung are fragmented and mix with soil during passage through the worm-gut and in corporation into the soil profile. This is an important step in the cycling of organic matter (Satchell, 1958).

Reports are available on the stimulatory effect of earthworm population and their secretions in pastures and some cereals crops (Atlavinyte and Vanagas, 1982; Atlavinyte *et al.*, 1971; [illegible] *et al.*, 1990; Tomati *et al.*, 1990).

Grappelli *et al.* (1985) and Tomati *et al.* (1988) have shown that application of earthworms can increase the plant growth. The beneficial effects of vermicompost on crops like maize, wheat (Sharma and Madan, 1988), soybean, sugarcane (Shuxcin *et al.*, 1991), wheat (Van Rhee, 1965), barley and oats (Edwards and Lofty, 1978), have shown that chemical fertilizer application along with vermicompost increased the yield and nutrient uptake by the crops.

In present investigation an attempt has been made (i) to standardize the method to collect the exudates of the body fluid of earthworm, *Eisenia foetida* without killing the worms and (ii) to use the exudate as spray on *Tagetes erecta* to study their effect on growth and development.

Materials and Method

The plant material used for study was developed by sowing the Pusa Narangi variety of *Tagetes erecta*. The uniform plants were selected and transplanted for treatment.

Clitellated earthworms *Eisenia foetida* were collected from the culture beds of organic waste. One kg earthworms, were placed into a dry enamel tray for 15 to 20 minutes to clear out the cast then were carefully separated from their excreta (castings) then 500 ml. luke warm (37–40°C) distilled water were added. The worms were agitated for 3–5 minutes and placed into another beaker containing 500 ml of water at room temperature (25–27°C) to rinse them thoroughly to collect the exudate adhering to its body wall before releasing back to the culture beds. The contents of two beakers were mixed to use as the spray. The syrupy, light yellow fluid were collected, and their spray were carried out at different intervals. The freshly prepared vermiwash was applied at fortnightly interval for 3 months in the forenoon with a fine nozzled jet hand sprayer. The vegetative growth, number or flowers and total yield were recorded.

The current study was carried out at D.S. College, Aligarh (U.P.).

Results and Discussion

The plant height at different intervals were varied by application of different concentration of vermiwash. At the intervals of 15 days the maximum plant height was reported in T_2 while after 30 days the result was same in T_1 and T_2. After 45 days the maximum height were reported in T_2 (Figure 30.1).

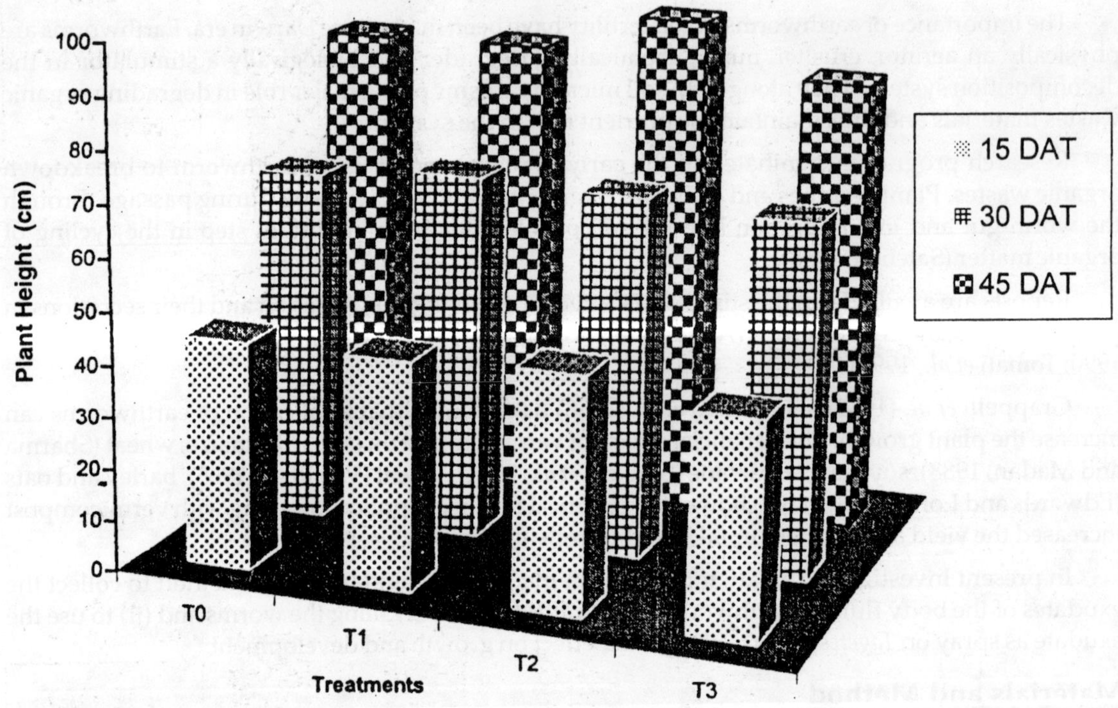

Figure 30.1: Histogram Showing Plant Height of *Tagetes erecta* Under Different Treatment of Earthworm Exudates during Different Time Intervals

After 15 days maximum number of branches was found in T_2 and least in T_3 while after 45 days, the rate of branching becomes enhanced in T_1. The budding rate was also higher in T_2 and lower in T_3. However the maximum yield were reported in T_2 and least in T_0 (Figures 30.2–30.5).

The minimum disturbance will stimulate the earthworms to pour out the coelomic fluid throughout their dorsal pores. This fluid is a source of phosphate, sulphates and chlorides of potassium, sodium and magnesium, sulphur, iron and other organic acids are also found in small quantities. All these salts being in soluble form they can diffuse into the plant system through the stomatal openings. So, vermiwash definitely show a positive effect on growth and total yield, but it should apply in low concentration if it is applied in higher concentration, the rate of growth and yield declines. This shows improved absorption efficiency of the leaves on reducing the concentration of the components.

The results of this study has now made the way to further investigate the bench life of the extract.

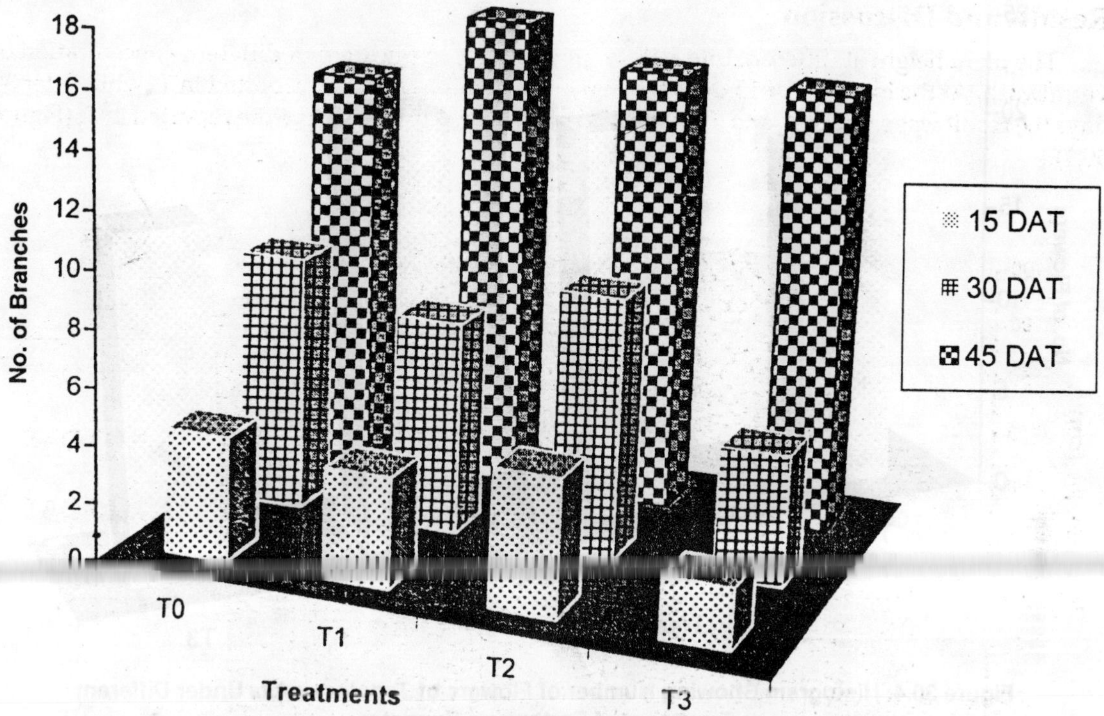

Figure 30.2: Histogram Showing Number of Branches of *Tagetes erecta* Under Different Treatment of Earthworm Exudates during Different Time Intervals

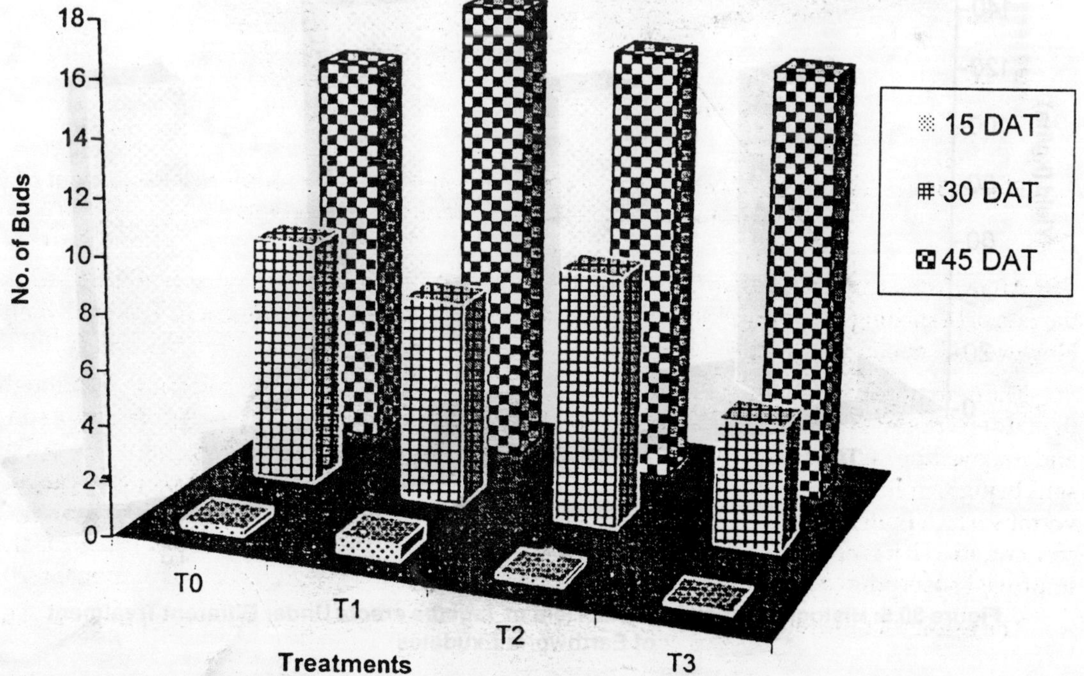

Figure 30.3: Histogram Showing Number of Buds of *Tagetes erecta* Under Different Treatment of Earthworm Exudates during Different Time Intervals

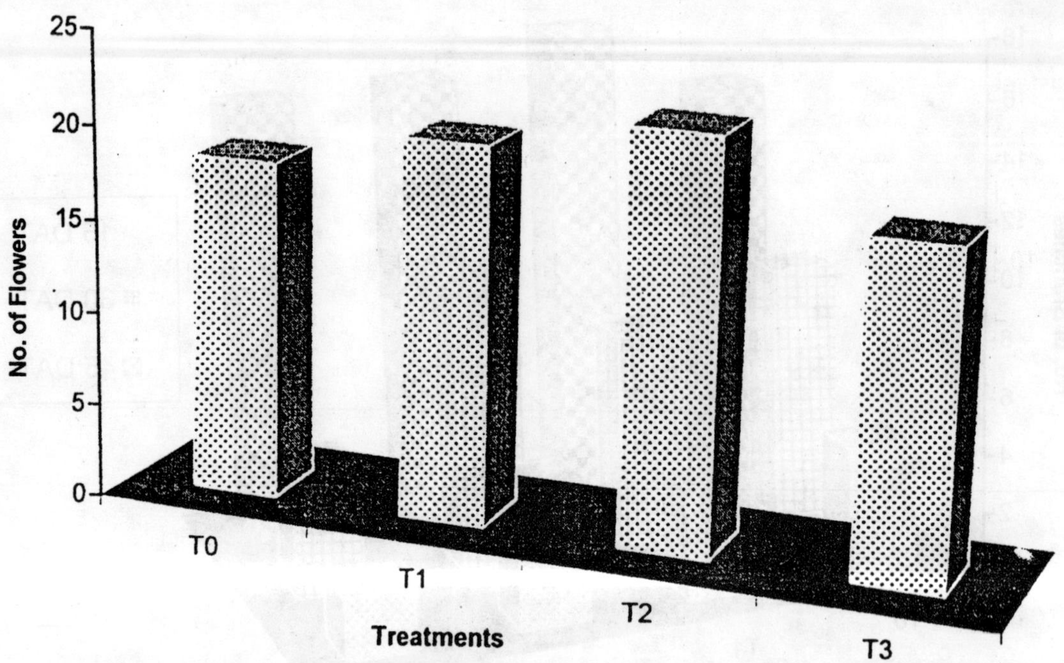

Figure 30.4: Histogram Showing Number of Flowers of *Tagetes erecta* Under Different Treatment of Earthworm Exudates

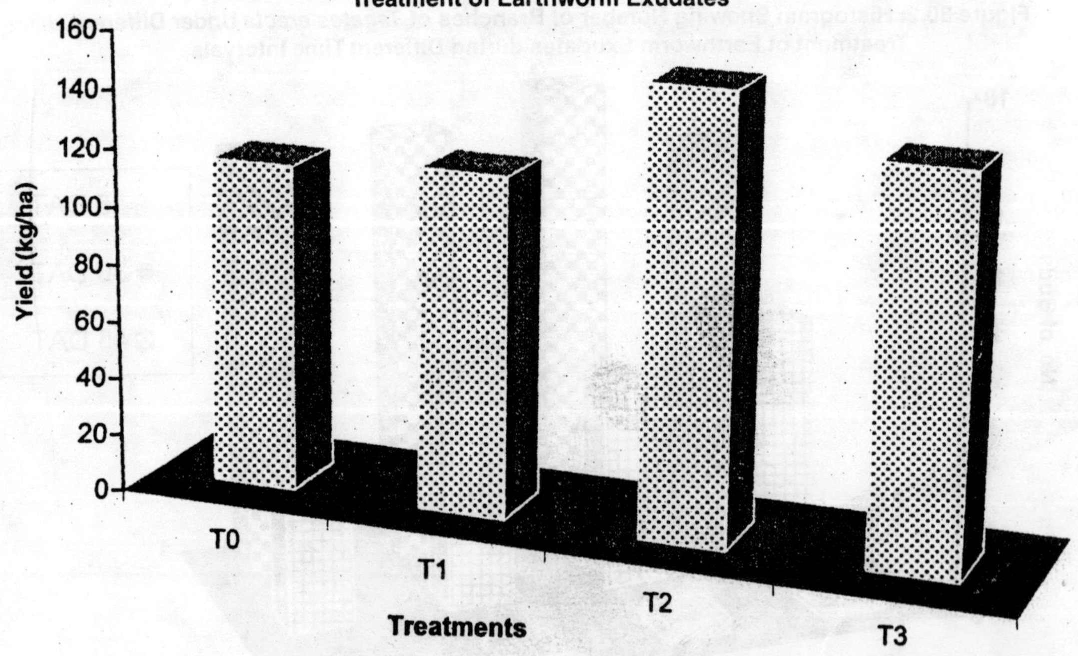

Figure 30.5: Histogram Showing Total Yield of *Tagetes erecta* Under Different Treatment of Earthworm Exudates

Acknowledgement

The authors are thankful to the University Grants Commission, New Delhi for providing financial assistance.

References

Atlavinyte, O. and Vanagas, J. 1982. The effect of earthworms on the quality of barley and rye grain. *Pedobiologia,* 23: 256–262.

Atlavinyte, O., Daciulyte, J. and Lugauskas, A. 1971. Correlations between the number of earthworms, microorganisms, and vitamin B_{12} in soil fertilized with straw. *Liet. T.SR Mokslu. Akad. Darb., Ser. B,* 3: 43–56.

Dhingra, S.N. and Dhingra, D.R. 1956. *Perfum. Essent. Oil Rec.,* 47: 391–394.

Edwards, C.A. and Lofty, J.R. 1978. The influence of arthropods and earthworms upon root growth of direct drilled cereals. *J. App. Ecol.,* 15: 789–795.

Galli, E., Tomati, V., Grappelli, A. and Delina, G. 1990. Effect of earthworm cast in protein synthesis in *Agaricus bisporus. Biol. Fertil. Soil,* 9: 1–2.

Grappelli, A., Tomati, U. and Palma, G. 1985. Plant consisting of modular elements for degrading organic wastes by means of earthworms. Patent entry at Consiglio Nazionale dell Ricerche, Rome (Italy).

Satchell, J.E. 1958. Earthworm biology and soil fertility. *Soils Fert.,* 21: 209–219.

Sharma, M.L. *et al.,* 1961. *Perfum. Essent. Oil Rec.,* 52: 561–562.

Sharma, N. and Madan, M. 1988. Effects of various organic wastes alone and with earthworms on the total dry matter of wheat and maize. *Biol. Wastes,* 25: 33–40.

Shuxcin, L., Dezhong, V. and Debing, W. 1991. Studies on the effect of earthworms on the fertility of red-arid soil. In: *Advances in Management and Conservation of Soil Fauna,* (Eds.) Veeresh, G.K., Rajgopal, D. and Vitakkamath, C.A. Oxford and IBH, p. 543–546.

Tomati, U., Grappelli, A. and Galli, E. 1988. The hormone like effect of earthworm casts on plant growth. *Biol. Fert. Soils,* 5: 288–294.

Tomati, U., Galli, E., Grappelli, A. and Dilena, G. 1990. Effects of earthworm cast on protein synthesis in radish (*Raphanus sativum*) and lettuce (*Lactuga sativa*) seedlings. *Biol. Fert. Soils,* 9: 1–2.

Van Rhee, J.A. 1965. Earthworm activity and plant growth in artificial cultures. *Plant and Soil,* 22: 45–48.

Chapter 31

Population Dynamics and Carrying Capacity of Thoubal District

S.R. Singh, P. Rukamani Devi,* N.B. Devi,* W.K. Devi* and N.S. Devi***

**Post Graduate Studies Centre, HRDRI, Canchipur – 795003*

***Department of Botany, Imphal College, Imphal, Manipur, India*

ABSTRACT

The present article deals with carrying capacity, a finite number of people who can be supported without degrading the natural environment of a place and population dynamics of Thoubal district, Manipur. Carrying capacity is never simple and direct and its potential decreases from 61.71 to 20.5 when based on fruits, from 39.32 to 4 based on vegetables and from 1.5 to 1.4 based on rice from 2001 to 2002. The carrying capacity of the place is influenced by the impact of population growth, the dynamics, the changes. The population dynamics in lagarithmic scale rises from 11.28 to 14.37 and 79208 to 3960722 in arithmetic scale from 1951 to 2001. These basic component of developmental planning and programme an area without improvement to human environment have analysed.

Keywords: Carrying capacity, Population dynamics, Population potential, Human environment, Inherent property.

Introduction

A sense of the future is an essential component in the mix factors of today. Regarding environmental issues (*i.e.*, some circumstances), only a few factors with respect to time may need to be considered. In case of wastewater treatment plant operator may need to predict the growth rate of bacteria in a digester over a period of hours or days. With relation to human population, the designer of the plant probably needed to estimate the population growth ratio over the next decade or two in order to size the facility. At the global environmental issues, to make even the crudest estimate of future carbon emissions and their effect on global warming, scientists need to forecast population and economic growth rates, anticipated improvement in energy efficiency, the global fossil fuel resources base and the consumption rate of individual fuels, the rate of deforestation or reforestation that could be anticipated. Such estimates need to be made for periods of time it may be measured in hundreds of years. On view of Niels Bohr, it cannot expect to make accurate predictions of the future, especially when the required time horizon may be extremely long. But we can make simple estimates that are robust enough that the insight they provide is most certainty that world population growth at today's rates cannot continue for hundred years. We can also use simple estimates from the different districts (*i.e.*, from the smallest unit of calculating population growth rate). The relationship between population dynamics and carrying capacity is indispensable. They are not related each other but also regulate each other at a higher degree. Carrying capacity (*i.e.*, environment) has become a major worldwide concern today. As there is a two-way relationship between both of them likely to influence each other at a higher degree. So, now-a-days the population dynamics and carrying capacity issues have drawn the attention of different programme personnel and Policy makers. With connection to population and carrying capacity (*i.e.*, environmental factors) it has risen to the top of the international agenda and has brought in its wake a renewed attention to population issues. The centre of this consensus was the realization of the global nature of the most salient environmental issues (Ashok Kumar and Dibya Lochan Mohanta, 2003). In 1994, the International Conference on Population and Development adopt the programme of Action of the International Conference on Population and Development, which recommended for integrating population, environment and development.

So far, such works have not yet been undertaken in Thoubal district Manipur. Thus execute the present research investigation with the following objectives:

1. To determine the population dynamics
2. To analyse the capacity of carrying capacity
3. To determine the consequence of population dynamics and carrying capacity to the environment.

Materials and Methods

Investigation on the population studies on human being was conducted during five decades *i.e.*, from 1951 upto 2001 for determination of the population dynamics in Thoubal district an area of 405 sq. km of the Manipur valley. It is extended between 24°15'N and 24°4'N latitude and 93°50'E to 94°10'E longitudes, which is centrally located in the state of Manipur. For this study, 300 houses were selected randomly and house-to-house census has been undertaken. The information of census data have based on the census of Manipur as well as census of Thoubal district upto 2001 (Statistical abstract of Manipur, 2001). In certain cases method of questionnaire and schedule, direct personnel interview etc. were adopted. For calculating carrying capacity all required data being taken from Agriculture Department and Horticulture Department.

For estimating the population dynamics two types of method has been formulated such as:

Arithmetic Scale

$$N_t = N_0 e^{rt}$$

where,

N_t = the number at time t

N_0 = initial amount/number at time zero

e = the base of natural logarithms

r = growth rate.

Logarithmic Scale

$$\ln N_t = \ln N_0 + rt$$

$$r = (\ln N_t - \ln N_0)/t$$

where,

N_0 = the number at time zero,

N_t = the number at time t

e = the base of natural logarithms

For the purpose of calculating carrying capacity, calculate as per the following formula:

Carrying Capacity (C.C.) = Maximum food production/Food requirement for a person

An estimation of potential crop land/arable land has been computed for three point scale. (Murak *et al.*, 1990) and base on FAO (1987) Statistics, the existing and potential agricultural productivity in calories were calculated.

The potential habitability of Thoubal district was estimated with the information on the amount of food energy in Kcal that is necessary to support one person.

Results and Discussion

Table 31.1 depicts the analysis result of population dynamics of Thoubal district during 1951 to 2001. In 1951, population was observed with values of 76,782 with corresponding logarithmic scale as 11.2799 and arithmetic scale as 79208. On the next decade *i.e.*, 1961 population was increased with 58142 peoples (*i.e.*, 7678 to 134924) the corresponding logarithmic scale as 12.1249 and arithmetic scale as 792083 respectively. In 1971 population was observed as 181771 it was increased 46847 numbers of persons from the previous decades *i.e.*, 134924 to 181771 persons. The corresponding logarithmic scale accorded as 12.7355 and arithmetic as 1584243 respectively. In 1981, population was observed as 231781 persons with corresponding to 13.8412 in Logarithmic scale and 2376402 in Arithmetic scale. In this observation period rate of increment of population were recorded as 50010 persons *i.e.*, 181771 to 231781 persons. During 1991 population of Thoubal district were accorded as 293958 numbers of persons it was observed that 62177 numbers of persons were increased from the previous decade (*i.e.*, 231781 to 293958) and corresponding logarithmic scale as 13.81112 and Arithmetic scale as 3168562 respectively. In 2001, population was accorded as 366341 with an increment of rate of population as 72383 (*i.e.*, 293958 to 366341), the corresponding Logarithmic scale of 14.3738 and Arithmetic scale as 3960722 respectively.

The inherent property of an organism to reproduce to survive *i.e.*, to increase in numbers (Chapman, 1928) and the algebraic sum of the number of young produced at each reproduction (Graham, 1952) was observed with highest rate in 2001 with a increment rate of population as 72383. The lowest increment rate of population was accorded as 46847 numbers of persons. The increment rate of population in the area was so different observed with a tremendous jump from 1981 upto 2001. The values were graphically presented and the growth is plotted as Logarithms or on semi logarithmic the log of population number plotted against time gives somewhat diverted form of straight line (Lotka, 1925; Pearl, 1930) (Figure 31.2). The present finding clearly showed the population expansion in a high speed. It might be due to less obstruction to the entities of population growth in the area. The.population growth in nature often exhibits exponential growth when there is ample food and no crowding effects, enemies and so forth. Under these conditions the population as a whole is expanding at a terrific rate. The finding was in agreement with that of previous workers (Slobodkin's 1962; Wangersky, 1956; Cunnings, 1967).

Table 31.1: Population of Thoubal District During 1951–2001 with Relation to Logarithmic Scale and Arithmetic Scale

Year	Population	Rate of Increment of Population	Coefficient of Population Growth	Logarithmic Scale $lnN_t = lnN_0 + rt$	Arithmetic Scale $N_t = N_0e^{rt}$
1951	76782			11.2799	79208
1961	134924	58142	0.056374	12.1249	792083
1971	181771	46847	0.02981	12.7355	1584243
1981	231781	50010	0.0243	13.2910	1376402
1991	293958	62177	0.0237	13.8412	3168562
2001	366341	72383	0.02201	14.3738	3960722

Figure 31.1 revealed the growth curve represented with population in numbers (N) in an arithmetic scale. The Figure depicts the hypothetical population of Thoubal district experiencing 5 decades of exponential growth. The population curve represented on an arithmetic scale, the part of the growth curve that represents population increase and it takes the form of 'I' not in 'J' shape (Figure 31.1). The present finding shape is deflected from the shape of previous workers (Bertataffy, 1957; Nicholson 1954). It might be due to the period of studies that is starting from 1951 not from the origination *i.e.*, zero population of the investigating area.

Table 31.2 displayed the analysis results of potential carrying capacity based on three levels of consumption during 2001 and 2002 respectively. During 2001, accorded cultivated area as 23.33 ('000 ha), production as 60.46 ('000 tonnes) yield as 2591.51 kg/ha and population potential as 1.5 (10^5 people). In vegetable it were observed as 17.05 ('000 ha) in arable land, production as 13.240 ('000 tonnes), yield as 7765.39 kg/ha and population potential as 39.32 (10^5 people). Regarding fruit, it was accorded as 37.38 ('000 ha) under arable land, production as 20.909 ('000 tonnes) and population potential as 61.71 (10^5 people) respectively.

In 2002, population potential factors for rice were observed as 22.51 ('000 ha) in arable land, 67.50 ('000 tonnes) in production, 2998.66 kg/ha yield and 1.8 (10^5 people) in population respectively. With relation to vegetable it was accorded arable land as 16.80 ('000 ha), production as 12.951 ('000 tonnes), yield with values of 7768.92 (kg/ha) and population potential as 4–0 (10^5 people) respectively. In fruit, arable land was observed as 24.647 ('000 ha), production as 69.223 ('000 tonnes), yield as 2808.57 kg/ha and population potential as 20.5 (10^5 people) respectively. The present investigation showed the importance of population dynamic studies with relation to planning and programme of developmental schemes in an area. In this connection Strong (1971) cited the needs and essentialities of components of urban development environments.

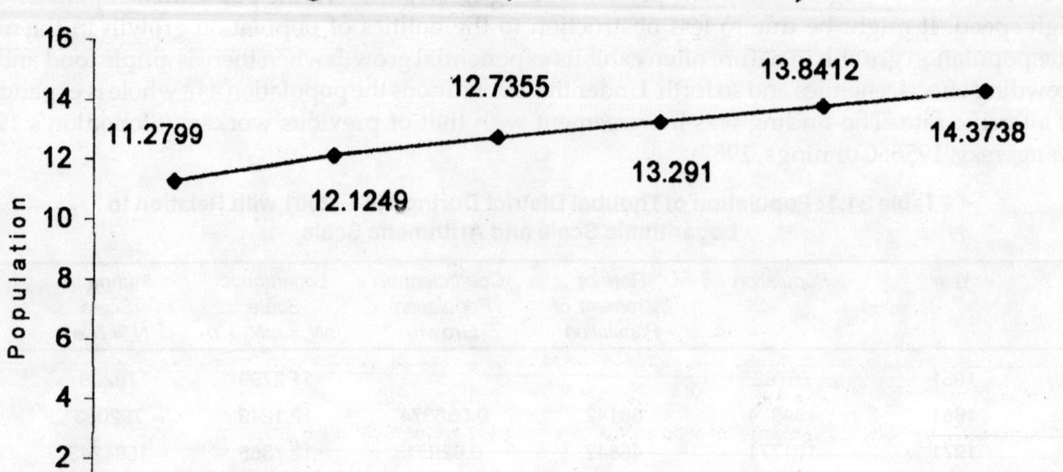

Figure 31.1: Representation of Population Growth Curve of Thoubal District with Relation to Time (Yrs.) in Arithmetic Scale for 1951–2001

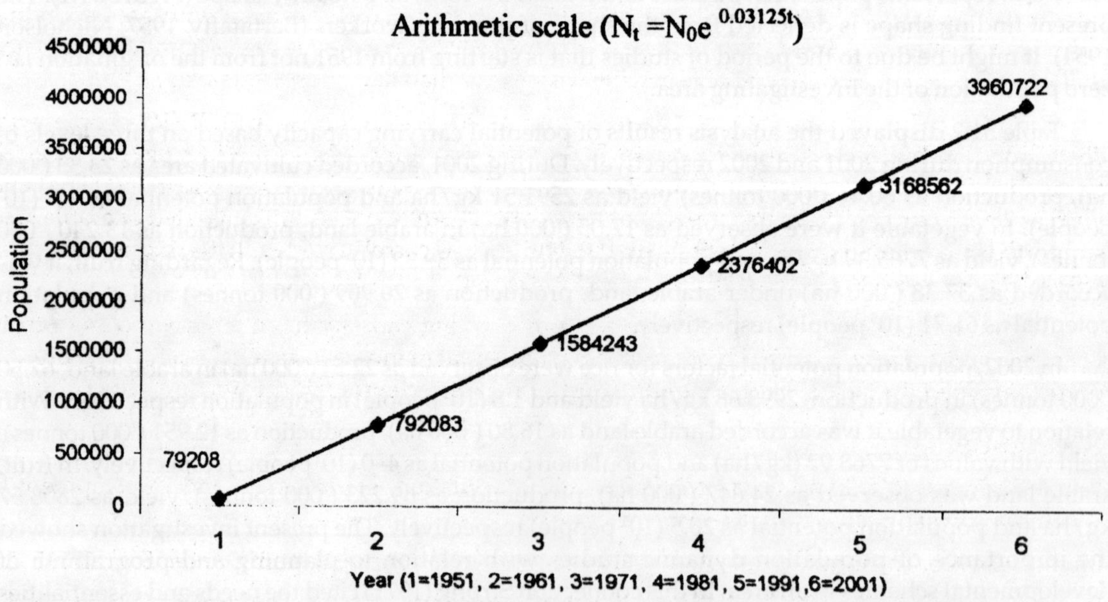

Figure 31.2: Displayed of Population Growth Curved of Thoubal District with Relation to Time (Yrs.) Logarithmic Scale for 1951–2001

Table 31.2: Potential Carrying Capacity of Thoubal District, Manipur Based on Three Consumables *viz.* Rice, Vegetables, Fruits (10^5 people)

Sl.No.	Food Items*	Area ('000 ha)	Prod. ('000 tonnes)	Yield (kg/ha)	Carrying Capcity**#
			Y_1 (2001)		
1.	Rice	23.33	60.46	2591.51	1.5
2.	Vegetable	17.05	13.240	7765.39	39.32
3.	Fruit	37.38	20.909	5593.63	61.71
			Y_2 (2001)		
1.	Rice	22.51	60.20	2448.60	1.4
2.	Vegetable	16.80	12.951	7768.92	4.0
3.	Fruit	24.647	69.223	2808.57	20.5

*: Murak *et al.*, 1990; **: FAO, 1987.

#: Carrying capacity (C.C.) = Maximum food production/Food requirement for a person.

Table 31.2 depicted population potential of rice as high as 1.5 (10^5 people) with corresponding cultivated/arable land area 23.33 ('000 ha) in 2001. Population potential of rice in 2002 was accorded as 1.4 (10^5 people) with corresponding arable land 22.51 and yield as 2448.60 kg/ha. With comparison between these two food items during two consecutive years showed that maximum arable land supported more population potential. The present finding was in an agreement with that of previous workers (Brown *et al.*, 1994, Mc Connell and Robert, 1995, Pimentel *et al.*, 1989; Rees and William, 1992; Wackernagel *et al.*, 1993). In case of vegetable, population potential was accorded as 39.32 (10^5 people) with corresponding cultivated arable land area 17.05 ('000 ha) in 2001 but in case of 2002, population potential with relation to vegetable was recorded as high as 4.0 (10^5 people) with corresponding cultivated/arable land 16.80 ('000 ha) and yield as 776892 (kg/ha). The present finding indicated that in 2002 population potential of vegetable was higher than that of previous year *i.e.*, 2001.

Potential carrying capacity based on fruit during 2001 was accorded as 61.71 (10^5 people) with corresponding cultivated/arable land as 37.38 ('000 ha) and yield as 5593.63 kg/ha. In 2002, population potential of fruit was recorded as 20.5 (10^5 people) with coincides 24.647 ('000 ha) and yield as 2808.57 kg/ha. From the present analysis it picturised as maximum cultivated or arable land supported maximum population potential. The finding was corroborated with the finding of Kessler (1994), Hayama and Rutlan (1971) and Cohen (1995).

The investigation credited the prime tasks of carrying capacity in all environments of fruit, vegetable and rice yield and production for developmental planning in an area. Further, the present finding justified the lack of basic continuing research based scientific studies on the subject.

References

Ashok Kumar and Dibya Lochan Mohanta. 2003. Population, Environment and Development in India. *J. Hum. Ecol.*, 14(5): 383–392.

Birch, L.C. 1948. The intrinsic rate of natural increase of an insect population. *J. Anim. Ecol.*, 17: 15–26.

Brown, L.R. and Kane. 1994. *Full House: Reassessing the Earth's Population Carrying Capacity*. W.W. Noreton, New York.

Cohen, J. 1995. *How Many People Can the Earth Support*? W.W. Norton and Co., New York.

Cummings, K.W. 1967. Calorific equivalents for studies in ecological energetics. Mimeographed, Pymatuning Lab. Ecol. University of Pittsburgh, Pittsburgh, pp. 52

FAO. 1987. *Production Yearbook*. Food and Agriculture Organisation of the United Nations, Rome.

Graham, S.A. 1952. *Principles of Forest Entomology*, Revised ed. McGraw-Hill Book Co., Inc., New York.

Hayama, Y. and Rutlan, V. 1971. *Agriculture Development: An International Perspective.* Md: John Hopkins Univ. Press, Baltimore.

Kessler, J.J. 1994. Usefulness of the human carrying capacity concept in assessing ecological sustainability of land–use in semi-arid regions. *Agriculture Ecosystems and Environment*, 48: 273–284.

Leslic, P.H., and Ranson, R.M. 1940. The mortality, fertility, and rate of natural increase of the Vole (Microtus agretis) as observed in the laboratory. *J. Anim. Ecol.*, 9:27–52.

Lotka, A.J. 1925. *Elements of Physical Biology*. Williams and Wilkins, Baltimore, pp. 460, Reprinted by Dover Publ., New York, 1956.

McConnell, Robert, L. 1995. The human population carrying capacity of the Chesapeake Bay Watershed: a preliminary analysis. *Population and Environment, A Journal of Interdisciplinary Studies*, 16: 335–351.

Murai, S., Y. Honda, K. Asakura and S. Goto. 1990. An Analysis of Global Environment by Satellite Remote Sensing: What population can the Earth Feed? Institute of Industrial Science, University of Tokyo.

Pimentel, David and Carl W. Hall (Eds.). 1989. *Food and Natural Resources*. Academic Press, Toronto, pp. 512.

Rees, William, E. 1992. Ecological footprints and appropriated carrying capacity: What Urban economics levels out. *Environment and Urbanisation*, 4: 121–129.

Solobadkin, L.B. 1962. Energy in animal ecology. In: *Advances in Ecological Research*, (Ed.) J.B. Cragg. 1: 69–101.

Statistical abstract of Manipur, 2001. Directorate of Economics Statistics. Government of Manipur.

Strong, A.L. 1971. *Planned Urban Environments*. Md. Johns Hopkins Univ. Press, Baltimore.

Terman, C.R. 1968. Population dynamics. In: *Biology of Peromyscus (Rodentia)*, (Ed.) King, J.A. The American Society of Mammalogists, special publication No. 2, Stellwater, OK.

Wackernagel, Mathes and William, E. Rees. 1993. How big is our ecological footprint: a handbook for estimating a community's appropriated carrying capacity discussion draft. Vancouver: UBC Department of Family Practice, pp. 106.

Nangersky, P.J. and Cunningham, W.J. 1956. On time lags in equations of growth. *Proc. Natl. Acad. Sci. (Wash.)*, 42: 699–702.

Chapter 32

Pesticide Induced Impairment on the Carbohydrate Metabolism in the Fish *Mystus vittatus*

R. Sornaraj*, A.J.A. Ranjit Singh** and A. Pushparaj*

**Reader, Research Department of Zoology, Kamaraj College, Thoothukudi*

***Reader, Department of Biology, S.P.K. College, Alwarkurichi*

ABSTRACT

Pesticides stress on the carbohydrate metabolism in the fish *Mystus vittatus* was studied by treating the fish with different sublethal concentrations of Parrysulfan (0.1672 and 0.3345 ppm), a organo chlorine pesticide and Sicocil (0.3358 and 0.6717 ppm), a organophosphate pesticide. The total glycogen level in the liver and muscle tissues and also the SDH level and lactic acid level showed significant changes when the fish was exposed to the pesticides for a long duration. This fish exposed to the concentration 0.3345 ppm of parrysulfan for 60 days showed 60 per cent and 50 per cent reduction in glycogen level in the muscle and over respectively. Similarly the fish exposed to 0.6717 ppm concentration of sicocil showed 59 per cent and 48 per cent reduction in the level of glycogen in muscle and liver respectively. The same trend is noted in the presence of SDH also. There is 55 per cent and 59 per cent reduction observed in the muscle and liver of the fish exposed to parrysulfan for 60 days and 48 per cent and 54 per cent reduction in the muscle and liver of the fish exposed for 60 days in sicocil. The presence of lactic acid in the tissues of fish exposed to pesticide showed an increasing trend. The fish exposed to the parrysalfan for 60 days in the 0.3345 ppm concentration showed 69 per cent and 60 per cent increase in lactic acid accumulation in the muscle and liver respectively where as the fish exposed to sicocil at the concentration 0.6717 ppm for 60 days mowed 68 per cent and 46 per cent increase in the lactic acid accumulation in the muscle and liver respectively. With regard to carbohydrate metabolism both the pesticides create a massive stress and inflict physiological disturbances to the fish *Mystus vittatus*.

Keywords: Pesticide, Sicocil, Parrysalfan, Glycogen, Organophosphate.

Introduction

In recent years, pesticides are widely used to control the various pests of our crops. Pesticides used ultimately reach the nearby water bodies through rain, irrigation surface runoff, etc. and affect the entire ecosystem. The dominating group of the aquatic system, fish, is the most affected animal. Much emphasis is being made to evaluate the effect of pesticides on the general organization of fishes. The pesticide that is reaching the aquatic system gets entrenched into the aquatic food chain through bio-accumulation bio-concentration and biomagnification process. Ramasami (1990) stated that the pesticides accumulated in the tissues of fishes alters the biochemical constitution of the fish. The effect of thiodon, dithane and nuvan on biochemical consituents in muscle, liver and kidney of cyprinid fish *Barilus benedelisis* was studied by Deoray (1987).

Rao and Rao (1981) and Babu *et al.* (1988) had reported a decrease in glycogen, total carbohydrates and lipid contents in tissues of *Sarotherodon mossambicus* exposed to pesticides. Ghosh (1987) reported the toxic effects of three organophosphate pesticides on the carbohydrate metabolism of the freshwater fish, *Channa punctatus.* Ramasamy and Maheswari (1993) reported lactic acidosis in fishes exposed to pesticides.

Shrivastava and Singh (1997) reported enhancement of glucose level in, the intestinal tissues and liver during early period of exposure of the fish, *H. fossilis,* to the pesticide endosulfan. But the trend was completely altered when the exposure duration was extended.

From the available literature it was found that the pesticide exposure impairs the biochemical constituents in the experimental animal, fish. Though much work has been done to trace the quantity of the biochemical lesions in several fishes, less attention has been focussed on the cat fish, *Mystus vittatus,* although they inhabit all freshwater system and are highly edible. Hence the present biochemical investigation was carried out on the toxicity of this chemicals on the fish, *Mystus vittatus.*

Material and Methods

The fish, *Mystus vittatus,* (weight 5–6 gms) were collected from the local pond, brought to the laboratory and acclimatized in the laboratory condition for two weeks under natural photoperiod and temperature (29±2°C) in 50 L glass aquaria.

From the 96h LC_{50} value, two sublethal concentrations *viz.*, $1/4^{th}$ and $1/8^{th}$ of parrysulfan (0.1672 ppm and 0.3345 ppm) and $1/4^{th}$ and $1/8^{th}$ of sicocil (0.3358 ppm and 0.6717 ppm) were chosen to expose the fish for biochemical studies. The experiment was run for 60 days keeping control separately. All fish were regularly fed *ad libitum* with nodules prepared from groundnut oil cakes and rice bran during the experimental period. After 10, 30 and 60 days of exposure the fish were taken out and the following tissues *viz.*, muscle and liver, were taken out under aseptic condition. The biochemical parameters *viz.*, total glycogen, SDH and lactic acid were estimated using standard procedures.

Results

Carbohydrate Metabolism

Glycogen

Due to the continuous exposure to the sublethal concentrations of parrysulfan and sicocil, the fish, *Mystus vittatus,* has developed a stress and reduction in food intake resulted in the depletion of muscle and liver glycogen from the beginning to the termination of the experiment. The percentage reduction in the glycogen level in both the liver and muscle tissues of parrsulfan and sicocil exposed fish was high with the increase of exposure duration (Table 32.1).

Table 32.1: Changes in the Level of Glycogen (mg/g wet tissue) in the Tissues of *Mystus vittatus* Exposed to Two Sublethal Concentrations of *Parrysulfan* and *Sicocil*. Values in parentheses indicate percentage change over control.

Pesticides	*Tissue*	*Concentration ppm*	*Sublethal Exposure During (days)* 10	30	60	*'t' Value*
Parrysulfan	Muscle	Control	0.94±0.07	0.92±0.05	0.90±0.04	
		0.1672	0.76±0.03	0.58±0.06	0.45±0.03	3.5682
			(–19.15)	(–36.96)	(–50.00)	P < 0.05 S
		0.3345	0.69±0.08	0.52±0.08	0.36±0.04	4.1330
			(–26.60)	(–43.48)	(–60.00)	P < 0.05 S
	Liver	Control	9.94±0.53	9.81±0.35	9.85±0.64	
		0.1672	7.80±1.08	6.04±0.34	5.11±0.34	4.4954
			(–21.52)	(–38.43)	(–48.12)	P < 0.05 S
		0.3345	7.01±0.74	5.71±0.27	4.96±0.41	6.6214
			(–29.48)	(–41.79)	(–49.64)	P < 0.05 S
Sicocil	Muscle	Control	0.94±0.07	0.92±0.05	0.90±0.04	
		0.3358	0.87±0.01	0.73±0.01	0.48±0.06	1.9770
			(–7.45)	(–20.65)	(–46.67)	P > 0.05 NS
		0.6717	0.81±0.05	0.61±0.04	0.37±0.14	2.5317
			(–13.83)	(–33.69)	(–58.87)	P > 0.05 NS
	Liver	Control	9.94±0.53	9.81±0.35	9.85±0.64	
		0.3358	8.41±0.94	7.01±0.53	5.91±0.23	3.8052
			(–15.09)	(–28.54)	(–40.00)	P < 0.05 S
		0.6717	7.93±0.68	6.83±0.11	5.11±0.43	3.9481
			(–20.22)	(–30.38)	(–48.12)	P < 0.05 S

Student 't' analysis revealed that the muscle glycogen content of the fish exposed to sicocil do not show any significant change.

Succinic Dehydrogenase (SDH)

The SDH level in the tissues of both parrysulfan and sicocil exposed fish was considerably reduced with the increase in the duration of exposure. The degree of SDH decrease was comparatively less in sicocil exposed fish than parrysulfan exposed fish. A significant decrease in SDH level in the liver tissues of both parrysulfan and sicocil exposed fish was also observed (Table 32.2).

Lactic Acid

Lactic acid level in the muscle and liver of the pesticide exposed fish registered an increasing trend. The increasing trend was more pronounced in parrysulfan exposed than sicocil exposed fish. The increment in lactic acid level was higher when the fish was exposed for 60 days. The level of lactic acid was almost double in both muscle and liver tissues after 60 days of exposure. The increase in lactic acid level in the muscle tissues of both parrysulfan and sicocil exposed fish was not statistically significant whereas the increase in lactic acid level of liver was statistically significant in both the pesticide exposed fish (Table 32.3).

Table 32.2: Changes in the Level of SDH (μ mole formazon/mg protein/hr) in the Tissues of *Mystus vittatus* Exposed to Two Sublethal Concentrations of *Parrysulfan* and *Sicocil*. Values in parentheses indicate percentage change over control.

Pesticides	*Tissue*	*Concentration ppm*	*Sublethal Exposure Duration (days)*			*'t' Value*
			10	*30*	*60*	
Parrysulfan	Muscle	Control	0.63±0.04	0.65±0.03	0.64±0.07	
		0.1612	0.51±0.02	0.46±0.08	0.36±0.04	4.4222
			(−19.05)	(−29.23)	(−43.75)	P < 0.05 S
		0.3345	0.49±0.01	0.41±0.09	0.29±0.06	4.1663
			(−22.22)	(−36.92)	(−54.69)	P < 0.05 S
	Liver	Control	0.89±0.03	0.88±0.05	0.84±0.03	
		0.1672	0.70±0.04	0.55±0.05	0.42±0.03	3.8059
			(−21.35)	(−37.50)	(−50.00)	P < 0.05 S
		0.3345	0.61±0.03	0.50±0.05	0.35±0.08	4.9859
			(−31.46)	(−43.18)	(−58.33)	P < 0.05 S
Sicocil	Muscle	Control	0.63±0.04	0.65±0.03	0.64±0.07	
		0.3358	0.55±0.06	0.50±0.03	0.39±0.04	3.3607
			(−12.69)	(−23.08)	(−39.06)	P < 0.05 S
		0.6717	0.53±0.08	0.45±0.02	0.33±0.01	3.4815
			(−15.87)	(−30.77)	(−48.44)	P < 0.05 S
	Liver	Control	0.89±0.03	0.88±0.05	0.84±0.03	
		0.3358	0.75±0.03	0.57±0.08	0.45±0.04	3.1636
			(−15.73)	(−35.23)	(−46.42)	P < 0.05 S
		0.6717	0.64±0.01	0.51±0.07	0.39±0.04	4.8338
			(−28.09)	(−42.04)	(−53.57)	P < 0.05 S

Discussion

After exposure to the sublethal concentrations of parrysulfan and sicocil the fish, *Mystus vittatus*, developed alterations in the quantity of various biochemical constituents. The modulations in biochemical set up impose stress to the fish. It has been stated that (Ghosh, 1987) the normal physiological functioning of different tissues is greatly disturbed by the pesticide pollution. The relative potency of toxicity of pesticides differs in relation to tissues, biomolecules and duration of exposure (Sheela *et al.*, 1992; Ganathy *et al.*, 1994).

In the present investigation the organochlorine pesticide, parrysulfan, and the organophosphorous pesticides sicocile were found to have induced alterations in the cellular machinery and have altered the level of biochemical constituents.

Liver and muscle are two active sites where storage and metabolism of glycogen take place. The observed glycogen loss in the liver and muscle of the fish, *Mystus vittatus*, is attributed to the toxicogenic effects of parrysulfan and sicocil on tissue energy reserve. There are two possibilities about the mode of action of the pollutant: either, the parrysulfan and sicocil are responsible for cellular disorganization of the internal tissue system affecting the storage metabolism as reported by Ghosh, 1986; Sambasiva

Rao *et al.*, 1984; or it forced the exposed fish to heavy physical exercise in the form of erratic and rapid movement and high respiration rate, thus increasing utilization of reserve glycogen of the liver to meet the glycogen hunger of the body followed by less glycogenesis and increased glocogenolysis. Corbett (1974), Matsumera *et al.* (1972) have added that the glycogen loss in muscle may be due to accumulation of the pesticide in the pancreatic islets of the fish, which selectively damages the insulin producing β cells. Mazeaud *et al.* (1977), Anil *et al.*,(1980) had reported that the pesticide methyl parathion stimulated rapid secretion of both glucocorticoides and catecholamines from the adrenal tissues of the fish, *Channa punctatus*, resulted in hyperglycemia.

Table 32.3: Changes in the Level of Lactic Acid (mg/g wet tissue) in the Tissues of *Mystus vittatus* Exposed to Two Sublethal Concentrations of *Parrysulfan* and *Sicocil*. Values in parentheses indicate percentage change over control.

Pesticides	*Tissue*	*Concentration ppm*	*Sublethal Exposure Duration (days)* 10	30	60	*'t' Value*
Parrysulfan	Muscle	Control	11.7±0.38	11.6±0.34	11.7±0.44	
		0.1672	12.4±0.86	13.8±0.48	18.1±0.34	1.7276
			(5.98)	(18.96)	(59.82)	P > 0.05 NS
		0.3345	13.1±0.98	17.5±0.41	19.8±0.34	2.6113
			(11.96)	(50.86)	(69.23)	P > 0.05 NS
	Liver	Control	6.8±0.35	6.8±0.38	6.8±0.49	
		0.1672	7.9±0.34	8.9±0.44	10.4±0.44	3.1201
			(16.18)	(30.88)	(52.94)	P < 0.05 S
		0.3345	8.2±0.74	9.9±0.38	10.9±0.98	3.6374
			(20.59)	(45.59)	(60.29)	P < 0.05 S
Sicocil	Muscle	Control	11.7±0.38	11.6±0.34	11.7±0.44	
		0.3358	12.2±0.64	13.2±0.47	14.8±0.86	2.2870
			(4.27)	(13.79)	(26.49)	P > 0.05 NS
		0.6717	12.9±0.38	16.3±0.78	19.6±0.81	2.3779
			(10.26)	(40.52)	(67.52)	P > 0.05 NS
	Liver	Control	6.8±0.35	6.8±0.38	6.8±0.49	
		0.3358	7.4±0.41	8.5±0.84	9.7±0.44	2.6098
			(8.82)	(25.00)	(42.65)	P > 0.05 NS
		0.6717	7.9±0.48	9.5±0.86	9.9±0.64	3.7643
			(16.18)	(39.71)	(45.59)	P < 0.05 S

The succinic dehydrogenase (SDH) level in the muscle and liver of *Mystus vittatus* showed a decreasing trend when exposed to parrysulfan and sicocil. The degree of decrease was correlated to the concentrations of the pesticides and exposure duration. As SDH is one of the enzymes in the TCA cycle, the decrease in SDH activity indicates depression of oxidative metabolism at the mitochondrial level, leading to an overall depression of TCA cycle (Radhaiah *et al.*, 1987). It has been reported that energy crisis induced by the pesticides alters the metabolism in the tissue. In such cases, a shift of aerobic metabolism towards anaerobic side could be presumed with the increase of glycolytic pathway.

Joyce *et al.* (1989, 1990) reported SDH reduction results in the switch over of aerobic to anaerobic pathway in *Clarius batrachus* exposed to organophosphorous pesticide. Increase in the glycolytic path way was observed in the present study. The fall in SDH indicates the oxygen stress and energy crisis and to rome extent mitochondrial disturbances (Azhar *et al.*, 1991).

In both muscle and liver, significantly higher levels of lactic acid had persisted in all the three intervals in both concentrations of parrysulfan and sicocil when compared to that of control. Increase of lactic acid in the muscle and liver tissues of pesticide treated fishes was due to the inadequacy of oxygen supply in cells to cope up with the complete break down of carbohydrate into carbondioxide and water (Seskin and Levine, 1947). Hence the persistence of tissue lactate in fishes exposed to parrysulfan and sicocil in due present study implies hypoxia. Significant decrease in oxygen consumption of the whole animal and corresponding decrease in succinic dehydrogenase (SDH) enzyme activity in liver and muscle of parrysulfan and sicocil exposed fish further support the above implication of tissue hypoxia.

Moreover the persistence of higher levels of lactic acid in both muscle and liver of parrysulfan and sicocil treated fishes may be due to the impairment in the diffusion of lactic acid from the two tissues studied. Such impairment in the diffusion of tissue lactate in fishes was reported by Heath and Pritchard (1965).

References

Azharbaig, M.D., K. Vijay Joseph and K. Jayantha Rao. 1991. Heptachlor induced changes in the enzymes associated with carbohydrate metabolism in functionally different muscles of *Channa punctatus*. *J. Ecotoxicol. Environment*, 1(2): 156–161.

Anil, K. Srivastava and N.N. Singh. 1980. Observations on hyperglycemia in the murrel *Channa punctatus* after acute exposure to methyl parathion. *Comp. Physiol. Ecol.*, 5(2): 100–101.

Babu, P.R.A., G.R. Reddy, G.R.V. Babu and C.S. Chetty. 1988. Glycolytic oxidation in freshwater fish, *Sarotherodon mossambicus* during benthicarp exposure. *Curr. Sci.*, 57(11): 591–594.

Corbett, J.R. 1974. *Biochemical Mode of Action of Pesticides*. Academic Press, New York, USA.

Deorary, B.M. 1987. Effects of pesticides on freshwater cyprinid fish, *Barilus benedelisis* (Ham.) *Ph.D. Thesis*, Marathwada University.

Ganathy, V., D. Srinivasa Reddy, S.L.N. Reddy and K. Shankaraiah. 1994. Effect of hexachlorocyclohexane on biochemical composition of the fish. *Channa punctatus*. *J. Ecotoxicol. Envi. Monit.*, 4(1): 15–20.

Ghosh, T.K 1986. Effect of dimethoate on tissue glycogen content of some freshwater fishes. *J. Environ. and Ecol.*, 4(4): 554–557.

Ghosh, T.K. 1987. Toxic impact of three organophosphate pesticides on carbohydrate metabolism in a freshwater fish, *Channa punctatus*. *Adv. Biol. Sci.*, 6: 20.

Heath, A.G. and A.W. Pritchard 1965. Effect of severe hypoxia on carbohydrate energy stores and metabolism in two species of freshwater fish. *Physiol. Zool.*, 38: 325–334.

Joyce Shobha Rani, V., P. Venkateshwaralu and C. Janaiah. 1989. Changes in carbohydrate metabolism of *Clarias batrachus* when exposed to two organophosphorus insecticides. *J. Environ. Biol.*, 10(2-supp.): 197–204.

Joyce Shobha Rani, V., P. Venkateshwaralu and C. Janaiah. 1990. Impact of sub-lethal of concentration of malathion on certain aspects of metabolism in freshwater fish in *Clarias batrachus. Comp. Physiol. Ecol.*, 15(1): 13–16.

Matsumera, F., G.M. Boush, and J. Misato. 1972. *Environmental Toxicology of Pesticides*. Academic Press, New York, USA.

Mazeaud, M.M., F. Mazeaud and E.M. Donaldson. 1977. Primary and secondary effects of stress in fish; some new data with general review. *Trans. Am. Fish. Soc.*, 106: 210–212.

Radhaiah, V., M. Girija and K. Jayantha Rao. 1987. Changes in selected biochemical parameters in the kidney and blood of fish, *Tilapia mossambica. Bull. Env. Cont. Toxicol.*, 30: 1006–1011.

Ramasami, M. 1990. Adaptive trends in lipid levels of liver and muscles of *Sarotherodon mossambicus* exposed to sevin. *J. Ecobiol.*, 2(1): 56–61.

Ramasami, M. and R.V. Maheswari. 1993. Comparative lactic acidosis in fishes following pesticide stress. *Proc. 2nd European Conf. on Ecotoxicol.*, 1(2): 877–885.

Rao, P.S.K. and P.KV. Rao. 1981. Lipid derivatives in tissues of the freshwater teleost fish, *Sarotheroden mossambicus.* Effect of methyl parathion. *Proc. IInd. Nat. Sci. Acad.*, 47: 53– 57.

[illegible]

Sheela, M., R. Mathivannan and S. Muniandi. 1992. Impact of Fenvalarate and phosphamidon on food intake, growth and conversion efficiency in the fish *Channa striactus. J. Env. and Ecol.*, 10(3): 593–596.

Sambasiva Rao, K.R.S., K. Sivaprasada Rao and K.V. Ramana Rao. 1984. Impact of technical and commercial grade phenthoate on some selected parameters of oxidative metabolism in the fish *Channa punctatus* (Bloch). *Ind. J. Ecol.*, 11: 6–11.

Srivastava, A.K. and Singh, N.N. 1997. Effect of endosulfan on carbohydrate metabolism in Indian cat fish. *Ecotoxicol. Env. Safety*, 5: 412–417.

Chapter 33

The Studies on Fisheries of Tilapia-Dominated Perennial Tank

A. Madhusudhan Rao and Ravi Shankar Piska

Department of Zoology, University College of Science, Osmania University, Hyderabad – 500 007, India

ABSTRACT

The present work deals with the studies on fisheries of a Tilapia-dominated perennial tank, Julur, Nalgonda District, Andhra Pradesh during three years from 2000–01 to 2002–03. The water-spread area of the tank is 14.06 ha and water is useful for irrigation and fisheries. The tank is in eutrophic nature. The fishes grouped into major carps, Tilapia, Murrels, catfishes and other food fishes, out of which Tilapia dominated with 61.04 per cent and followed by major carps (36.39 per cent). The mean fish production was 737.50 kg/ha/yr ($p>0.05$). Tilapia production reduced from 596.44 to 432.13 and further decreased to 387.85 kg/ha/yr during the above period. The major carp production increased from 215.46 to 251.76 and further to 318.99 kg/ha/yr (mean production 262.03 kg/ha/yr) during the above period ($p > 0.05$). Only major carp seed (25–30 mm total length) stocked every year (4267, 7113 and 8534 fry/ha) and other fishes depend on self-recruitment. The major carp production increased with increase of stocking densities. The number of fry for production of 1 kg fish was 25.33. The fish productions per day and per fisherman were 106 and 53.34 kg respectively. The growth rate of major carps reduced significantly and attained 1 kg only during third year. Tilapia was responsible for decreased growth rate and major carp yield, due to the competition with other fishes because of voracious feeding and prolific breeding.

Keywords: *Fisheries, Tilapia, Major carps, Fish production, Growth, Perennial tank.*

Introduction

Reservoir constitute the single largest inland fishery resource, both in terms of size and fish production potential. These man made ecosystems offer enough scope for stock manipulation through ecological maneuvering, paving the way for production hikes at relatively low capital investment. Reservoir fisheries development is labour productive and ensure employment for weaker section of our society. 3.15 million ha of water spread area is available in India for reservoir fisheries. Experience has revealed that these water bodies of immense potential for fish husbandry through extensive aquaculture techniques. Reservoirs in India are considered the growing resource with enormous potential for fish production augmentation through capture fisheries. Reservoir fisheries is essentially a stocking-cum-capture system.

The exotic fish, Tilapia, *Oreochromis mosambicus* is common fish in Indian waters and is uneconomical and has less demand in Indian markets (Mitra, *et al.*, 1989; Piska and Rao, 2002; Piska, 2005; Narayana and Kohli, 2004 and Sakhare, 2005). It is dominant in some of the water bodies and effecting fisherman economy. The main objective of present work is to observe the impact of Tilapia-dominance on fish growth and production in a perennial tank.

Materials and Methods

The present work conducted in a perennial tank, Julur, Nalgonda District, Andhra Pradesh during three years period from 2000–01 to 2002–03. The water spread area is 14.06 ha and water is useful for agriculture and fisheries.

The fishes caught with the help of gill nets, cast nets, drag nets, etc. in holidays during monsoon and post-monsoon months and other than holidays during pre-monsoon months. Since Tilapia population was more, its population reduced by repeated netting. The carnivores and trash fishes also reduced by repeated netting.

Major carps seed (25–30 mm total length) stocked every year during monsoon months, except common carp seed during first year. The over all stocking of major carps was 60,000, 100,000 and 120,000 during 2000–01, 2001–02 and 2002–03 respectively. The stocking rates were 4267, 7113, and 8534 no/ha with the mean of 6639 no/ha. The stocking percentages of the catla, rohu and common carp were 42.86 per cent, 42.86 per cent, 14.29 per cent. The growth rate of fishes analysed every fortnight, then calculated for every season. Number of fry for the production of 1 kg fish calculated with the help of stocking densities and fish production. ANOVA test used according to Bailey (1959).

Results

The mean fish production was 737.50 kg/ha/yr (F 0.9619; df 2, 4; $p > 0.05$). The fish production decreased from 11829 kg (841.32 kg/ha/yr) to 9910 kg (704.84 kg/ha/yr) and further to 9375 kg (666.79 kg/ha/yr) during 2000–01, 2001–02 and 2002–03 respectively (Table 33.1).

Tilapia dominated with a mean of 61.04 per cent. Tilapia production declined from 596.44 to 337.84 kg/ha/yr during 2000–01 to 2002–03 with mean production of 444.19 kg/ha/yr (Table 33.1). The landing size of Tilapia was 100–150 gr. The ratio between major carps and Tilapia production was 1 : 1.75 and the ratio between other fishes and Tilapia was 1 : 23.27. The dominance ratio of Tilapia declined from 1 : 2.78 to 1 : 1.07 with the major carps during first year to third year. Major carps occupied the second place in the fish landings with a mean of 262.03 kg/ha/yr (F 0.9619; df 2, 4; $p > 0.05$). The major carps productivity increased from 215.46 to 318.99 kg/ha/yr (3030 to 4485 kg).

Cat fishes occupied the third place in the fish landings with 10.86 kg/ha/yr. Murrels occupied the fourth place with 6.78 kg/ha/yr. The contribution of the other food fishes landings was least with 2.39 kg/ha/yr.

Table 33.1: Total Fish Landing in Julur Tank During 2000–01 to 2002–03

Sl.No.	Type of Fish	2000–2001			2001–2002			2002–2003			Total Fishes Landing (kg/ha/yr)
		Total Quantity (kg)	Fish Landing (kg/ha/yr)	Percent-age (%)	Total Quantity (kg)	Fish Landing (kg/ha/yr)	Percent-age (%)	Total Quantity (kg)	Fish Landing (kg/ha/yr)	Percent-age (%)	
1.	Major carps	3030	215.46	25.61	3541	251.76	35.73	4485	318.99	47.84	2000–01:841.32
2	Murrels	149	10.60	1.26	96	6.83	0.97	41	2.92	0.44	2001–02:704.84
3.	Tilapia	8386	596.44	70.9	6077	432.13	61.32	4750	337.84	50.67	2002–03:666.79
	Catfishes	210	14.94	1.78	166	11.81	1.68	82	5.83	0.87	Total landings 31114
	Other fishes	54	3.84	0.45	30	2.13	0.3	17	1.21	0.18	Mean:737.50 (2000-01 to 2002-03)
	Total	**11829**	**841.32**	–	**9910**	**704.84**	–	**9375**	**666.79**	–	

Catla dominated among the major carp landing with the production of 382.22 kg/ha/yr (Table 33.2). Next dominated fish after catla among major carps was rohu with 286.20 kg/ha/yr. Least dominated fish in major carp landings was common carp with 117.90 kg/ha/yr. Catla landing slightly decreased from 1769 to 1581 kg during 2000–01 to 2001–02, in the following year catla landings increased to 2024 kg (Table 33.2). Similar trend observed in case rohu. The major carp landings increased from 2000–01 to 2002–03 (3030 to 4485 kgs).

Table 33.2: Major Carp Landings in Julur Tank During 2000–01 to 2002–03

Sl.No.	Type of Fish	2000–2001			2001–2002			2002–2003			Total Fishes	Fish
		Total Fish Catchs (kg)	Fish Catch (kg/ha/yr)	Percent-age (%)	Total Fish Catchs (kg)	Fish Catch (kg/ha/yr)	Percent-age (%)	Total Fish Catchs (kg)	Fish Catch (kg/ha/yr)	Percent-age (%)	Catch (kg)	Catch (kg/ha/yr)
1.	Catla	1769	125.81	58.39	1581	112.45	44.65	2024	143.95	45.13	5374	382.22
2.	Rohu	1261	89.69	41.61	1202	85.49	33.95	1561	111.02	34.81	4024	286.20
3.	Common carp	–	–	–	758	53.91	21.4	900	64.01	20.06	1658	117.9
	Total	**3030**	**215.46**		**3541**	**251.76**		**4485**	**318.99**		**11056**	**786.34**

The average numbers of fishing days/year were 46.33. Maximum number of days engaged in fishing during pre-monsoon months in three years. The average number of fishermen per year were 108.66. The average catch per day per month was 106 kg. The average catch of fish/fisherman was 53.34 kg (Figure 33.1).

Only rohu and catla stocked during July 2000.The average growth rate of rohu was 19.83,46.09 and 45.23 g/month for 1st, 2nd 3rd years respectively (Table 33.3). The average growth rate of catla per month were 23.95 g, 46.0 g and 61.90 g for above period. Rohu and catla and common carp seed

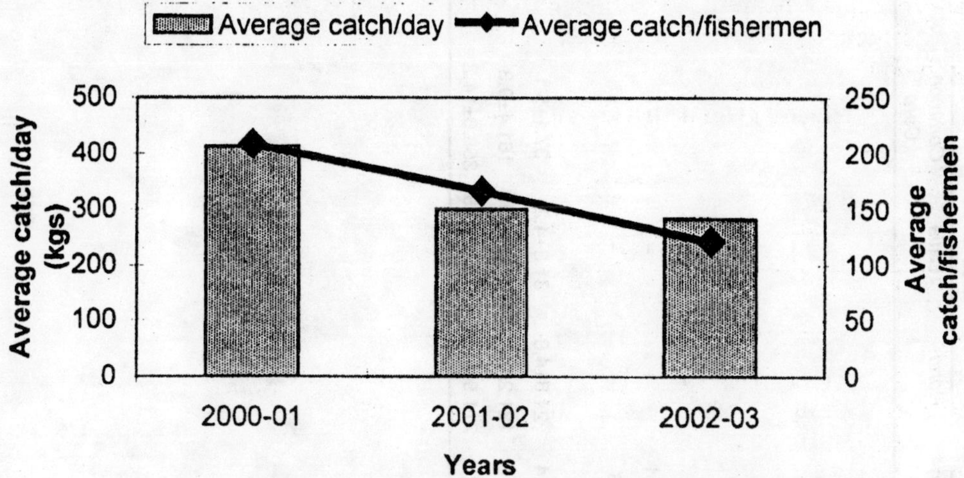

Figure 33.1: Fishing Efforts in Julur Tank During 2000–01 to 2002–03

stocked during July, 2001 and the average growth rate of rohu per month was 19.35 g and 47.35 g during first and second years respectively. The average growth rate of catla was 23.83 g/month in first and 48.28 g/month during second years respectively. The common carp growth rate per month was 25.16 g and 52.90 g during first and second years respectively. The fishes also stocked during July 2002 and the average growth rates were 21.39 g/month (rohu), 27.14 g/month (catla) and 28.46 g/month (common carp) (Table 33.3).

The number of fry for production of 1 kg fish was 19.80, 28.25 and 26.75 during the above period with a mean of 25.33. During the present study 262.03-kg/ha/yr major carp production achieved with the stocking density of 6638 fry/ha (Table 33.4; Figure 33.2).

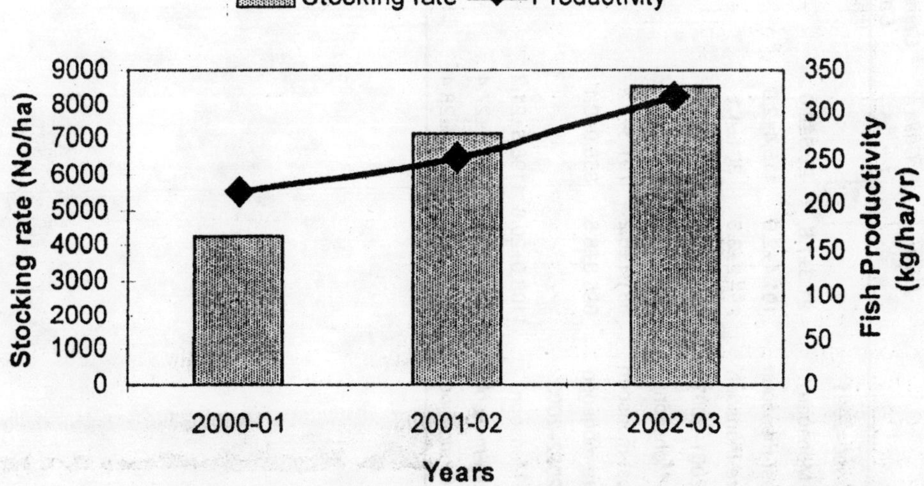

Figure 33.2: Stocking Densities and Production of Major carps from Julur Tank During 2000–01 to 2002–03

Table 33.3: Growth Rate (g) of Major Carps Stocked During 2000–01 in Julur Tank

Season and Year	2000–2001			2001–2002			2002–2003		
	Rohu	*Catla*	*Common Carp*	*Rohu*	*Catla*	*Common Carp*	*Rohu*	*Catla*	*Common Carp*
2000–2001									
Monsoon	21.3±1.8	28.5±2.0	–	–	–	–	–	–	–
Post-monsoon	101.4±2.6	115.4±2.9	–	–	–	–	–	–	–
Pre-monsoon	252.2±3.6	304±4.1	–	–	–	–	–	–	–
2001–2002									
Monsoon	482.3±3.8	502.1±4.2	–	20.6±0.2	26.3±1.8	31±2.1	–	–	–
Post-monsoon	681.1±6.2	701.3±7.2	–	98.8±6.1	113.4±4.8	142.2±3.1	–	–	–
Pre-monsoon	805.3±8.5	856.0±8.1	–	246.2±9.8	301.4±8.6	318±3.6	–	–	–
2002–2003									
Monsoon	1108.0±20.6	1196.5±21.3	–	502.2±4.9	526.1±5.3	551.3±8.4	24.8±4.8	31.3±4.2	37.1±5.1
Post-monsoon	1236.1±22.4	1371.1±27.4	–	698.1±9.2	722.4±8.4	794.1±6.8	121.2±12.3	135.5±8.1	161.4±9.3
Pre-monsoon	1348.1±28.2	1598.8±28.4	–	821.4±14.1	880.3±12.5	952.8±12.1	271.5±13.4	342.8±14.5	358.6±14.2

Table 33.4: Stocking Density and Production of Major Carps from Julur Tank During 2000–01 to 2002–03

Sl.No.	*Year*	*Stock Rate No/ha*	*Productivity (kg/ha/yr)*	*No. of Fish for Present Production of 1 kg Fish*
1.	2000–2001	4267	215.46	19.80
2.	2001–2002	7113	251.76	28.25
3.	2002–2003	8534	318.99	26.75
	Average	6638	262.03	25.33

Discussion

The fish productivity of the Julur tank was much higher when compared to the average Indian production of 29.70 kg/ha/yr (Dehadrai, 2001). The present figures were much higher than small reservoirs of India–49.9 kg/ha/yr. (Piska. 2000). The present productivities were higher than other minor reservoirs like Baghla–106 kg/ha/yr, Bachra–139 kg/ha/yr and Gularia–100 kg/ha/yr which managed by scientific methods (Jhingran and Sugunan, 1990). Devi (1997) and Piska (2000) recorded the productivity of 445 kg/ha/yr and 528 kg/ha/yr in Ibrahimbagh and Shanthamrai reservoirs Andhra Pradesh. Chary (2003) reported that productivity of Durgamcheruvu as [illegible] kg/ha/yr. Mahapatra (2003) recorded only 15.6 kg/ha/yr in Hirakud reservoir and 5–10 kg/ha/yr in other major reservoirs in Orissa. Higher productivity of fish found in Bidikiyar lake (825 kg/ha/yr) and Ambala municipal lake (500 kg/ha/yr) in Haryana (Agarwal, 1990).

Saxena (1990) reported the potential yield in Ukai (220 kg/ha/yr). Nagarjunasagar (119 kg/ha/yr). Bhavanisagar (54 kg/ha/yr). Rihand (54 kg/ha/yr) and Gobindsagar (263 kg/ha/yr) reservoirs. These figures were much less than the potential yield of present study. Burli and Sontakki (2002) reported that low fish production (4–6 kg/ha/yr) noticed in Malaprabha and Ghataprabha reservoirs of Karnataka. Tamot (2002) studied the fish production of irrigation reservoirs of Madhya Pradesh 70 kg/ha/yr. Bandyopodhyay *et al.* (2003) reported that after the introduction of Indian major carps and exotic carps in Manipur waters, dramatic change in the fish production was observed. Gowda (2002) reported the productivity of reservoirs of Karnataka as 30 kg/ha/yr with a total water spread area of 2.10 lakh ha (73 reservoirs).

Jhingran and Sugunan (1990) reported that the major carps contributed 90.54–94.76 per cent in Tillaiya and 57.44–76.61 per cent in Konar reservoirs. Devi (1997) reported that the contribution of major carps was 91.42 per cent in Ibrahimbagh and 96.40 per cent in Shathamraj reservoirs of Andhra Pradesh and Chary (2003) also reported the dominance of major carps in Durgamcheruvu of Andhra Pradesh.

In the present study Tilapia production was in declining trend, where as major carps production was in increasing trend, *i.e.*, the production levels of the above were in inversely proportion to each other. Tilapia and other fishes production levels were in decreasing trend during the above period, *i.e.*, the production levels of Tilapia and other fishes were directly proportional to each other.

Importance of Tilapia on fish production was reported by Rao and Shakuntala (1998) in Nelligudda reservoir, Karnataka and Piska and Rao (2002) in perennial tank, Nalgonda. They reported inverse relationship between major carps and Tilapia production. They also reported that Tilapia competes with major carps for natural food and space. When Tilapia was dominated there was scarcity of natural food in reservoir, which reflects the growth of carps and resulted low yields. Mitra *et al.* (1998),

Narayanan and Kohli (2004) and Sakhare (2005) reported that Tilapia was responsible for reduction of fish production in reservoirs due to its competition for food with other fishes.

This increase of major carps production was due to the stock manipulation and reduction of Tilapia and predators like murrels, catfishes and other fishes. This indicated that the major carp percentage was dependent upon the other fishes, especially on Tilapia and predators. The major carps especially in the younger stages easily fell as pray to the predatory fishes. The present results indicated that minor carps and trash fishes never played an important role in increase or decline of major carps, due to their less number. Most of the minor carps and trash fish were herbivorous.

The average catch/fisherman/month was 36.4 kgs in Karnal lake in Haryana (Agarwal, 1990). This value was very less when compared to the values of the present study. This may be due to more number of fishermen engaged in Karnal lake. Chary (2003) reported that 3.29 kg per fishermen per day in Durgamcheruvu. Agarwal (1990), Devi (1997) and Piska *et al.* (2000) stated that the fish production increased with increase of number of fishermen engaged in fishing. They concluded that the fish production from the natural water bodies could be increased by proper utilization of manpower and material for fishing.

In the present study stocked fishes attained 1 kg only during third year. This indicated that the growth rates of major carps were very slow. Usually major carps attained 1 kg weight during first year in ponds (Piska, 1999, 2000). Where as fishes reared in natural waters fish attained 1 kg either during first year or early part of second year. (Devi, 1997; Piska 2001 and Chary, 2003).

Piska (2000) reported that rohu grow to 401.35 g, catla to 407.32 g and common carp to 653.46 g in six months in a seasonal rain-fed tank. The growth rates in present study were less when compared earlier studies. Venugopal *et al.* (1988) reported the growth of common carp was more than 600 g and catla 400 g and rohu 300 g in 7 months rearing in a seasonal rain-fed tank. These growth rates were also more when compared to that of those in the present work.

During the present study the growth rates of rohu, catla and common carp were low. The growth rates were poor due to the competition for food and space with Tilapia. Due to the voracious feeding, omnivorous nature early maturity and prolific breeding Tilapia domination responsible for competition for food and space. Due to the presence of more number of Tilapia the natural food scarcity was found in the tank. This was indicated by analysis of plankton in the tank. During the first and second years the growth of major carps was poor when compared third year. This was due to more number of Tilapia in first two years. Tilapia population was reduced by the management and the growth of major carps was slightly improved.

Different authors suggested different stocking densities in Indian reservoirs. Mohanty (1984) suggested 1000 seed/ha in reservoirs of Orissa. Agarwal (1990) recommended the stocking rates in reservoirs of Haryana of minor reservoirs (10–100 ha) as 2000 fry/ha/yr and small reservoirs (below 10 ha) as 10000 fry/ha/yr. The present water body comes under minor reservoir category. Mathew and Mohan (1990) reported the stocking rate was 841 fingerlings/ha in Kerala and they never considered the stocking rate was low. Piska (2003) suggested the stocking density of 2000 seed/ha in reservoirs and tanks and 5000 seed/ha in the ponds in Andhra Pradesh. He also stated that the stocking density of 2000/ha was slightly high, which was due to compensate the mortality of sensitive fry in the open waters.

Sinha (1998) studied the stocking density and size of fingerlings in Indian reservoirs. Sreenivasan (1999) reported the relation to stocking, recapture of fish. Devi (1997) and Piska (2000) studied the stocking relationship with fish yield rates of Shathamraj and Ibrahimbagh reservoirs. Das *et al.* (2002),

Mahapatra *et al.* (2002) and Chary (2003) studied the stocking and fish yield of various reservoirs of India. Some of the reservoirs also stocked with advanced fry or fingerlings (Mathew and Mohan, 1990).

Different authors reported different figures for number of fry for production of 1 kg fish in different reservoirs of India–5–35 fry in reservoirs of Haryana (Agarwal, 1990), 9–12 fry in reservoirs of Andhra Pradesh (Devi, 1997; Chary, 2003).

The results indicated that the fish production can be increased by stock manipulation. The similar reports were given by Devi (1997), Velayndhan and Siddarthan (1998), Piska *et al.* (2000), Piska (2000) and Chary (2003).

Acknowledgement

We are thankful to Prof. B. Raghavendra Rao and Prof. P. Umamaheshwer Reddy for their suggestion during the present work.

References

Agarwal, S.C. 1990. *Fishery Management.* Ashish Publishing House, New Delhi.

Bailay, N.T.J. 1959. *Statistical Methods in Biology.* The ELBS and English University Press Limited, Great Britain.

Bandyopadhyay, M.K., Singh, S.K. and Sarkar, B. 2003. Possibility of raising Indian major carp seed in cases for stocking larger water bodies. CA–1, *Nat. Sem. Fish and Shellfish Culture,* pp. 37.

Burli, A.S. and Sontakki, B.S. 2002. A comparative study of fisheries management in Malaprabha and Ghataprabha reservoirs of Karnataka, AR-16. *The Sixth Indian Fisheries Forum,* pp. 54.

Chary, K.D. 2003. Present status, management and economics of fisheries of a minor reservoir, Durgamcheruvu of Rangareddy District. *Ph.D. Thesis,* Osmania University.

Das, A.K., Krishna Rao, D.S., Sukumaran. P.K. and Karthikeyan, M. 2002. Strategies for management of Karnataka reservoirs. *Fishing Chimes,* 22(2): 17–21

Dehadrai. 2001. Reservoir fisheries in India. *Proc. Nat. Sem. Riverine and Reservoir Fisheries,* p. 97–104.

Devi. B. S.1997. Present status, potentialities, management and economics of fisheries to two minor reservoir of Hyderabad. *Ph.D. Thesis,* Osmania University.

Gowda, H.S.V. 2002. Status of fisheries development in Karnataka. *Fishing Chimes,* 22(1): 86–94.

Jhingran, A.G. and Sugunan, V.V. 1990. General guidelines and planning criteria for small reservoir fisheries management. *Proc. Nat. Workshop Reservoir Fish,* p. 1–8.

Mahapatra, B.K., Vinod, K. and Mandal, B.K. 2002. Stocking of Indian major carps in Umiam reservoir of Meghalaya: An approach for culture based fisheries. *Proc. Nat. Symp. Fisheries. Enhanc. Inland Waters,* p. 43–44.

Mahapatra, D.K. 2003. Present status of fisheries of Hirakud reservoirs, Orissa. *Fishing Chimes,* 22(10&11): 76–79.

Mathew, P.M. and Mohan, M.U. 1990. Reservoir fisheries in Kerala status and scope for development. *Proc. Net. Workshop Reservoir Fish,* p. 111–117.

Mitra, A., T. Nayak and S.K. Sarkar. 1989. Social acceptance and economic returns of Tilapia culture in W.B. Exotic aquatic species in India sp.pub. *J. Asian Fish. Soc., Ind. Br.,* 101–104.

Mohanty, S.K. 1984. An approach for the development of reservoir fisheries in Orissa. *Proc. Freshwater Fish. Rural Devel. Sec.*, 1: 40–46.

Narayanan, A. and M.P.S Kohli. 2004. Biotype imbalances due to unscientific introduction of exotic fishes: the Indian experience. *Fishing Chimes*, 24(3): 39–41.

Piska, R.S. 1999. *Fisheries and Aquaculture.* Lahari Publications, Hyderabad.

Piska, R.S. 2000. Impact of stocking densities of major can seed on fish production in a minor reservoir. *Fishing Chimes*, 20(9): 39–41.

Piska, R.S., Devi, B.S. and Chary, K.D. 2000. The present status of fisheries of Ibrahimbagh, a minor reservoir of Hyderabad. *Fishing Chimes*, 20(2): 41–43.

Piska, R.S. 2001. Fisheries of minor reservoir, Shathamraj, Hyderabad. *Ind. J. Fish*, 48(2): 119–122.

Piska, R.S. and Rao, A.M. 2002. Phenomenal fish production in the Tilapia-dominated tank. *Aquacult.*, 3(2): 171–176.

Piska, R.S. 2003. Carp seed productions scenario in Andhra Pradesh. *Fishing Chimes*, 23(1): 122–123.

Piska, R.S. 2005. Planktonic diversity in a Tilapia dominated minor reservoir. *Trends Life Sci.* (Press).

Rao, D.S.K. and K. Shakunthala. 1998. Phenomenal Fish Production in the dominated Nelligudda reservoir in Karnataka. *Fishing Chimes*, 48(4): 5–16.

Sakhare, V.B. 2005. Impact of Tilapia on indigenous fishes. *Fishing Chimes*, 24(11): 55–56.

Saxena, B.S. 1990. Potentialities production and problems of Indian reservoir fisheries. *Proc. Nat. Workshop Reservoir Fish*, p. 34– 38.

Sinha, M. 1998. Policy options for integrated development of reservoir fisheries from production to Marketing. *Fishing Chimes*, 18(1): 54–59.

Sreenivasan, A. 1999. The role of small reservoir in fish production. *Fishing Chimes*, 18(10): 29–31.

Tamot, S. 2002. Fisheries Potential in Madhya Pradesh. *AR–15 The Sixth Indian Fisheries Forum*, pp. 96.

Velayudhan, T.D. and Sidharthan, T. 1998. Influence of stocking on fish yield in Chit tore Reservoir and its prospect on rural economy. *Fishing Chimes*, 18(4): 17–20.

Venugopal, G., H.G. Hingorani, K., Venkateshvaran and J.P. George. 1998. Culture of carps in a seasonal tank. *Comp. Physiol., Eco.*, 13(1): 20–23.

Chapter 34

Study on Soil Respiration in the Rainy Season for Subtropical Pine Forest Stand, Manipur

Ujala Devi and E.J. Singh
Department of Botany (P.G.), Ecology Research Laboratory, D.M. College of Science, Imphal

ABSTRACT

The present study deals on the parameters like soil respiration, soil temperature and soil moisture in both disturbed and undisturbed pine forest stand in the rainy season. Soil respiration in the undisturbed site ranged from 73.6 mg $CO_2 m^{-2}h^{-1}$ (July) to 236.5 mg $CO_2 m^{-2}h^{-1}$ (August). The maximum soil temperature and soil moisture percentage was recorded to be 26.2°C and 28.7 in the month of August. However in the month of October the minimum soil temperature and soil moisture percentage was found to be 20.3°C and 24.6. In the disturbed pine forest site the soil respiration varied from 86.0 mg $CO_2 m^{-2}h^{-1}$ (May) to 172.9 mg $CO_2 m^{-2}h^{-1}$ (August). However soil temperature ranged from 22.0°C (October) to 26.3°C (August) and the soil moisture percentage also range from 24.9 (June) to 31.0 (August). The highest value of soil respiration in both the sites was correlated with the highest value of soil temperature and soil moisture content in the month of August.

Keywords: Soil respiration, Soil temperature, Soil moisture percentage, Rainy season and Pine forest.

Introduction

Soil respiration is defined as the total CO_2 production in intact soils resulting from the respiration of soil organisms, roots and mycorrhizae. It includes three biological processes namely, microbial respiration, root respiration and faunal respiration and non biological process *i.e.*, chemical oxidation which may be particularly pronounced at high temperatures (Bunt and Rovira, 1954). This activity is sustained by organic matter inputs to the soils from above ground and below ground roots. Soil respiration is a useful parameter for studying soil biological activity, carbon cycling and energy flow in an ecosystem and is also considered as an important index of the decomposition subsystem (Singh and Gupta 1977). Soil temperature and soil moisture explain most of the variation in soil respiration but additional factors such as forest type, soil fertility and soil texture can be also important for site variation in soil respiration. Bhupinderpal Singh (2003) reported on the tree root and soil heterotrophic respiration in the boreal Scotspine forest at Aheden in northern Sweden. Makoto Suwa *et al.* (2004) studied the impact of elevated atmospheric CO_2 on forest floor respiration in a temperate pine forest. Pekka vanhala (2002) studied seasonal variation in the soil respiration rate in coniferous forest soils. Raich (1992) reported the global carbondioxide flux in soil respiration of *Pinus regida* forest and its relationship of vegetation and climate. Werner Borken *et al.* (2002) studied variation of soil respiration in European beech, Norway spruce and Scots pine forest of Solling and Unterlub in Germany. Several workers have studied soil respiration in different ecosystems in the world.(Mall *et al.*, 1991; James and Christopher,1995; Sundaravalli *et al.*, 2001; Monika *et al.*, 2002; Smith, 2003). Christoph *et al.* (2002) studied on soil respiration in the Sweet Chestnut forest of southern Switzerland after firing. However in India, soil respiration were conducted in different habitates by many workers (Singh and Gupta, 1977; Gupta and Singh, 1981; Jasbir and Ramakrishnan, 1982; Malkanian and Singh 1987). In Manipur, no work has been reported in soil respiration in the pine forest except the work of Laishram *et al.* (2002) in mix-Oak forest ecosystems at Shiroy Hill.

The present study deals on the correlation of soil respiration with the monthly soil temperature and soil moisture percentage in the rainy season for both disturbed and undisturbed pine forest stand.

Description of the Study Site

The pine forest of Engel Ching (Engel Hill) is an undisturbed site located near Tendongyan village and it falls at 24°55/N latitude and 93°53/E longitude at an altitude of 895 m. from the mean sea level, at the distance of 16 km from Imphal city.

The another disturbed pine forest is extending on the Langol hill ranges near the Khamaran village which lies at 24°52/N latitude and 93°54/E longitude at an altitude of 960 m. from the mean sea level; at the distance of 12 kms from Imphal.

The climate of the both study sites is monsoonic. The annual rainfall at the study site is about 1537.56mm. The mean maximum temperature ranges between 28.0°C.and 30.0°C.and the minimum temperature ranges between 16.1°C.and 28.0°C.

Materials and Methods

Soil respiration was measured by an alkali absorption method using aluminium cylinder of 25 cm tall and 13 cm diameter with surface area of 132.6 cm^2 (Anderson, 1982). In each cylinder, there was a 2 cm hole to facilitate the drainage in order to avoid rain water in the enclosed area. The cylinders are inserted into the soil upto 15 cm depth with three replications in both disturbed and undisturbed forest sites. Rate of soil respiration was taken at regular intervals of one month for one year. During the experiment, all the vegetations within the cylinder were clipped. Plastic vials containing 50 ml of 0.25 N NaOH solution were kept inside the aluminium cylinder and the cylinders were also covered by aluminium tin with air tight. The joining of aluminium cylinder along with the hole (2 cm) were sealed for airtight and placed it for 24 hrs. in the field. The absorbed CO_2 by NaOH was measured by titration with 0.25N HCL solution using phenolphthalein as an indicator. The advantage of using NaOH is that it can absorbed 99.6 per cent of CO_2 during the process of soil respiration in the soil (Witkamp and Drift, 1961 and Boulen *et al.*, 1984). A blank experiment was also estimated on each site for the control. Soil moisture was determined by Soil moisture meter in depth wise (*i.e.*, 0–10 cm, 10–20 cm and 20–30 cm).

Results and Discussion

In the forest site I (undisturbed) during the rainy season the rate of soil, respiration was found highest in the hill base in the month of September (367.3 mg $CO_2m^{-2}h^{-1}$) followed by slope (240.0 mg $CO_2m^{-2}h^{-1}$) and top (238.3 mg $CO_2m^{-2}h^{-1}$) in the month of August and June respectively (Figure 34.1). It may be due to the approach of warm temperature and high moisture content. During this period, the microflora and microfauna present in the soil becomes much active in their respiratory metabolic activities thereby enhancing the liberation of CO_2 gas from the forest floor (Behera and Padi 1980). However, the minimum rate of CO_2 was found in the month of July in both base (127.7 mg $CO_2m^{-2}h^{-1}$) and top portion (31.0 mg $CO_2m^{-2}h^{-1}$) whereas in the slope the minimum was observed in the month of October (33.2 mg $CO_2m^{-2}h^{-1}$).

Similarly in the rainy season an attempt has been made to correlate mean monthly soil respiration rate with soil temperature and soil moisture content in the forest site I (Figures 34.2 and 34.3). The maximum monthly soil respiration (236.5 mg $CO_2m^{-2}h^{-1}$) is highly correlated with the soil temperature (26.2°C) and soil moisture percentage (28.7) in the month of August. It is obvious that the soil temperature and the soil moisture are the most important factors for influencing the soil respiration in the forest ecosystems (Wiant, 1967).

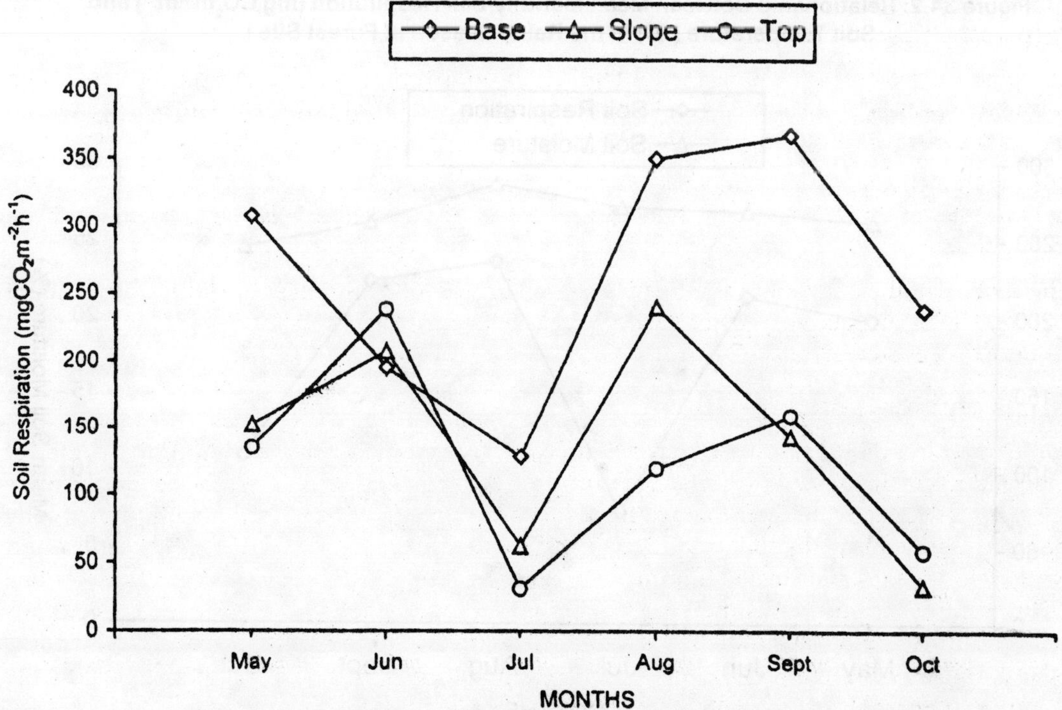

Figure 34.1: Monthly Variation of Soil Respiration (mg $CO_2m^{-2}m^{-1}$) Rate in the Rainy Season at Forest Site I

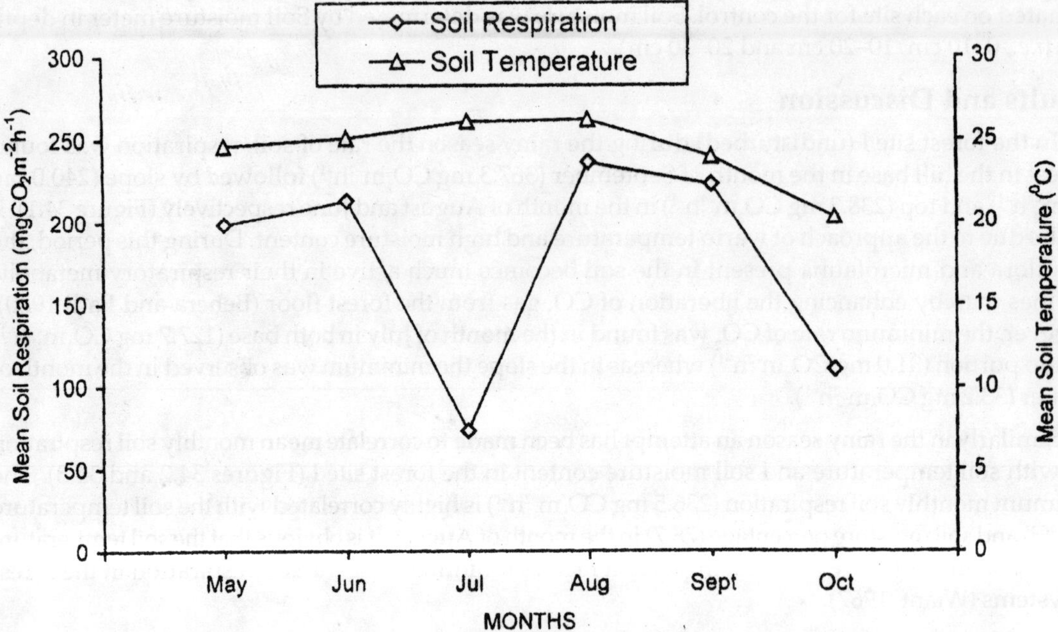

Figure 34.2: Relationship Between Mean Monthly Soil Respiration (mg $CO_2m^{-2}m^{-1}$) and Soil Temperature (°C) in the Rainy Season at Forest Site I

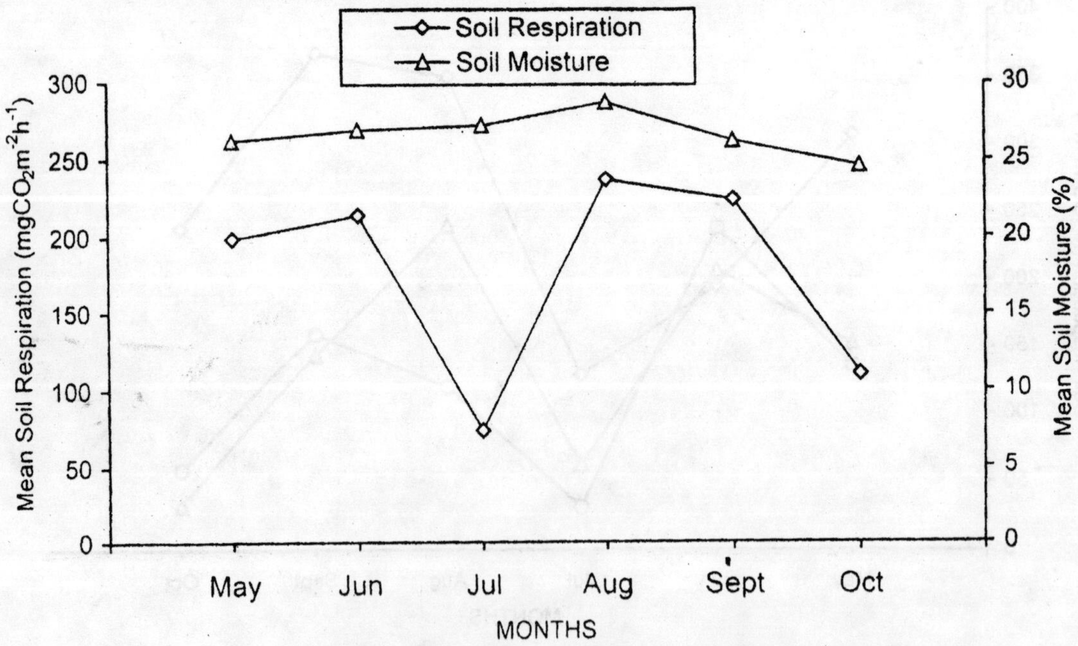

Figure 34.3: Relationship Between Mean Monthly Soil Respiration (mg $CO_2m^{-2}m^{-1}$) and Soil Moisture in the Rainy Season at Forest Site I

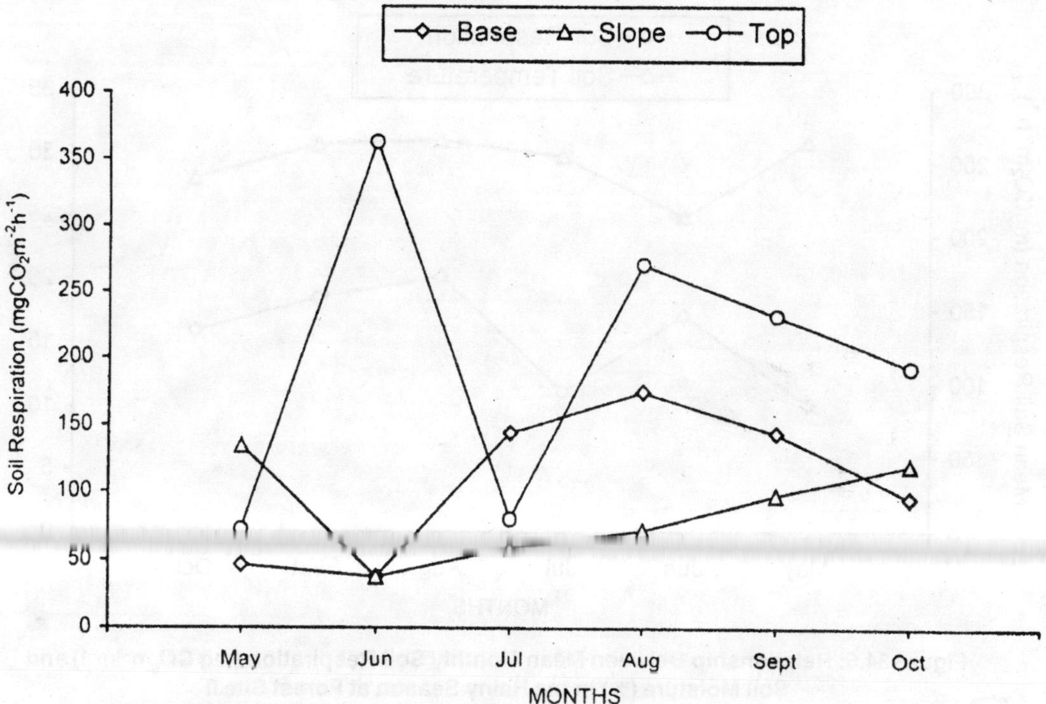

Figure 34.4: Monthly Variation of Soil Respiration (mg $CO_2m^{-2}m^{-1}$) Rate in the Rainy Season at Forest Site II

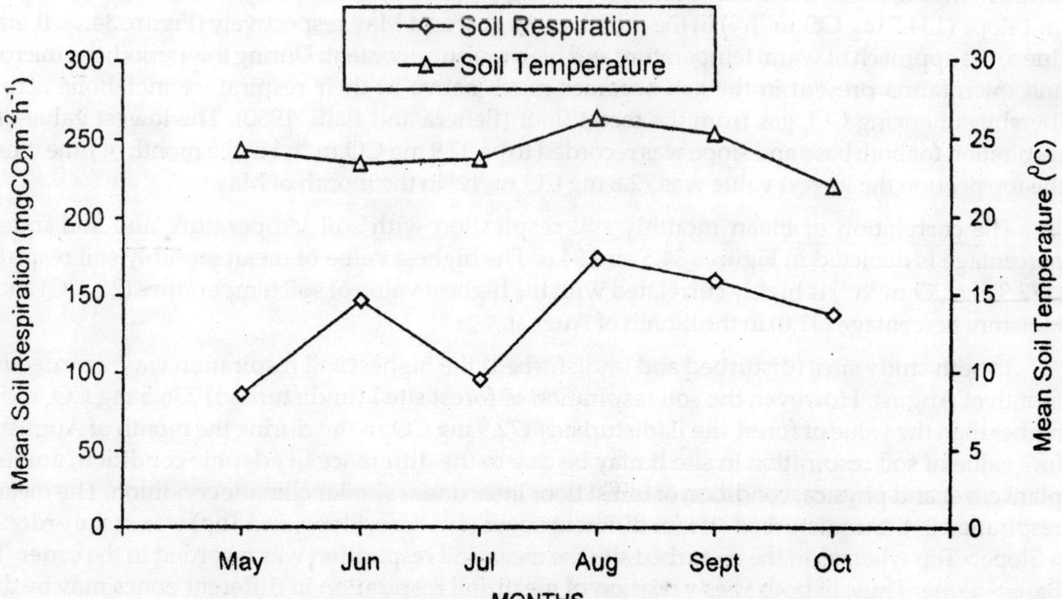

Figure 34.5: Relationship Between Mean Monthly Soil Respiration (mg $CO_2m^{-2}m^{-1}$) and Soil Temperature (°C) in the Rainy Season at Forest Site II

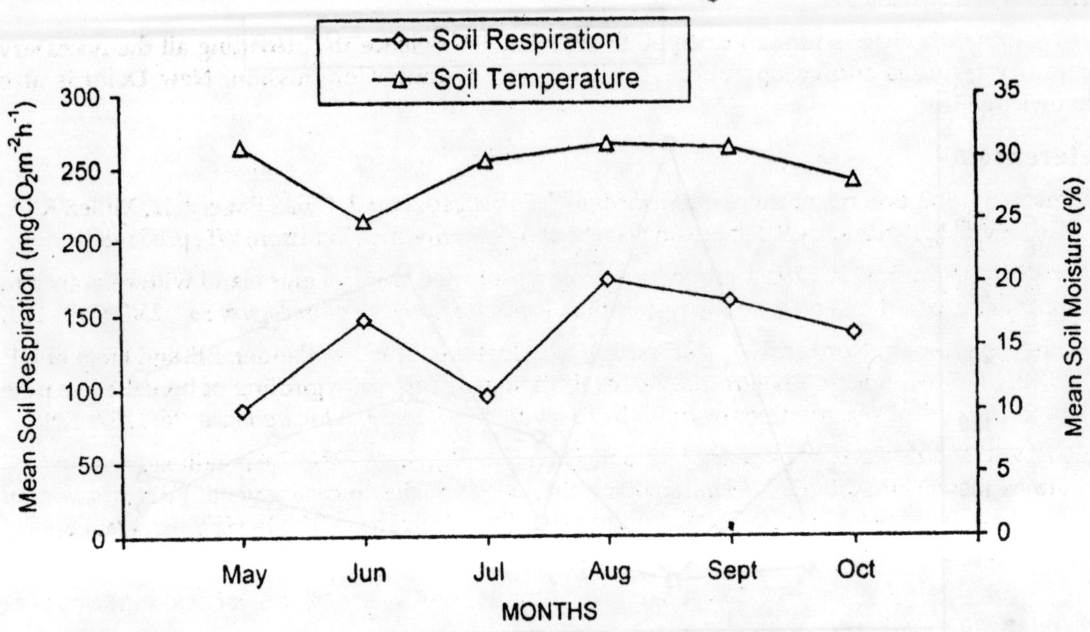

Figure 34.6: Relationship Between Mean Monthly Soil Respiration (mg $CO_2m^{-2}m^{-1}$) and Soil Moisture (%) in the Rainy Season at Forest Site II

In the forest site II (disturbed) the maximum rate of soil respiration during the rainy season was found in the hill top in the month of June (362.6 mg $CO_2m^{-2}h^{-1}$) followed by base (174.8 mg $CO_2m^{-2}h^{-1}$) and slope (134.7 mg $CO_2m^{-2}h^{-1}$) in the month of August and May respectively (Figure 34.4). It may be due to the approach of warm temperature and high moisture content. During this period, the microflora and microfauna present in the soil becomes much active in their respiratory metabolic activities thereby enhancing CO_2 gas from the forest floor (Behera and Padi, 1980). The lowest value of soil respiration for both base and slope was recorded to be 37.9 mg $CO_2m^{-2}h^{-1}$ in the month of June whereas the top portion the lowest value was 72.8 mg $CO_2m^{-2}h^{-1}$ in the month of May.

The correlation of mean monthly soil respiration with soil temperature and soil moisture percentage is depicted in Figures 34.5 and 34.6. The highest value of mean monthly soil respiration (172.9 mg $CO_2m^{-2}h^{-1}$) is highly correlated with the highest value of soil temperature (26.3°C) and soil moisture percentage (31.0) in the month of August.

In both study sites (disturbed and undisturbed) the highest soil respiration was recorded in the month of August. However, the soil respiration of forest site I (undisturbed) 236.5 mg CO_2 $m^{-2}h^{-1}$ is higher than the value of forest site II (disturbed) 172.9 mg $CO_2m^{-2}h^{-1}$ during the month of August. The low value of soil respiration in site II may be due to the difference in edaphic condition, amount of plant cover and physical condition of forest floor litter under similar climatic condition. The mean soil respiration in the undisturbed site for different zones (*i.e.*, Base, Slope, and Top) was in the order: Base > Slope > Top whereas in the disturbed site the mean soil respiration was recorded in the order: Top > Base > Slope. Thus, in both sites variation of mean soil respiration in different zones may be due to variation of moisture content and soil temperature.

Acknowledgement

We gratefuly thanks to the Principal, D.M College of Science for providing all the necessary laboratory facilities and co-operation. The University Grants Commission, New Delhi is also acknowledged.

References

Anderson, J. 1982. Soil respiration. In: *Methods of Soil Analysis, Part 2*, (Eds.) Page, A.L., Miller, R.H., Keeney, D.R. Agronomy 9. American Society of Agronomy, Inc., Madison, WI., p. 831–871.

Behera, N. and Padi, D.P. 1980. Carbon budget of a protected tropical grass land with reference to primary production and total soil respiration. *Review of Ecology and Biology of Soil*, 23(2): 167–181.

Bhupinderpal, Singh; Nordgren, A., Ottosson, L.M., Hogberg, M.N., Mellander, P.E. and Hogberg, P. 2003. Tree root and soil heterotrophic respiration as revealed by girdling of boreal Scots pine forest extending observations beyond the first year. *Plant Cell and Environment*, 26: 1287–1296.

Boulen, W.De., Vre, F.M., Verstraten, J.M. and Duysinge, J.J.H.M. 1984. Carbondioxide in the soil atmosphere: Simulation model parameter estimation from field measurements. In: *Hydrochemical Balance of Freshwater Systems*, (Ed.) E. Eriksson. IAHS Publication No. 150. IAHS Press. Wallingford, Oxford Shire, p. 25–30.

Bunt, J.S. and Rovira, A.D. 1954. Oxygen uptake and carbondioxide evolution of heat sterilised soil. *Nature*, 173: 1242.

Christoph, W., Daniel, S., Markus, W., Peter, M. and Marco, C. 2002. Soil respiration and soil microbial biomass after fire in a Sweet chestnut forest in southern Switzerland. *Catena*, 48: 201–215.

Gupta, S.R. and Singh, J.S. 1981. Soil respiration in a tropical grassland. *Soil Biology and Biochemistry*, 13: 261–268.

James, W.R. and Christopher, S.P. 1995. Global patterns of carbondioxide emission from soils. *Global Biogeochemical Cycles*, 9: 23–36.

Jasbir, S. and Ramakrishnan, P.S. 1982. Structure and function of a subtropical humid forest of Meghalaya (II). Litter dynamics and nutrient cycling. *Proc. Indian Acad. Sci. (Plant Sci.)*, 91(3): 255–268.

Laishram, I.D. Yadava, P.S. and Kakati, L.N. 2002. Soil respiration in a Mixed Oak forest Ecosystem of Shiroy Hills, Manipur in North Eastern India. *International Journal of Ecology and Environmental Sciences*, 28: 133–137.

Makoto, S., Gabriel, G.K., Ram, O., Jeft, A., Jeft, P., Anthony, M. and William, H.S. 2004. Impact of elevated atmospheric CO_2 on forest floor respiration in a temperate pine forest. *Global Biogeochemical Cycles*, 18: 1–2.

Malkanian, U. and Singh, J.S. 1987. Studies on soil respiration under disturbed and protected forests in Central Himalaya, India. In: *Proc. 9th Symp. Trop. Ecol. Ecological management of tropical ecosystems and International Conference on Rehabitation of Disturbed Ecosystem: A global issue*, pp. 327.

Mall, L.P., Singh, V.P. and Garge, A. 1991. Study of biomass, litterfall, litter decomposition and soil respiration in monogeneric Mangrove and mixed Mangrove forests of Andaman Islands. *Tropical Ecology*, 32: 144–152.

Monika, R., Shalini, S. and Pathak, H. 2002. Emission of carbondioxide from soil. *Current Science*, 82: 510–517.

Pekka Vanhala. 2002. Seasonal variation in the soil respiration rate in coniferous forest soils. *Soil Biology and Biochemistry*, 34: 1375–1379.

Raich, J.W. and Schlesinger, W.H. 1992. The global carbondioxide flux in soil respiration and its relationship to vegetation and climate. *Tellus*, 44:81–99.

Singh, J.S. and Gupta S.R. 1977. Plant decomposition and soil respiration in terrestrial ecosystem. *Botanical Review*, 43: 443–523.

Smith, V.R. 2003. Soil respiration and its determinants on a sub-Antarctic island. *Soil Biology and Biochemistry*, 35: 77–91.

Sundaravalli, V.M., Kailash, P. and Illakiam, D. 2001. Leaf litter decomposition and soil respiration in a semi-arid ecosystem near Madurai, South India. *International Journal of Ecology and Environmental Sciences*, 27: 221–224.

Werner, B., Yi-Jun Xu, Eric, A.D. and Friedrich, B. 2002. Site and temporal variation of soil respiration in European beach, Norway spruce and Scots pine forests. *Global Change Biology*, 8: 1205–1216.

Wiant, H.V. Jr. 1967. Influence of temperature on the rate of respiration. *Journal of Forestry* 65: 489–490.

Witkamp, M. and Drift, J.V. 1961. Breakdown of forest litter in relation to environmental factors. *Plant and Soil*, 15: 295–311.

Chapter 35

Pesticidal Stress Influenced Respiratory Alterations in the Freshwater Fish, *Mystus vittatus*

R. Sonaraj,* A.J.A. Ranjit Singh, ** A. Pushparaj,* and G. Ramathilagam*

**Reader, Research Department of Zoology, Kamaraj College, Thoothukudi*

***Reader, Department of Biology, S.P.K. College, Alwarkurichi*

ABSTRACT

The fish *Mystus vittatus* exposed to two sub-lethal concentrations of parrysulfan (0.1672 ppm and 0.3345 ppm) ana organochlorine and two sub-lethal concentrations of sicocil (0.335 ppm and 0.6717 ppm) and organophosphate pesticide showed an elevated consumption of oxygen on the very first day. The elevated change in parrysulfan medium was 15.85 per cent and 18.23 per cent in 0.1672 ppm and 0.3345 ppm respectively and 11.52 per cent and 15.19 per cent in the 0.3358 ppm and 0.6717 ppm in sicocil medium respectively. As the duration of exposure extends beyond 24 hrs, the rate of oxygen consumption showed a decreasing trend. The percentage decline in oxygen consumption in the parrysulfan exposed fish was –8.74, –24.91, and –0.28 after 10, 30 and 60 days of exposure in the concentration of 0.1672 ppm and –10.44, –26.06 and –36.35 after 10, 30 and 60 days of exposure in the 0.3345 ppm respectively. The sane trend was noticed in sococil exposed fish also and the percent reduction in oxygen consumption was –7.52, –22.34 and –26.14 after 10, 30 and 60 days of exposure in 0.3358 ppm and –9.52, –23.28 and –28.45 after 10, 30 and 60 days of exposure in 0.6717 ppm concentration. The decline in oxygen consumption was higher in the fish exposed to parrysulfan medium than to sicocil added medium.

Keywords: Parrysulfan, Sicocil, Oxygen consumption, Mystus vittatus.

Introduction

Among different environments aquatic environment is the serious sufferer of the damage caused by manmade and natural chemicals. The fauna of aquatic environment affected by pesticides is a subject of great concern. The pesticides are drained into the nearby pond and riverine system during monsoon and the land run off The indiscriminate and widespread use of pesticides and their subsequent passing into the aquatic system question the very survival and the behavioural, physiological and biochemical functioning of aquatic animals, particularly fishes of commercial import. The extent and degree of their harmful effect on fish can be gauged to a greater extent by experimental studies in the laboratory.

Studies on oxygen consumption form a useful tool in the assessment of toxicant stress on the aquatic organisms and give an index of energy expenditure mechanisms for environmental variations. Hence, in the present study, an attempt is made to study the oxygen consumption of a catfish, *Mystus vittatus,* exposed to two pesticides commercially called Parrysulfan (Organochlorine) and Sicocil (Organophosphate).

Material and Methods

The fish, *Mystus vittatus.* of the size 5–6 gms were collected from the laboratory acclimatized stock. The fish were fed on *ad libitum* with pellets prepared from groundnut oil cake and rice bran. The excreta of fishes were removed periodically (every one hour) without disturbing the fish.

Respiratory measurements were made by the method of Ganapati and Prasada Rao (1960) and dissolved oxygen was estimated by Winkler's method (Welsh and Smith, 1953). The fishes exposed to two different sublethal concentrations of Parrysulfan (0.1672 ppm and 0.3345 ppm) and Sicocil (0.3358 and 0.6717 ppm). The oxygen consumption of the exposed fish to different sublethal medium was analysed on 1st, 10th, 30th and 60th days of exposure.

The oxygen consumption was measured initially in the pesticide free medium and test concentrations separately. Prior to each experiment, the medium was renewed and then the oxygen consumption of the fish was estimated. The fish were not disturbed during the course of the experiment. A total of 5 fish was exposed separately in each of the different concentrations to obtain the average oxygen consumption. Triplicates were maintained All the experiments were carried out at temperature of 29±1°C. The results were presented as ml of oxygen consumed/g/hr. A test of significance on the difference in the mean values between the experimental fish and control fish, using correlation coefficient and regression equation was made.

Results

The effect of oxygen consumption on the fish, *Mystus vittatus,* exposed to various sublethal concentrations of Parrysulfan and Sicocil is presented in Table 35.1. The rate of oxygen consumption got elevated on the 1st day of release of fish into the pesticide medium. The elevated change in Parrysulfan medium was 15.85 per cent and 18.23 per cent in 0.1672 ppm and 0.3345 ppm respectively and 11.52 per cent and 15.19 per cent in the 0.3358 ppm and 0.6717 ppm of Sicocil medium respectively. As the duration of exposure extends beyond 24 hrs a negative trend was observed in oxygen consumption. The percentage decline in oxygen consumption in the Parrysulfan exposed fish was –8.74, –24.91 and –30.28 after 10, 30 and 60 days of exposure in the concentration of 0.1672 ppm and –10.44, –26.06 and –36.35 after 10, 30 and 60 days of exposure in the 0.3345 ppm respectively. In the Sicocil exposed medium the percentage reduction was –7.52, –22.34 and –26.14 after 10, 30 and 60 days of exposure in 0.3358 ppm mid -9.52, –23.28 and –28.45 after 10, 30 and 60 days of exposure in the 0.6717 ppm concentration respectively. The decline in oxygen consumption was higher in the fish exposed to Parrysulfan than to Sicocil.

Table 35.1: Changes in the Level of Oxygen Consumption (ml O_2 consumed/g/h) of the Fish *Mystus vittatus* Exposed to Two Sublet Concentrations of *Parrysulfan* and *Sicocil.* Values in tne parentheses indicate percentage of change over cont. Correlation coefficient 'r' and regression equation are also indicated.

Pesticide	*Concentration ppm*	*Oxygen Consumption/ml/g/r* *1 Day*	*10 Days*	*30 Days*	*60 Days*
Parrysulfan	Control	0.3555±0.0053	0.3499±0.0024	0.3415±0.0048	0.3352±0.0038
	0.1672	0.4119±0.0062	0.3193±0.0042	0.2564±0.0050	0.2323±0.0057
		(+15.85)	(–8.74)	(–24.91)	(–30.28)
		r=0.762	r=0.7957	r=0.7633	r=0.3716
		Y=0.0944+0.8930X	Y=0.1665–1.3884X	Y=0.0165–0.7992X	Y=0.0437+0.5627X
		t= 13.8024, p<0.05 S	t=12.6558, p<0.05 S	t=24.6581, p<0.005 S	t=30.2442, p<0.005 S
	0.3345	0.4203±0.0106	0.3134±0.0022	0.2524±0.0043	0.2120±0.0064
		(+18.25)	(–10.44)	(–26.06)	(–36.35)
		r=0.7643	r=0.8276	r=0.7112	r=0.4000
		Y=0.1223–1.5262X	Y=0.5816–0.7665X	Y=0.0311+0.6481X	Y=0.8533–1.9131X
		t=10.881, p<0.05 S	t=22.3054, p<0.05 S	t=27.5981, p<0.05 S	t=14.6767, p<0.05 S
Silocil	0.3358	0.3964±0.0068	0.3236±0.0033	0.2652±0.0075	0.2461±0.0063
		(+11.52)	(–7.52)	(–22.34)	(–26.14)
		r=0.3988	r=–0.0354	r=–0.4065	r=0.1656
		Y–0.2143+0.5124X	Y=0.3395–0.0455X	Y=0.0484+0.6350X	Y=0.1522+0.2801X
		t =9.4423, p<0.05 S	t=12.9981, p<0.05 S	t=17.2457, p<0.05 S	t=24.2022, p<0.02 S
	0.6717	0.4095±0.0067	0.3166±0.0041	0.262±0.0041	0.2384±0.0052
		(+15.19)	(–9.52)	(–23.28)	(–28.45)
		r=0.7080	r=0.6757	r=–0.5260	r=–0.5494
		Y=0.0951+0.8824X	Y=0.818–1.1384X	Y=0.4160–0.4507X	Y=0.4945–0.7640X
		T=12.6386, p<0.05S	t=14.1558, p<0.05S	t=25.3132, p<0.05S	t=30.1476, p <0.05S

The correlation coefficient 'r' estimated for oxygen consumption as a function, of duration of exposure confirmed that the decrease in oxygen consumption was statistically significant, in all cases except for the 1st day of exposure. The 't' value estimated confirmed that there was a significant difference ($p < 0.05$) in the intake of oxygen with that of the control value in all concentrations and duration of exposure. The regression equations derived indicated that there was an inverse relationship between the oxygen consumption and duration of exposure to the pesticide irrespective of the concentration (Table 35.1).

Discussion

Oxygen intake by an organism is an important physiological event. The quantum of intake of oxygen determined the physiological status and metabolic activity of the animal. The oxygen consumption not only indicates the metabolic rate, but it also provides an index of stress condition (Mount, 1962; Ferguson and Bingham, 1966; and Hutner *et al.*, 1968). The intimate contact of the gills with the polluted water may lead to alteration in normal respiratory mechanism, lower the diffusion mechanism through gill and thereby oxygen consumption is reduced, which creates a physiological imbalance to the organism (Finney and Berman, 1976). Accelerated rate of oxygen consumption of *Anabas scandens* after exposure to sumithion was reported by Natarajan (1980). Rao *et al.* (1980) showed that in case of *L. rohita* oxygen consumption increased progressively with increase in the concentration upto certain level and decreased with further increase of toxicant until death ensured. Roy and Datta (1988) reported that the *Cirrhinus mrigala* exposed to malathion showed a decreasing trend in oxygen intake in relation to concentration and duration of exposure. Rafia and Uma (1995) stated that the oxygen consumption of *Mystus gulio* was altered due to exposure to different concentrations of copper sulphate and zinc sulphate. According to Saxena and Chauhan (1996) the reduction in oxygen consumption in *Labeo rohita* exposed to effluent was because of coagulation of gill mucus, which has resulted in asphyxiation as well as inhibition of enzymes at mitochonmial level due to the toxic components in the effluent Devi (2000) reported that the oxygen consumption in *O. mossambicus* exposed to endosulfan showed an initial increase in the sublethal and medium lethal concentrations followed by a decline in the subsequent hours.

In the present study, the oxygen consumption of the fish, *Mystus vittatus,* exposed to two sublethal concentrations of Parrysulfan and Sicocil showed an increasing trend during the first day of exposure and a declining trend as a function of the duration of the exposure. The increase in oxygen intake during the first day of exposure in the present study might be due to behavioural manifestation like hyperactivity and escape behaviour, which was observed after exposure to Parrysulfan and Sicocil stress. The decrease in the rate of oxygen consumption after exposure to Parrysulfan and Sicocil for 10, 30 and 60 days might be due to the sluggishness of the fish as a result of the pesticide stress and also the secretion of excessive mucus which formed a thin film over the gill thereby preventing absorption of oxygen during the process of gaseous exchange.

References

Devi Swetharanyam. 2000. The effects of endosulfan on oxygen consumption of the fish *O. mossambicus*. *J. Ecotoxicol. Environ. Monit.*, 10(1): 21–24.

Ferguson, D.E. and C.R. Bingham. 1966. The effect of combination of pesticides on susceptible and resistant mosquito fish. *Bull. Environ. Contam. Toxicol.*, 1: 97.

Finney, L. and E.R. Berman. 1976. Studies in oxygen uptake. *Invest. Opthalmol.*, 15: 789.

Ganapati, P.N. and D.G.V. Prasada Rao. 1960. Studies on the respiration of barnacles. Oxygen uptake and metabolic rate in relation to body size in *Balanus amphitrite communis* (Darwin). *J. Anim. Morph. Physiol.m* 7: 27–31.

Hutner, J.Y., B.F. Dowden and H.J. Benett. 1968. The effect of endrin on the oxygen consumption of the blue gill sun fish *Lepomis macrocharus. Proc. Ind. Acad. Sci.*, 90: 80–86.

Mount, D.I. 1962. Chronic effect of endrin on blunt nose minnows and guppies. *U.S. Fish Wild. Ser. Res. Resp.*, 58: 1–38.

Natarajan, G.W. 1980. Effect of sumithion on oxygen consumption of *Anabas scandens. Geobios*, 7: 128–130.

Rafia Sultana and V. Umadevi. 1995. Oxyen consumption in a cat fish, *Mystus gulio* (Ham.) exposed to heavy metals. *J. Environ Bio.*, 16(3): 107–210.

Rao, D.M.R., A.P.Devi and A.S. Murthy. 1980. Relative toxicity of endosulfan, its isomers and formulated production to the fresh water fish *Labeo rohita. J. Toxicol. Environ. Hlth.*, 6: 323.

Roy, P.K. and J.S. Dattamunshi. 1988. Oxygen consumption and ventilation rate of a freshwater major carp *Cirrhinus mrigala* (Ham) in fresh and malathion treated waters. *J. Environ. Biol.*, 9(1): 5–13.

Sexana, K.K. and R.R.S. Chauhan. 1996. Oxygen consumption in fish *Labeo rohita* (Ham) caused by distillery emuent. *Poll. Res.*, 15(3): 277–280.

Welsh, J.H and R.I. Smith. 1953. *Laboratory Exercises in the Invertebrate Physiology*. Burgess Publishing Co., Minneapolis, pp. 126.

Chapter 36

Acute Toxicity of Curacron (Profenofos) and Karate (*Lambda cyhalothrin* to *Cyprinus carpio*, Linn.)

C. Radhkrishnan Nair*# and A. Palavesam**

**Reader, P.G. Department of Advanced Zoology and Biotechnology, S.T. Hindu College, Nagercoil – 629 002, Kanyakumari District, Tamil Nadu, India*

***Reader, Centre for Marine Science and Technology, Marine Campus, Rajakkamangalam, Kanyakumari District, Tamil Nadu, India*

ABSTRACT

Static bioassay experiments conducted under laboratory conditions on *Cyprinus carpio* using organophosphate pesticide, curacron (Profenofos) and pyrethroid, karate (*Lambda cyhalothrin*) revealed the toxicity. The percentage mortality and the LC_{50} values of curacron and karate were 0.194, 0.154, 0.148, 1.140, 0.134, 0.129 ppm and 0.136, 0.131, 0.139, 0.127, 0.125 and 0.122 ppm, respectively for 24, 48, 72, 96 and 120 hours were determined. The results indicated that the mortality for any fixed period increased with the increase of exposure period. LC_{50} values decreased with increased exposure period.

Keywords: Pesticides, Curacron and Karate toxicity, Mortality, Cyprinus carpio.

Introduction

Rapid industrialization and successful green revolution have introduced a large variety of chemicals into our environment. These chemical pollutants reach the aquatic ecosystem in considerable

Corresponding Author.

amounts via industrial effluents, agricultural runoff, contaminated groundwater, bottom sediments, urban runoff, manufacturing and formulating plant wastes and atmospheric fallout, rain etc., (Anderson *et al.*, 1978). As far as aquatic ecosystem is concerned, it has been recognized as an important global problem and thus, stimulated the eco-toxicologists to determine the lethal and sub-lethal impact on fish and other aquatic organisms (Omoregie *et al.*, 1998 and Nineawe, 2000).

Pesticides have, no doubt, been a boon to human civilization by sustaining agricultural revolution, but at the same time, they have bared their ravaging force on humanity by pushing to a point of almost no return. Accumulation of pesticide pollutants in aquatic environment leads to adverse effect on non-target organisms, particularly on fishes. One of the reasons for large-scale fish mortality, and low fish landing has been attributed to the increasing concentration of pesticides in natural water bodies (Pandey *et al.*, 2000 and Alam, 2000).

Unfortunately, most farmers apply these pesticides indiscriminately without much consideration of the hazards they are bound to cause. They may cause contamination of the food chain through the built-up of toxic residues, and the pollution of the ecosystem, eventually.

Majority of the studies concerning the effects of pesticides have been confined to the acute toxicity tests with the death of fish as an end point. Acute toxicity tests aimed at estimating the effects of toxicants on organisms in a short period of time have been conducted.

The objective of toxicity test is to determine the concentration of a test material or the level of an agent that produces a deleterious effect on a group of test animals during a short-term exposure under controlled conditions. Besides, toxicity tests with aquatic animals can be conducted by applying toxicant indirectly to the animal or application of toxicant directly through intravenous injection.

Studies on acute toxicity impact of pesticide, curacron on certain thermo conformers have been few and far between. Therefore, the present study was initiated to focusing on acute toxic effect of curacron on *Cyprinus carpio*.

Materials and Methods

For the present study the freshwater fish *Cyprinus carpio* and organophosphate pesticide curacron (Profenofos) and pyrethroid, karate (Lambda cyhalothrin) were used for the evaluation of toxicity, Healthy and unifonn size (7 + 1 g) fishes were procured from private firm. They were acclimated to laboratory conditions at 28±1°C in glass aquaria for 15 days.

The experiments were conducted in 10 liter tanks with 10 fishes each starved for 24 hrs prior to the experiment, following the recommendations of Florin-Muller (1917) for the maintenance of bioassay. The experimental medium was renewed daily till the end of the experiment. The mortality of fishes in different concentrations was noted at 12, 24, 48, 72, 96 and 120 hrs and the dead animals were removed immediately. Then 12, 24, 48, 72, 96 and 120 hrs LC_{50} values were computed using software by transforming mortalities (percentage values) in to probit scale (Finney, 1971). The experiment was repeated three times and mean values recorded separately for each test fishes.

Results and Discussion

The mortality of *C. carpio* increased with the increase in the concentration of pesticides, curacron and karate, but was not observed in control. The minimum mortality rate (20 per cent) was recorded at lowest concentration (0.110 ppm) of curacron when *C. carpio* reared to 120 hrs exposure whereas, at highest concentration (0.190 ppm) of curacron produced cent percent (100 per cent) mortality at 24 hrs exposure (Table 36.1).

Table 36.1: Percentage Mortality of *C. carpio* Exposed to Different Concentrations of Curacron for Different Duration

Concentrations of Curacron (ppm)	*Exposure Duration*					
	12	*24*	*48*	*72*	*96*	*120*
Control	0	0	0	0	0	0
0.100	0	0	0	0	0	0
0.110	0	0	0	0	10	20
0.120	0	0	0	8	20	30
0.130	0	20	30	40	50	50
0.140	10	30	40	50	60	70
0.150	10	40	50	60	70	80
0.160	20	50	60	80	90	100
0.170	30	60	80	100		
0.180	40	100				
0.190	40	100				
0.200	60					

The LC_{50} values of curacron to *C. carpio* ranged from 0.129 (120 hrs) to 0.194 ppm (12 hrs) and are provided in Table 36.2. Based on the regression equation fitted for log-dose and probit mortality, the "a" values (Y intercept) ranged between–0.013 (12 hrs) and –0.021 (96 hrs); while "b" values (regression coefffficient) ranged between 3.656 (12 hrs) and 4.587 (120 hrs). The lower and upper fiducial limits were obtained for each LC_{50} value and the lowest deviation was noticed in 72 hrs exposure (0.131 and 0.150). The deviation between empirical and expected probabilities was tested using chi-square and the values were compared with critical values at appropriate degrees of freedom at the probability level of 0.05 and in the majority of cases it was statistically not significant (Figure 36.1).

Table 36.2: Toxicity Parameters of Curacron to *C. carpio* (log-dose/probit regression analysis)

Exposure Period (hours)	*LC_{50} ppm*	*Regression Equation*	*Regression Coefficient (b)*	*95% Fiducial Limits*		*Chi-square Test*	
				Lower (LFL)	*Upper (UFL)*	*X^2*	*Critical Value*
12	0.194	Y = –0.013+3.656 X	3.656	0.174	0.215	0.45	7.81
24	0.154	Y = –0.021+4.232 X	4.232	0.145	0.162	6.81	9.49
48	0.148	Y = –0.011+4.309 X	4.309	0.136	0.159	0.31	11.07
72	0.140	Y = –0.015+4.559X	4.559	0.131	0.150	2.08	11.07
96	0.134	Y = –0.021+4.459 X	4.459	0.127	0.141	0.63	14.07
120	0.129	Y = –0.021+4.587X	4.587	0.121	0.136	1.48	12.59

Curacron exerted lethal effect on *C. carpio* with the increase in exposure period, the mortality rates also increased; while the LC_{50} values decreased. The decrease in LC_{50} value suggested decreasing resistance of the fish with increasing exposure duration, as signified by the co-efficient of the fitted regression equation. Singh and Kumar (2000) have also reported exactly similar findings in toxicity studies involving organophosphorous pesticides with *C. carpio.*

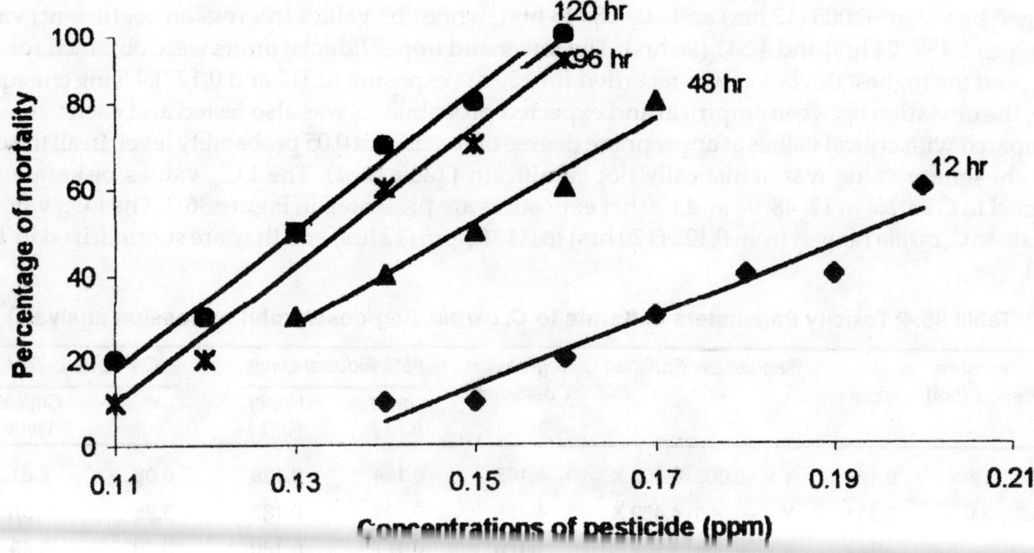

Figure 36.1: LC_{50} Values of Curacron to *C. carpio*

In *C. carpio* exposed to different concentrations of karate, the percentage of mortality was found to be increasing with the increase in the concentration of pesticide, against the nil mortality for those fishes reared in control medium. At the lowest concentration (0.090 ppm) karate produced 10 per cent mortality at 120 hrs exposure, whereas, 100 per cent mortality was recorded at the highest concentration (0.136 ppm) during 24 hrs exposure (Table 36.3).

Table 36.3: Percentage Mortality of *C. carpio* Exposed to Different Concentrations of Karate for Various Exposure Period

Concentrations of Curacron (ppm)	*Exposure Period (hrs)*					
	12	*24*	*48*	*72*	*96*	*120*
Control	0	0	0	0	0	0
0.080	0	0	0	0	0	0
0.090	0	0	0	0	0	10
0.100	0	0	0	0	10	10
0.120	0	0	0	10	10	20
0.122	0	0	0	10	20	30
0.124	0	0	10	20	30	40
0.126	0	10	30	40	50	60
0.128	0	30	40	50	60	80
0.130	10	40	50	70	80	100
0.132	20	50	70	80	100	
0.134	30	70	90	100		
0.136	50	100				
0.138	60	100				

Based on the regression equation fitted for log dose and probit mortality, "a" values (Y intercept) ranged between –0.005 (12 hrs) and –0.008 (48 hrs), while "b" values (regression coefficient) varied between 4.459 (24 hrs) and 4.541 (96 hrs). The lower and upper fiducial limits were obtained for each LC_{50} and the highest deviation was recorded for 120 hrs exposure (0.116 and 0.127). Using chi-square test, the deviation between empirical and expected probabilities was also tested and each value was compared with critical values at appropriate degree of freedom at 0.05 probability level. In all the cases the chi-square value was statistically not significant (Table 36.4). The LC_{50} values of karate with respect to *C. carpio* in 12, 48, 96 and 120 hrs exposures are presented in Figure 36.2. The LC_{50} values of karate to *C. carpio* ranged from 0.122 (120 hrs) to 0.136 ppm (12 hrs) and they are summarized in Table 36.4.

Table 36.4: Toxicity Parameters of Karate to *C. carpio.* (log-dose/probit regression analysis)

Exposure Period (hrs)	*LC_{50} ppm*	*Regression Equation*	*Regression Coefficient*	*95% Fiducial Limits*		*Chi-square Test*	
				Lower (LFL)	*Upper (UFL)*	*x^2*	*Critical Value*
12	0.136	Y = –0.005+4.105 X	4.105	0.134	0.138	0.08	7.81
24	0.131	Y = –0.008+4.459 X	4.459	0.129	0.132	3.25	11.07
48	0.129	Y = –0.008+4.473 X	4.473	0.127	0.130	0.70	9.49
72	0.127	Y = –0.009+4.465 X	4.465	0.125	0.128	1.89	12.59
96	0.125	Y = –0.010+4.541 X	4.541	0.122	0.128	2.098	12.59
120	0.122	Y = –0.016+4.583 X	4.583	0.116	0.127	2.116	12.59

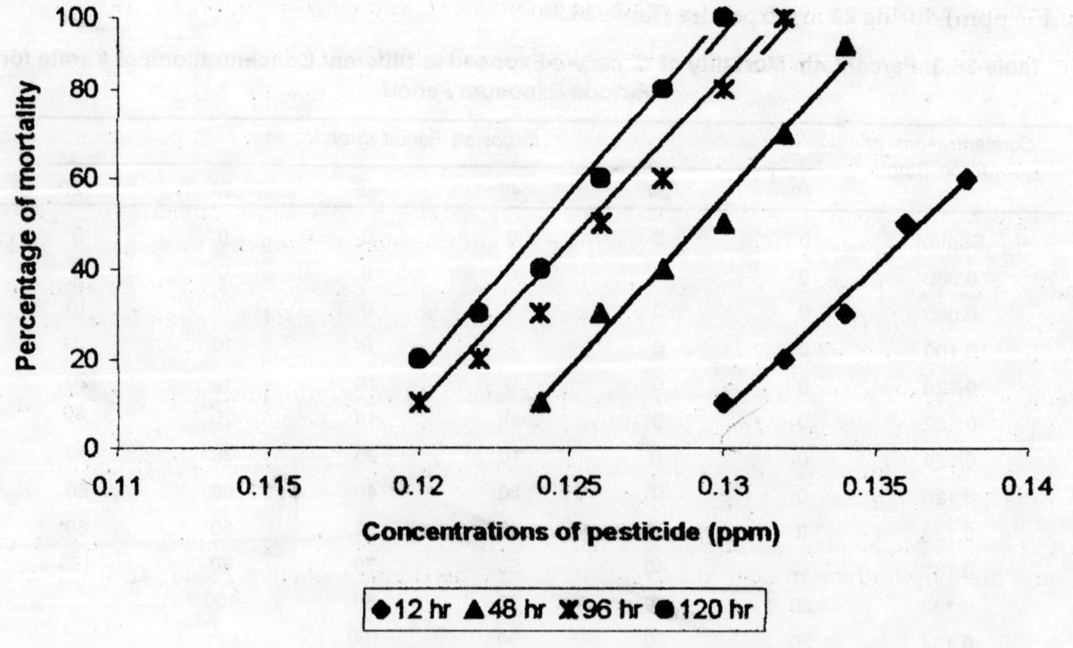

Figure 36.2: LC_{50} Values of Karate to *C. carpio*

Tested for acute toxicity in the present study, *C. carpio* was highly susceptible to karate, compared to curacron pesticide as indicated by 96 hrs LC_{50} values.

Earlier studies on *C. carpio* revealed the high susceptibility of the species to different pesticides. The 96 hrs LC_{50} value for methyl parathion, was 0.8 ppm (Thiruvalluvan *et al.*, 1997), for profenofos (curacron) to rainbow trout was 0.025 ppm and blugill sunfish 0.3 ppm (EPA, 1999), for sicocil to *Mystus vittatus* was 2.687 ppm (Sornaraj, 2000) and for malathion to *Clarius batrachus* was 0.04 ppm (Aruna *et al.*, 2000). The 96 hrs LC_{50} values of karate to blue gill sunfish was 0.21 ppm, for rainbow traout it was 0.24 ppm and 0.807 ppm for sheephead minnow.

It has been pointed out that the toxicity of a pesticide can be modified by temperature, pH, alkalinity and hardness also influence the toxicity (Saxena, *et al.*, 1996).

To sum up, these findings recommend that farmers should avoid indiscriminate use of both pesticides curacron and karate. Since karate is more toxic, it is advisable not to use karate on wetlands. But, however, farmers should take great care in making the agricultural land drainage harmless enough letting the run off into the nearby aquatic environment to ensure safety for all organisms.

References

Alam, M.N. 2000. Toxicity of metacid 50 to a paddy field fish, *Channa punctatus* (*Bloch*). *Indian J. Environ. and Ecoplan.*, 3(3): 701–705.

Anderson, P.D. and S.D. Apollonia. 1978. *Principle of Ecotoxicology*. John Wiley and Sons, New York, USA.

Aruna, K., S. Singh and K. Srivastava. 2000. Malathion induced biochemical changes in the kidney of freshwater fish, *Clarias batrachus*. *J. Ecotoxicol. Environ. Monit.*, 10(1): 11–14.

Environmental Protein Agency. 1988. Fact sheet Number 171, Karate, Washington, DC, pp. 321.

Environmental Protection Agency. 1999. Executive Summary for Profenofos.

Finney, D.J. 1971. *Probit Analysis*. Cambridge University Press, Cambridge, pp. 333.

Florin, J. and B. Muller. 1970. Toxicity testes with fish. Swis federal office or water protection. FAO–EIFAc Doc, *Unit. Res.*, 10618, (1).

Omoregie, E., P.C. Ofojekwu and E.I. Amali. 1998. Effects of sub–lethal concentrations of formalin on weight gain in the Nile Tilapia, *Oreochromis niloticus* (Trewavas). *Asian Fisheries Sci.*, 10: 323–327.

Pandey, A.C., A.K. Pandey and P. Das. 2000. Fish and fisheries in relation to aquatic pollution. In: *Environmental Issues and Management*, (Eds.) S.R. Verma, A.K. Gupta and P. Das. Nature Conservators, Muzaffarnagar, p. 87– 112.

Saxena, V.S. Trivedi, and D.N. Saksena. 1996. Acute toxicity of two organophosphorus pesticides to a freshwater murrel, *Channa orientalis* (Schneider). *J. Environ. Poll.*, 4: 79–83.

Singh, M. and S. Kumar. 2000. Effect of sub lethal concentration of dimethate and malathion on *Catla catla* (Ham.). *Uttar Pradesh J. Zool.*, 20(2): 131–135.

Thiruvalluvan, M., N. Nagendran and A. Charles Manoharan. 1997. Bioaccumulation of cadmium and methylparathion in *Cyprinus carpio* Var. communis (Linn). *J. Environ. Pollut.*, 4(3): 221–224.

Chapter 37

Impact of Textile Mill Effluent on Seed Germination and Seedling Growth of *Lablab purpureus* L.

M. Rajasekara Pandian, G. Sharmila Banu*, G. Kumar** and K. H. Smila**

**Centre for Biotechnology, Muthayammal College of Arts and Sciences, Kakkaveri, Rasipuram, Namakkal, Tamil Nadu, India*

***Department of Biochemistry, Selvamm Arts and Science College, Namakkal, Tamil Nadu, India*

ABSTRACT

The impact of various concentrations *viz.*, 0, 25, 50, 75 and 100 per cent (v/v) of the effluent of textile mill on seed germination, seedling growth and pigment content of *Lablab purpureus* were investigated. The effluent caused toxic effects to the seedlings at higher concentrations. But good growth of the seedling (75.33 per cent) was observed at 25 per cent concentration. Undiluted effluents had an inhibitory effect (28.93 per cent), where as 25 per cent effluent had a growth promoting effect (14.72 per cent), which was significantly better than control. Beyond 25 per cent effluent root length (3.27, 14.75 and 27.86 per cent) and shoot length (–11.76, 8.8 and 26.47 per cent) decreased. Thus the effluent of textile mill can be used safely for irrigation purpose at 25 per cent dilution for beneficial cultivation.

Keywords: Textile mill effluents, Seed Germination, Lablab purpureus.

Introduction

Seed germination and growth are vital for continuation of plant life. Seeds and seedlings are extremely vulnerable to environmental stress due to the presence of polluting agents in the environment especially during seed hydration period which is very important for initiating and triggering the intricate sequence of metabolism essential for germination and growth of seedlings. Industries produce various environmental problem in and around factory areas and hence it is necessary to understand the impact of industrial effluent on the environment.

Untreated disposal of industrial waste has become a major threat to the surrounding land areas bearing vegetation as it may render the soil infertile. The industrial effluents generally contain various contaminants in the form of organic and inorganic chemicals. These chemicals have the detrimental effects on plant development including seedling germination and growth. Previous reports have shown that effluents discharged from industries are capable of inhibiting seed germination and seedling growth. (Varadarajan *et al.*, 1970; Chaney, 1990; Shukula and Pandey, 1991). The present investigation was carried out to study the effect of textile mill effluent on seed germination and seedling growth in Groundnut (*Lablab purpureus* L).

Materials and Methods

The effluent from Madura textile mills, Vickramasingapuram was obtained from the main discharge pipelines and analyzed for its various physical and chemical properties following the method of APHA (1989).

Germination tests were carried out by keeping 25 seeds in sterilized Petridishes containing 50g of sterilized soil. In order to keep the soil constantly moist, each Petriplate was moistened daily with equal volume of different concentration (25, 50, 75, and 100 per cent) of effluent with 5 replicates. Control condition was maintained by using equal volume of distilled water (effluent free). Reading was taken at fixed time intervals and emergence of radical was taken as the criterion for germination. On the days of germination the growth parameters such as germination percentage, root length and shoot length were measured. Chlorophyll-α, and chlorophyll-β contents were measured by the method of Amon (1949) and the carotenoid content by that of Ikan (1969).

Results and Discussion

Groundnut treated with various concentration of effluent showed a different variation. The value increased at lower concentration (25 per cent) and then decreased continuously at higher concentration (100 per cent) (Table 37.2). The delay in seed germination in concentrated effluent might be due to inhibition of enzyme activity (Agarwal *et al.*, 1981). Neelam and Sahai (1988) have reported that the seed germination which prevent by enriching the salinity and conductivity of solute being absorbed by seed before germination.

A better seedling growth was recorded at 25 per cent concentration in comparison to control (Table 37.2). Vijayakumari (2003) also reported the same effect in Soya bean.

The higher concentration of the effluent inhibited the seedling growth and the germination percentage. Similarly, Vijayakumari and Kumudha (1990) and Madhappan (1993) have reported the same results in various crops.

Rao and Nandakumar (1983) confirmed the present findings, which reported that the relative reduction in root and shoot length of the seedlings treated with higher concentrations of effluents might be due to the physiological stresses because of high salinity. The retardation in plant growth was due to excess quantities of micronutrients, heavy metals and other toxic chemicals (Dolar *et al.*, 1972).

Table 37.1: Physico-chemical Characters of the Textile Mill Effluents

Parameters	*Values*	**IS–Tolerance Limit IS–2490–1982 Part–I*
Colour	Dirty white	Colourless
Temperature (°C)	45	< 40
pH	9.2	5.6–9
BOD	228	30
COD	731	250
Total suspended solids (mg/l)	870	200
Total dissolved solids (mg/l)	780	2100
Hardness (mg/l)	950	–
Chloride (mg/l)	600	600
Sulphate (mg/l)	15	1000
Nitrate (mg/l)	22	–

*Tolerance limits for industrial effluents discharged into inland surface waters prescribed by Bureau of Indian standards.

Table 37.2: Effect of Textile Mill Effluent on Seed Germination, Root and Shoot Length of *Lablab purpureus* L. in 7 Days

Concentration of the Effluents (%)	*Germination (%)*	*Root Length (cm)*	*Shoot Length (cm)*
0	65.66	5.8±0.3	9.5±0.5
25	75.33	6.1±0.3	6.8±0.07
50	60.33	5.9±0.5	7.6±0.3
75	55.33	5.2±0.6	6.2±0.7
100	46.66	4.4±0.1	5.0±0.8

Chlorophyll-α, β and Carotenoid gradually increased to 25 per cent effluent concentration and declined at higher concentration (Table 37.3). These findings are in agreement with those of Sahai *et al.* (1986) on *Oryza sativa* Linn and Sahai and Srivastavza (1986) on *Cajanas cajan*. The carotenoid content decreased gradually with increase in effluent concentration. The carotenoid synthesis decreased due to decrease in nitrogen and other inorganic elements (Cottenie 1973). The decline of these elements may be due to the accumulation of heavy metals in the roots, which may prevent the uptake of these nutrients by the plants.

Table 37.3: Effect of Textile Mill Effluent on Pigment Content of *Lablab purpureus* L.

Effluent Concentration	*Chlorophyll-a (mg/g)*	*Chlorophyll-β (mg/g)*	*Carotenoid (mg/g)*
0	2.18	2.95	1.35
25	2.34	3.25	1.40
50	1.82	2.73	1.32
75	1.32	2.46	1.24
100	1.02	2.12	1.16

The present investigation leads to conclude that the textile mill effluent at lower concentration (25 per cent) enhances growth. But the effluent was observed to be toxic when applied in undiluted form or at higher concentrations. So, the proper dilution may be carried out and then used for irrigation of crops.

References

Agarwal, P.K., Singh, V.P. and Dineshkumar. 1981. Bio-chemical changes in seedlings on *Brassica nigra* L. and *Linum usitatissimum* L. under the stress of Mansurpur distillery effluents. *J. Indian Bot. Soc.*, (Suppl.). 60: 95.

APHA. 1989. *Standard Methods for the Examination of Water and Wastewater*, 17th ed. APHA, AWWA, WPCF.

Arnon, D.I. 1949. Copper enzyme in isolated chloroplast. Polyphenoxidase in *Beta vulgaris*. *Pl. Physiol.*, 24: 1–15.

Chaney, R.L. 1990. Public health and Sludge utilization, Part 2. *Biocycle*, 31: 68–73.

Cottenie, A. 1973. Chemical fertilizers in relation to agricultural production Modedalingen Fakulteit land boun wetanschappen. *Genetics*, 38: 1722–1731.

Dollar, S.C., Boyle, J.R. and Keeney, A.D. 1972. Paper mill sludge disposal on soils: Effects of the yield and mineral nutrition of Oats (*Avena sativa* L). *J. Environ. Qual.*, 1: 405–409.

Ikan, R. 1969. *Natural Products: A Laboratory Guides*. Academic Press, New York, USA.

Madhappan, K. 1993. Impact of tannery effluent on seed germination, morphological characters and pigment concentration of *Phaseolus mungo* L. and *Phaseolus aureus* L. *Poll. Res.*, 12: 159–163.

Neelam, S. and Sahai, R. 1988. Effect of Fertilizer factory effluent on seed germination, seedling growth pigment content and biomass of *Sesamum indicum* Linn. *J. Environ. Biol.*, 9: 45–50.

Rao, M.G. and Nandakumar, N.V. 1983. Impact of effluent on seed germinability and chlorophyll content in *Cicer arietinum* L. *Poll. Res. J.*, 2: 33.

Sahai, R., Jabeen, S. and Saxena, P.K. 1983. Effect of distillery effluent on seed germination, seedling growth and pigment content of L. *Oryza sativa*. *Ind. J. Eco1.*, 10: 7–10.

Sahai, R. and Srivastava, N. 1986. Effect of distillery waste on the seed germination, seedling growth and pigment content of *Cajana cajan* L. *J. Ind. Bot. Soc.*, 65: 208–221.

Shukla, N. and Pandey, G.S. 1991. Oxalic acid manufacturing plant wastewaters: Effect on germination and seedling height in selected cereals. *J. Environ. Biol.*, 12: 149–151.

Varadarajan, S., Govinda Iyer, T.A. Gopalaswamy, A. and Premanathan, S. 1970. Influence of tannery effluents on soils and crops and proper disposal of effluents. *Madras Agri. J.*, 57: 353–360.

Vijayakumari, B. 2003. Impact of textile dyeing effluent on growth of Soya bean (*Glycine max* L.). *J. Ecotoxicol. Environ. Monit.*, 13(1): 59–64.

Vijayakumari, K. and Kumudha, P. 1990. Effect of distillery effluent on seed crop plant. *Geobios*, 17: 206–211.

Chapter 38

Problems Related to Processing of Manganese Ore Fines

V. Rajendra Prasad*, Y. Prasanna Kumar, P. King and V.S.R.K. Prasad

Environmental Pollution Control Engineering Laboratory, Department of Chemical Engineering, College of Engineering, Andhra University, Visakhapatnam – 530 003, A.P., India

Introduction

India has extensive deposits of manganese ore, a vital raw material in steel manufacture and has got wide use in non-metallurgical industries also. The specifications of the ore are, however, stringent and the value of the ore depends to a great extent on the chemical composition. Impurities such as Phosphorus, Iron, and Silica in manufacture of Ferro manganese and particularly Iron in manganese dioxide ores used for dry batteries should not exceed certain limits.

Ores that can be directly employed for the manufacture of ferro manganese are available only to a limited extent in India. Presently, manganese ores below–6 mm size generally do not find market for use in Ferro manganese and pig iron production. However, there is no restriction in case of Dioxide ores used in dry batteries but due to their lower grades and high impurities even here processing has become necessary to yield a product which meets the requirements of battery grade. Owing to these stringent physical and chemical specifications, the manganese ore mining in India has become highly selective in nature. Due to selective mining, rejected ores, which do not meet grade and size specifications have grown to enormous proportions at mine sites. More emphasis has to be laid on beneficiation of these ores for their better utilization in metallurgical and non-metallurgical industries. The practice of beneficiation is necessary not only for the conservation and proper utilization of our mineral wealth, but also for all round development of our mineral wealth.

* Corresponding Author, E-mail: rpvasanta@rediffmail.com.

Occurrence and Uses of Manganese Ores

Manganese deposits in India occur mostly in Maharashtra, Madhya Pradesh, Bihar, Orissa, Karnataka, Goa and Gujarat. The main deposits are in the Nagpur–Bhandara Balaghat belt of Maharashtra and Madhya Pradesh. Other occurrences are in Bellary, Chitradurga and North Kanara district of Karnataka, Sundergarh, Bolangir and Keonjhar of Orissa, Srikakulam district of Andhra Pradesh, Bansawara of Rajasthan and Panch Mahal and Baroda of Gujarat.

Manganese ore is vital raw material in the production of steel. It is one of the most important deoxidisers suppressing pinholes in the ingot, or casting and oxysulphide formation; it helps to control the undesirable effects of residual sulphur. It provides essential alloying elements that improve the mechanical properties of steel. Manganese ore is used in metallurgical industries for making Ferro Manganese (Mn 74–78 per cent, Fe 6–8 per cent), Silica manganese (Mn 70–74 per cent, Si 16–20 per cent), Spiedgelesion (Mn 15–25 per cent, C 6–5 per cent), Chrome Manganese (Cr 50–55 per cent, Mn 20–25 per cent) etc.

Manganese Ore Reserves of MOIL's Mines

Manganese Ore (India) Limited (MOIL) is presently operating 12 mines in the States of Maharashtra and Madhya Pradesh. It has total recoverable reserves of all grades to the tune of 33.09 million tonnes. The grade wise breakup is given in Table 38.1. Out of total 33.09 million tonnes of recoverable reserves, low phosphorous manganese ore reserves are about 7.16 million tonnes. 73 per cent of the total recoverable reserves belong to the category of +46 per cent Mn grade, 8.5 per cent to 46–44 per cent. Mn grade, 8.3 per cent to 44–35 per cent Mn grade and 10.2 per cent to 35 per cent Mn grade. 10 per cent of the total reserves mainly consist of waste rocks and generated fines during the mining operations. Generation of fines is less in most of the mines. However, in Balaghat and Ukwa mines where low phosphorous ores are produced, the generation of fines is much higher.

Table 38.1: Grade-wise Recoverable Reserves in the Mines of MOIL

Sl.No.	Mines	Total Reserves in Million Tonnes	Mn% +46	Mn% 46–44	Mn% 44–35	Mn% –35
	Madhya Pradesh					
1.	Balaghat	7.5920	7.0940	0.4213	0.0767	–
2.	Tirodi	0.9180	0.5289	0.0898	–	0.2993
3.	Ukwa	7.6200	5.4102	1.8654	–	0.3444
	Maharashtra					
1.	Beldongri	0.2410	0.2208	–	0.0202	–
2.	Chikla	3.4190	1.4203	0.2164	–	1.7823
3.	Dongri Buzurg	10.1750	7.8612	0.1618	2.1520	–
4.	Gumgaon	0.6560	0.3060	0.0450	0.3050	–
5.	Kandri	1.4300	0.8095	–	0.1975	0.4230
6.	Munsar	1.0400	0.4888	–	–	0.5512
7.	Ramdongri	0.0020	0.0005	–	0.0015	–
Total in Million Tonnes		**33.0930** (100%)	**24.1402** (73%)	**2.7997** (8.5%)	**2.7529** (8.3%)	**3.4002** (10.2%)

Physical, Chemical and Mineralogical Composition of Fines

Generation of fines considerably depends on the nature of the ore, mining operations and handling systems. Particularly in Ukwa and Balaghat mines, the nature of ore is friable and generation of fines is considerable and is to the tune of 14 per cent and 12 per cent respectively. The percentage of manganese varies from 30–35 per cent in the fines. These fines are having size below–3 mm and these, even after beneficiation cannot be used directly in metallurgical operations. Mineralogical composition of the fines at Balaghat and Ukwa mines is similar in nature and consists of Hollandite, Braunite, Psilomelane, Hematite, Apatite and Quartz.

In other mines, the generation of fines is comparatively less and to the tune of 5 per cent on the average. The percentage of manganese in these fines ranges from 20–25 per cent and size will be –6 mm in some cases. Mineralogical constituents of fines available at other mines are listed Table 38.2.

Table 38.2

Sl.No.	*Mine*	*Minerological Composition*
1.	Adilabad (A.P.)	Pyrolusite, Psilimelane Braunite manganite, Quartz, Feldspar
2.	Balaghat	Hollandite, Braunite, Psilomelane, Hematite, Quartz and Apatite
3.	Chikla	Braunite, Pyrolusite, Psilomelane, Hematite, Goethite, Quartz, Feldspar, Apatite and Garnet
4.	Dongri Buzurg	Pyrolusite, Psilomelane, Jacobsite, Hematite, Goethite, Feldspar and Quartz
5.	Munsar and Kandri	Braunite, Jacobsite, Pyrolusite Psilomelane, Hematite Quartz and Apatite
6.	Tirodi	Braunite, Hollandite, Psilomelane, Quartz, Hematite and Apatite
7	Ukwa	Braunite, Hollandite, Pyrolusite Manganese Phyllite, Hematite, Quartz and Apatite

Problems Related to Manganese Ore Fines

Presently, ores below –6 mm size generally do not find market in Ferro manganese production. However, a small quantity of fine ores of higher grade is being utilized by Maharashtra Electrosmelt Ltd., for production of sinter. For the production of ferromanganese, the physical and chemical specification required in ores is given in Table 38.3.

Table 38.3

Physical		*Chemical*	
Size (mm)	*% Weight*	*Constituent*	*% Value*
–75 +10	83 (Min)	Mn	48% (Min)
–10 +6	12 (Max)	SiO_2	9% (Max)
–6 +3	3 (Max)	Fe	7.5% (Max)
–3	2 (Max)	P	0.16% (Max)

Size

As fines do not meet the physical and chemical specifications required by ferromanganese manufacturers processing and agglomeration has to be made before they are marketed for production of Ferro manganese. Processes have been developed to beneficiate fines to required grade. However the beneficiated fines need agglomeration (Pelletisation, sintering, briquetting or nodulising) to bring the high-grade fines to the required size to use them suitably in Ferro alloy production. The Pelletisation and sintering processes are successful in agglomerating the fines but due to the high investments and small scale of operations; the cost of production of pellets is higher. Presently, some percentage of sinters are used as a part of the feed to Ferro manganese furnaces at Maharashtra Electrosmelt Plant at Chandrapur and FACOR's Ferro Alloy Plant at Garividi (Andhra Pradesh). The conversion cost of fines to sinters is around Rs. 300 per tonne. Thus, the sinters produced may not be competitive by way of savings in the cost as compared to the standard grade lumps. Development of cheaper methods of agglomeration for utilization of upgraded fines has to be looked upon and especially improvements are sought in briquetting and nodulising processes.

Phosphorous

Phosphorus content of manganese ores poses a serious problem in their use for Ferro alloys production. Investigations have been made in minimizing the phosphorous content in the ore/fines at RRL. Bhubaneswar and IBM, Nagpur, on MOIL manganese ores. The process requires grinding and either leaching or flotation of ores for reduction of phosphorous. The concentrate thus produced is in the fines form and cannot be directly used for the production of Ferro manganese and needs to be agglomerated before use in the furnace.

Manganese Dioxide Ores

Manganese dioxide ore is used as depolariser in the manufacture of dry batteries. The suitability of the ore depends largely on the crystal structure and available oxygen in the ore. Generally, MnO_2 occurs in three crystal forms namely Alpha, Beta and Gamma. The ore having gamma crystal structure is battery active and is used directly as depolariser in dry cells, but the ore possessing Alpha and Beta crystal structures are used after blending with Synthetic Manganese Dioxide. Indian manganese ores do not possess Gamma crystal structures and cannot be exclusively used as depolariser in manufacture of dry batteries. Manganese dioxide ore is produced in Dongri Buzurg mine of the Company. The types of ores are produced at the mine are given in Table 38.4.

Table 38.4

Sl.No.	*Type*	*Size*	*MnO_2 %*
1.	Dioxide Ore	Lumpy	+80%
2.	DROM Lumpy	–75 +37 mm	74–75%
3.	DROM Small	–37 +3 mm	74–75%
4.	72/77	–3 +1.5 mm	72%
5.	Dried Sludge	–1.5 mm	60%

Dioxide ore with +80 per cent manganese dioxide is sold to dry battery manufacturers at higher price for use in production of dry cells. DROM Lumpy and Smalls are supplied to Ferro Manganese Producers for Ferro alloys production. The selling price of this ore is lesser than Dioxide ore. In India the reserve of dioxide ore is limited, and, therefore, it is necessary that manganese dioxide ore containing 74–75 per cent MnO_2 should be upgraded to +80 per cent MnO_2 with about 3 to.4 per cent Fe for use in dry battery industries. Present level of Fe is 7 to 8 per cent. This will increase the availability of dioxide ore used in manufacture of dry cells and this will be major step towards conservation of scarce dioxide ores. These ores were tested at Laboratories of IBM, Nagpur. The results showed that it is possible to upgrade these ores to 80 per cent MnO with 3–4 per cent Fe at –65 mesh liberation size with 60 per cent recovery by weight. However the process is not economically viable due to lower recovery as the feed material is of high value. A process has to be developed with much higher recovery to make the operation economically viable.

In Dongri Buzurg mine, about 1.00 lakh tonne of fines containing 53 per cent MnO_2, 11–12 per cent Fe are available. Investigations are under progress at ISM, Dhanbad and IBM, Nagpur to process them to 80 per cent MnO_2 and 3–4 per cent Fe to make them suitable for use in dry batteries. Tests indicated that these fines may be upgraded to 79 per cent MnO_2 and 6 per cent Fe at 27 per cent recovery by weight by the combination of gravity and magnetic separation tests. Further tests are proposed to be conducted to optimize parameters and to simplify flow sheet. MOIL proposes to put up a plant at Dongri Buzurg mine after successful establishment of the process for treatment of these fines and the fines likely to be produced in the future.

Investigations were conducted on Balaghat and Ukwa ROM ores and fines to study the improvement in the recovery of valuable minerals. Tests revealed that combination of heavy media separation, Dynawhirl process and shaking tables can produce a suitable grade with maximum recovery of fines. The company proposes to set up an integrated Ferromanganese plant consisting of beneficiation, sintering and manufacturing of Ferromanganese during 8th plan period.

Processes for Beneficiation of Manganese Fines

Conventional Methods

Gravity processes were commonly used for processing of low-grade manganese ores in which minerals of different specific gravities are separated by their relative movement in response to gravity and one or more other forces. Heavy media separation processes are normally employed for processing of coarser ores *i.e.*, upto 50 mm size. The process of jigging is used for upgrading the ores in the size range of –25 mm + 1 mm. The ores below 1 mm are presently treated by tables and spirals. However these processes have their limitations in treating fines. The other methods of separations used for manganese ores are magnetic separation and flotation. However, still in almost all manganese mines in India hand sorting and jigging of fines are practised.

Recent Developments: Gravity Processes

Gravity methods of various kinds have been used to recover minerals and new gravity devices were developed to meet the changing requirements, which have often been directed at finding out techniques for recovering finer and finer fractions of the given ores. From the beginning of the mineral technology, efforts have been made to extend the range of gravity concentration methods to recover finer size particles.

In the recent years, shaking tables along with hydro sizers and moving belt separators have been successfully employed for beneficiating manganese fines up to 200 mesh.

The Reichert cone concentrator was developed in Australia in the early 1960s primarily to treat titanium bearing beach sands and the success of cone circuits led to their application in many other fields. The single unit comprises several cone sections stacked vertically so as to permit several stages of upgrading. The cones are mounted in circular frames over 6 mm high. The feed pulp is distributed evenly around the periphery of the cone. As it flows towards the centre of the cone, the fine and heavy particles separate to the bottom of the film. This concentrate is removed by an annular slot in the bottom of the concentrating cone, the tailings will float over the slot. Reichert cones have high capacity and accept feed of upto 3 mm in size and can treat materials as fine as 30 microns. These cones may be efficiently applied for Upgradation of manganese ores also.

The heavy media process has long been recognized as the most efficient gravity separating method to concentrate minerals at low plant investment and operating cost. In recent years, the use of heavy media separation is extended to the treatment of intermediate and fine size ranges. The development in this direction is for the beneficiation of manganese fines in "Dyna Whirl Pool Process". The Dyna Whirl Pool process employs a unique method of creating and controlling cyclonic motion and of utilizing the separating characteristics of the resulting vortex. The sizes from 2" to 65 mesh can be successfully treated on commercial scale. The tests conducted on manganese fines at M/s. SALA International Laboratories have yielded promising results. The process was successfully employed for beneficiation of manganese fines (40 per cent Mn.) of Balaghat and Ukwa mines to 49 per cent Mn. With a recovery of 75–80 per cent (by weight) in the size range of –6 to 0.5 mm. Capacities of the Dyna Whirl Pool process separator vessels range from 10 TPH to 100 TPH of feed per unit. This processor permits compact plant design resulting in low capital investment, considerable space saving, easy supervision of plant operation and low operating cost.

The most recent development in this field is perhaps Bartles Mozley table, which consists of 40 smooth fiberglass decks mounted in stacks on a tilting frame. The fine feed is distributed to the forty decks where concentration is effected by the panning like motion given to the assembly from out of balance weight. At the end of each concentrating cycle (about one-half hour) the feed slurry is stopped and the concentrates washed from the table by tilting it. The Bartles Mozley separator can treat ores upto fine size range *i.e.*, 5–10 microns. This new apparatus may find considerable application in future processing plants where recovery of fines is desired. This process is yet to be established for beneficiating manganese fines.

Magnetic Processes

The common manganese minerals of India are feebly magnetic and hence can be separated from less magnetic minerals like quartz, barite, calcite, apatite etc. by high intensity magnetic separation. For concentration of manganese minerals, generally following types of magnetic separators are used:

1. Cross belt separators
2. Induced roll magnetic separators
3. Wet high intensity magnetic separators.

Cross Belt and Induced Roll Magnetic separators are applicable for dry operating conditions. Induced Roll Magnetic Separators are of higher capacities than Cross Belt Magnetic Separators but it requires clear and dry feed in the range of –8 mesh to 200 mesh. Recently "Permoll" high intensity magnetic separators have been developed where permanent magnets are used instead of electromagnets.

The application of wet high intensity separators eliminates the need for pre-drying and extends the range of separation for the recovery of particles finer than 200 mesh.

As the world's supply of mineral raw materials diminishes, lower grade fine ores will have to be sought as new sources of metals and minerals. Thus the need for continued and even expanded research in fine particle processing methods has to grow to fulfil the requirements.

Recent developments have shown that processed fines can be directly utilized in the production of Ferro manganese without agglomeration. Such developments are made in the name of 'Plasma Furnace Smelting Technology' for the production of Ferro-manganese at Tetronics Research and Development Co. Ltd., England. The process of production of ferromanganese is carried out in a transferred arc direct current plasma furnace. A plasma stabilizing gas flows between the electrodes between which very high power densities and temperatures can be generated. This system has shown much promise for utilization of manganese fines directly in the production of ferromanganese.

Chapter 39

Upgradation of Minerals through Bioleaching

V. Rajendra Prasad, Y. Prasanna Kumar, P. King and V.S.R.K. Prasad*

Environmental Pollution Control Engineering Laboratory, Department of Chemical Engineering, College of Engineering, Andhra University, Visakhapatnam – 530003, A.P., India

ABSTRACT

Minerals in general are complex materials not only for handling, processing and using in industries but also complex from characterization point of view. Minerals coming out of the mines and thereafter subjected to mechanical milling, chemical attrition and various leaching treatments are totally different in their physical properties thus influencing their behaviour during processing.

History is a witness to the fact that minerals have been the basis of classification of human civilization. A civilization undertakes development in consonance with a pattern of physical endowment of valuable minerals like Copper, Iron Ore, Aluminium, Zinc, Lead, Gold, Cobalt, Manganese, Graphite, Lime Stone, platinum etc. Since minerals resources are wealth hidden beneath the earth-curst and are wasting assets their conservation and planned exploitation is of paramount importance to national economic development and growth. As a matter of fact, mineral resources not only provide the basic inputs to domestic industry, but also account for sizeable export and foreign exchange earnings. Henceforth, the development and regulation of the mineral sector is concomitant to the optimal utilisation of resources which ensures all round development and growth of national economy.

Biohydrometallurgy is an interdisciplinary field involving metallurgy, chemistry, chemical engineering, microbiology and molecular biology. It has assumed tremendous practical significance since the last three decades. In a country like India, with its vast unexploited mineral potential Biohydrometallurgy assumes a great national significance.

* Corresponding Author, E-mail: rpvasanta@rediffmail.com.

Introduction

Increased demands for metals and thereby of their corresponding ores are causing fast depletion of high grade ore deposits. It is feared that the high grade ore deposits will be on the point of exhaustion within coming few decades. This will necessitate the use of low grade reserves of the world's mineral resources after upgradation through suitable technique. As such, without the development of new and more economic technologies for metal ore upgradation the industrial production of metals may cease in not far away future. This calls for putting special emphasis on biohydrometallurgical technology.

Biohydrometallurgy is an interdisciplinary field involving metallurgy, chemistry, chemical engineering, microbiology and molecular biology. It has assumed tremendous practical significance since the last three decades. It deals with the extraction of metals or upgradation of low grade minerals by application of various biological techniques and is considered as an extremely potent approach for the extraction of metals. Abetter understanding of these processes has allowed scientists to exploit microorganisms for the leaching of metals from ores and to adopt innovative microbe-based technologies for metal reclamation. Microbial technology is favoured as it requires less energy, chemicals, capital cost and it gives rise to very little pollution. Biohydrometallurgy may be the technique of choice for removing impurities from ores such as magnesite, dolomite, bauxite etc.

Occurrence of Silica

Several known minerals *e.g.*, magnesite, bauxite, dolomite, basalt, contain silica as common impurity. In nature silica occurs in polymerized forms and is almost insoluble. The presence of silica as impurity makes these minerals unsuitable for metal extraction. Compounds of silicates, aluminosilicates and the other anion radicals with Al, Fe, Mg, Mn, Ca etc., as cations constitute the most copious class of minerals.

Solubilization of Silica

Vernadskii, as early as 1934, thought about the possibility of solubilization of silicate by soil microorganisms to liberate various cations of silicate elements. He expressed that decomposition of aluminium silicate could take place through biological action. Later on, Aleksandrov and Zak (1950) isolated certain strains of bacteria capable of decomposing aluminosilicate, which they termed as "silicate bacteria."

Reports on depolymerization and solubilization of naturally occurring silica by biological action are available in the literature. It has been amply demonstrated that various plants and microorganisms including molds, diatoms, bacteria are capable of solubilizing silica and incorporate silicate ions from a silica contain medium. Okuda and Takahashi (1964) estimated that the concentration of soluble monomer silica in the bleeding sap exuded from cut rice stems vastly exceeds the concentration of the soil environment which contains high amount of polymer silica. They interpreted this as a depolymerization process, leading to the release of monomer silica from the polymerized material. Depolymerization of quartz suspension by molds and incorporation of monomer silica into the mycelia has been experimentally demonstrated (Holzapfel and Engel, 1954). Lauwers and Heinen (1974) reported that monomeric silica, released by mineralization of polymerized silica or quartz, was taken up by the higher plant *Equisetum* and the bacterium *Proteus mirabilis.*

For silicate minerals the state of the art concept is that chemical weathering leads to the destruction of their crystalline structure through hydrolysis which involves the ions H^+, OH^+, H_3O^+. In its general form the process of the destruction of heterogeneous silicate and aluminosilicate minerals is described as follows:

$$M^{+} \text{ mineral}^{-} + H^{+}OH^{-} \rightarrow H^{+} \text{ mineral}^{-} + M^{+} OH^{-}$$

where,

M^{+} mineral^{-} is the original mineral, H^{+} mineral is the weathered mineral; M^{+} are monovalent cations that are present in the mineral; M ions could be of other valences.

Mechanism of Silica Solubilization

Microorganisms interact with silicate materials in three major ways. They contribute to the decomposition and solubilization of silicate minerals. They utilize silica in a dissimilatory fashion by incorporating it into their cells or bodies in the dissolved form, releasing it as free silicic acid. They also assimilate silica by taking it up in dissolved form.

Microorganisms are capable of producing inorganic and organic acids which can react with minerals. By increasing the concentration of H^{+} ions in the medium, microorganisms activate the hydrolysis and release of metal cations.

Various microorganisms belonging to different systematic groups participate in the degradation of the same types of silicate minerals, thus stimulating the release of metal cations and the destruction of the siloxane and alumino-oxygen bonds of these crystalline minerals.

A siloxane bond is broken under the impact of biogenic processes of alkali formation. It was observed that an increase in release of silicon from quartz occurred as a result of the alkalification of the medium brought about by the growth of *Sarcina ureae.* The activity of photosynthesizing microorganisms also accelerates the degradation of quartz, since the pH of the medium rises with the fixation of CO_2. When it ceases the pH of the medium starts to shift towards acid range making it unfavourable for the degradation of siloxane bond.

Specific features of the destruction of silicate minerals and its mechanism are largely determined by the nature of the minerals and that of the microorganisms as well as the metabolites they produce. Acid producing organisms have been implicated in the weathering of rocks and minerals. Significant quantities of Si, Al, Fe, and Mg were reported to be solubilized by *Penicillum simplissimum* WB-28 from dunite, peridotite, basalt, granite and quartzite rocks (Silverman and Munoz, 1970) due to the production of citric acid. However, solubilization was not unique in the case of biogenic acid but was the result of acid attack in general. Solubilization was concomitant with alterations in the infra-red absorption spectra in the Si-O stretching vibration region of the rocks.

The release of Al, Mg and SiO_2 by fungal action from rocks has been documented by Henderson and Duff (1963). The destruction of apophyllite, olivine and Ca and Zn silicate by *Pseudomonas* and other soil organisms was found to be accompanied by the release of ketogluconic acid and other organic acids. Duff and Webley (1959) demonstrated that 2-ketogluconic acid was produced by similar type of soil organisms when grown in liquid culture containing insoluble phosphates or silicates. However, it does not seem that the production of 2-ketogluconic acid is an inducible process and the presence of an inducer is needed for their production. Many other metabolites of microorganisms such as a mino acids, di- and tri-carboxylic acids, oxyacids and other compounds as well have been found to participate in the destruction of silicate rocks and minerals forming soluble complexes with cations of metals and elements present in them. The role played by the process of complexation in weathering has been noted by several workers.

Stable soluble complexes of the chelate type lichen acids and oxyacids of bacterial origin form during the weathering of silicates. Fungi produce a large number of organic acids which are highly active as weathering agents. Organic acids formed by fungi have also been found to form complexes with Fe two to five times faster than lichen acids. Thus, in addition to hydrogen ions, organic compounds capable of forming complexes are capable of removing metalations from silicate materials.

Solubilization to fislica from diatom frustules has been found to be associated with bacterial activity. Solubilization in such instances is attributable, perhaps, to the production of hydrolytic enzymes (Patrick and Holding, 1985). It was also suggested that several different mechanisms might be responsible for solubilization and recycling of silica. Solubilization of silica by metabolic byproducts of microorganisms is considered to be one of these mechanisms. There are also some species of bacteria that elaborate voluminous extracellular slime capsules which attack silicate minerals when in intimated physical contact with them (Aleksandrov and Zak, 1950). 2-ketogluconic acid produced by may microorganisms, besides having acidifying effect, acts as a chelating agent and causes efficient solubilization of silicates or phosphates. Lewin (1966) suggested a similar mechanism involving chelating agents in silica solubilization from diatom frustules. From his observation it was also suggested that presence of metal ions could retard release of silica during frustules solubilization.

Biodegradation of different aluminosilicates for the recovery of aluminium has been widely studied and considerable information has been obtained (Silverman, 1979; Decey *et al.*, 1981). The aluminosilicate degradation caused by bacteria is related with the action of different metabolites secreted into their microenvironment. Acidolysis, complexolysis and alkalolysis are considered to be the acting mechanisms depending on the type of the metabolites secreted. These bacteria are in specific need of silicon and probably, also of aluminium for their optimum growth. The action of "silicate bacteria" on aluminosilicates has been connected with the formation of mucilaginous capsules as well as with the production of different metabolites such as organic and amino acids (Groudeva and Groudev, 1984). However, it has not been excluded that the bacterial action might be of an enzymatic nature and the bacteria are able to utilize energy released from biodegradation of aluminosilicate. The ability of heterotrophic bacteria, *B. mucilaginous,* to degrade silicate and aluminosilicate minerals has been used to develop a technological scheme for the dressing of low grade magnesite and bauxite.

Reports on bioleaching of silica from magnesite or by *Bacillus lichenoformis,* a bacterium isolated from magnesite ore deposit, is at hand (Mohanty *et al.* 1990). Concomitant mineralysis of silica and silicon uptake by the strain was observed. It was concluded that silicon uptake was restricted to adsorption onto bacterial cell surface rather than an internal uptake through the membrane. Phosphorus was found to be highly essential for the growth of the bacteria and silicon uptake, in agreement with the results of Scheleske *et al.* (1986), who observed that silicon uptake by diatom, is stimulated with increasing concentration of phosphate. Earlier, it was shown that some fungi and bacteria, isolated from Hawaiian bauxite were able to accumulate aluminium on the surface of cells, which they extract from aluminoorganic compounds during the mineralization of the latter.

Recently, Halder *et al.* (1993) reported different strains of root-nodule bacteria, *Rhizobium* and *Bradyrhizobium* capable of solubilizing silicates from different synthetic silicates. It was observed that the production of 2-ketogluconic acid and other organic acids during the growth of the organisms brought about by the release of silica from silicate compounds.

Application of Silica-solubilizing Microbes

Magnesite and bauxite are the basic raw materials for refractory industry. Most of the low silica refractory minerals are being imported for Indian industries causing drainage of a large sum of foreign exchange. After a successful laboratory scale experimentation to remove silica from magnesite by *Bacillus licheniformis*, the technology developed in the authors laboratory is being used for the first time for commissioning a 5 MT pilot plant at Salem Works of Burn Standard Co. Ltd., Tamil Nadu, India in collaboration with the Department of Biotechnology, Government of India to remove silica from magnesite ore available in the region. The plant consists mainly of three different parts, culture vessel, mixing vessel and reactor vessel with several accessories like steam boiler for sterilization, air compressor fitted with an air filter for supplying sterile air for the growth of bacteria, air heater for maintaining desired temperature etc. Results are quite promising in the sense that a removal of up to 60 per cent of silica present in magnesite ore has been possible. The pilot plant is working satisfactorily and is also being modifies to achieve maximum silica removal. This will, of course, reduce the import of enriched magnesite ore from developed countries.

Conclusion

In a country like India with its vast unexploited mineral potential, Biohydrometallurgy assumes a great national significance. Removal of silica from magnesite and bauxite ores for lowering the sintering temperature for magnesite and the upgradation of bauxite for aluminium production can be done by using microbes. Other ores, whether high or low grade, may be similarly upgraded with the help of appropriate microbes.

Chapter 40

Ambient Noise Quality Around Sensitive Areas in Asansol City, W.B.

D. Banerjee* and S.K. Chakraborty**

**Department of Environment and Water Management, B.B. College, Ushagram, Asansol – 713 303, Burdwan, West Bengal*

***Post Graduate Institute of Physics, B.B. College, Asansol – 713 303, Burdwan, West Bengal*

ABSTRACT

A comprehensive noise exposure study was conducted in and around some silence zones for summer and winter season from April 2003 to March 2004, in Asansol. A total of five sensitive locations including school, college, nursing home, hospital, court and a temple were selected for ambient noise monitoring. Sound level was monitored for day, evening and night time around the study site at various distances from them. Noise data from field was used for statistical analysis and generation of various noise parameters. Noise Contour Maps were also prepared using Noise Map, 2000 software. The study reveals that present noise level in all the locations exceeds the limit prescribed by Central Pollution Control Board. Mean Noise level during day, evening and night is above 65 dB(A). The results indicate significantly high Day-Night equivalent noise level in the range of 65 to 80 dB(A) for both summer and winter. The Community Noise Level is alarmingly high with average value of 75 dB(A). The mean Noise Pollution Level and Noise Climate in all locations were high. It can be concluded that the population in this industrial area, particularly around sensitive zones like educational institutions, hospitals, courts and places of worship are exposed to noise pollution ranging from moderate to severe level.

Keywords*: Sound Level, Sensitive areas, Noise Map, Asansol.*

Introduction

The Central Pollution Control Board (CPCB), in its notification on Ambient Air Quality (AAQ) standard for Noise, which has been included as an air pollutant under section 20 of Amended Air Act of 1987, as given in Table 5 (CPCB 1995). The permissible noise levels in Sensitive areas (Silence Zones) is 50 dB (A) during Daytime and 40 dB(A) at night. As per SCHEDULE [Rule 3(1) and 4(1)], CPCB, Day time shall mean from 6.00 a.m. to 10.00 p.m and Night time shall mean from 10.00 p.m. to 6.00 a.m. Silence zone is an area comprising not less than 100 metres around hospitals, educational institutions, courts, religious places or any other area which is declared as such by the competent authority. Use of horns, loudspeaker, heavy machinery, crackers, etc. is prohibited in and around such places.

The major sources of noise pollution in Asansol region is intense vehicular movement along cramped roads and to some extent industrial and commercial activities. The transportation of raw material and finished goods from the Iron and Steel Works located within the city premises creates high noise pollution throughout the day. The G.T. Road is the primary road that traverses across the region. The city has built up around this road. The road passes through the major portion of the city and joins the NH2 at a distance of about 8 km from the city centre. Secondary roads, lanes and by-lanes are all used for vehicular movement across the city. The roads are lined along side with residential, commercial and sensitive establishments. Many of the sensitive sites like the Asansol Court, along with the SDO and RTO office are located along one of the heavy traffic zone areas with hourly vehicular count of around 5000 during peak hours. Other sites like Schools, Colleges and Hospitals and Nursing Homes are mostly close to primary roads. The lack of well maintained roads, which are congested during peak hours, produces severe noise level within 100 meters of these silence zones.

The aim of the study was to investigate the present noise quality status and its seasonal variation in and around some silence zones including education institutions, health institutes, court and religious place. Development of Noise Contour maps of the city was also intended.

Methodology

Study Area and Sampling

Noise level around some sensitive locations were monitored between April 2003 and March 2004. The monitoring sites are represented in Figure 40.1 and details of the study sites are given in Table 40.1. A total of five sites were chosen susceptible sites ware preferred according to their immediacy to noise sources like industry, commercial activity and traffic movement. The sites are Asansol Court (ACO), Asansol Engineering College (EGC), Bari-Maidan-Temple and Church (BMD), Government Hospital (GHOS) and Raghunath Nursing Home (RGN).

A digital Sound Level Meter, Lutron, Type 2 with Frequency Weighting Network as per IEC651 specifications and frequency range of 31.5 Hz to 8,000 Hz was used for the study. A 'B and K' (Bruel and Kjaer) multi-function acoustic calibrator (Model: 4226) was used for calibration at 94.0dB (A) before and after sampling. All reading was taken on the 'A-Weighting' frequency network, at a height of 1.5 meters from ground level and on the 'Fast' range Time Weighting. The 'A' weighting characteristic and 'Fast' range is simulated as 'Human Ear Listening' response. This is typically required if making the environmental sound level measurement. Noise was monitored at each location during Morning, Evening and Night for summer and winter season from April 2003 to March 2004.

Table 40.1: Characterisation of Five Monitoring Stations

Sampling Station and Locations	*Zonal Activities*	*Major Sources of Noise Pollution in 100 m Radius*
Site 1: Raghunath Nursing Home	Residential/Commercial	Transportation Activities (major) Commercial Activities (less)
Site 2: Asansol Court	Commercial	Transportation Activities
Site 3: Engineering College	Industrial	Transportation Activities Cement Industry
Site 4: Government Hospital	Residential/Commercial	Transportation Activities Commercial Activities
Site 5: Bari-Maidan (Temple-Church)	Residential	Transportation Activities Commercial Activities

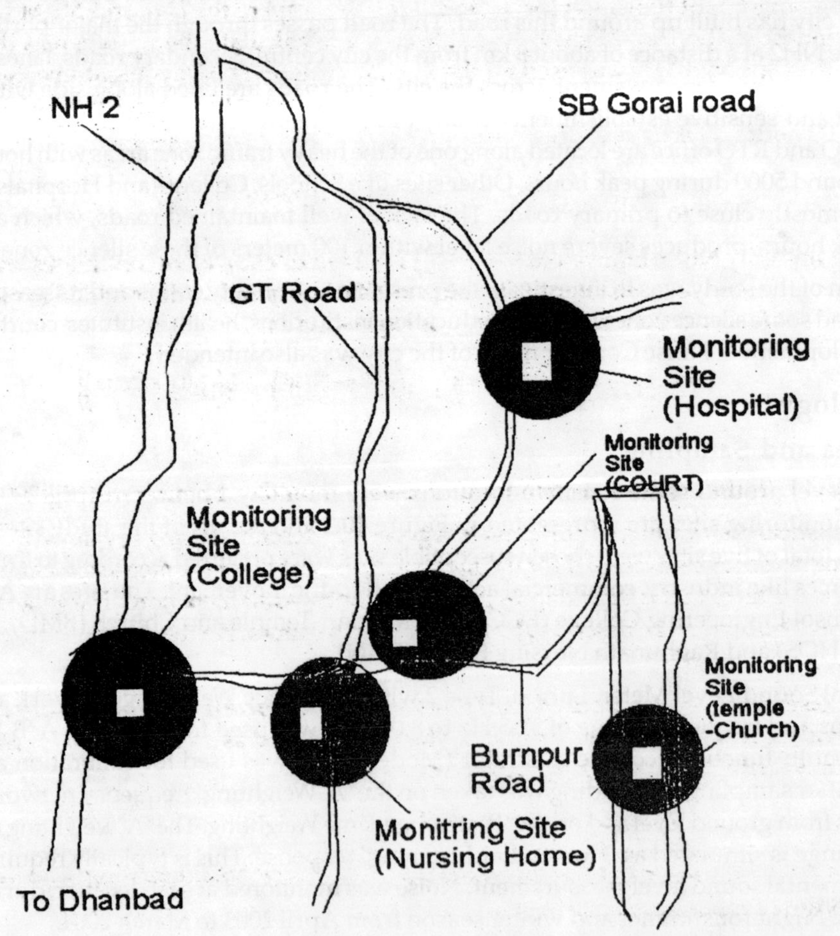

Figure 40.1: Map of Asansol Area Showing the Noise Monitoring Stations around Five Silence Zones. (1) RGN: Nursing Home, (2) ACO: Asansol Court, (3) EGC: Asansol Engineering College, (4) GHOS: Asansol Sub-Divisional (Govt.) Hospital, (5) BMD: Temple and Church.

Statistical Analysis and Mapping

Data collected from field was statistically analysed using MS Excel for determination of various noise parameters and other applications. Various parameters and indices estimated included Day-Night Equivalent Noise Level (L_{dn}), Community Noise Level (L_{den}), Noise Pollution Level (L_{np}), Percentile Noise Levels (L_{10}, and L_{90}) and Noise Climate (NC) were estimated. The sound level meter does not give a steady and consistent reading and it is quite difficult to assess the actual sound level over the entire monitoring period. To mitigate this shortcoming, the Continuous Equivalent Sound level, indicated by Leq, is used. It is defined as the level of that steady sound which over the same interval of time, contains the same-total energy (or dose) as the fluctuating sound. Equivalent sound level, 'Leq', can be obtained from variable sound pressure level, 'L', over a time period 'T' by using following equation

$$L_{eq} = 10\log_{10}\left[(1/T)\int_0^T 10^{L/10}\,dt\right] \quad \text{Eqn. 1}$$

Day-Evening-Night time monitoring was done to compare the Day and Night L_{eq} values with national standards. The Day-Night L_{eq} was also used to calculate the L_{dn} whereas the evening L_{eq} along with the Day-Night L_{eq} was used for the L_{den} calculation. The Day-Night Average Sound level, L_{dn}, is a 24-hour equivalent continuous level where 10 dB is added to night time noise levels from 22.00 hours to 0700 hours. The L_{dn} is expressed as

$$L_{dn} = 10\log(1/24)\left[16\left(10^{Ld/10}\right) + 8\left(10^{(Ln+10)/10}\right)\right] \quad \text{Eqn. 2}$$

The Community Noise Equivalent level, L_{den}, is a 24 hour equivalent continuous level where 5 dB is added to evening noise levels from 1900 hours to 2200 hours and 10 dB is added to night time noise levels from 2200 hours to 0700 hours. The L_{den} is expressed as

$$L_{den} = 10\log(1/24)\left[12\left(10^{Ld/10}\right) + 4\left(10^{(Le+5)/10}\right) + 8\left(10^{(Ln+10)/10}\right)\right] \quad \text{Eqn. 3}$$

where,

L_d, L_e and L_n is the equivalent continuous noise level for day, evening and night respectively.

The statistical distribution of noise levels is generally described using L-values (Percentile level), which are the noise levels exceeded for the indicated proportion of the measurement period[1]. Thus, the L_{10} level is the level that is exceeded 10 per cent of the measurement time. The general shape of the noise level distribution can be described by limited set of L-values. For the purpose of this study the L_{10}, L_{50} and L_{90} were estimated.

Other noise indices that were used for the study to characterize the noise exposure were the Noise Pollution level (L_{NP}) and Noise Climate (NC). The L_{NP} is defined as a variation on L_{eq} which accounts for short term variability in noise levels, expressed as

$$L_{NP} = L_{eq} + (L_{10} - L_{90}) \quad \text{Eqn. 4}$$

The Noise Climate is the range over which the sound levels are fluctuating in an interval of time and is given by

$$NC = (L_{10} - L_{90}) \quad \text{Eqn. 5}$$

where,

L_{10} and L_{90} are the noise level exceeded for 10 per cent and 90 per cent of monitoring time respectively.

Noise mapping was conducted and noise contour maps were developed for 18 hour Leq time period in the five study areas using 'Sound Plan' (Ver. 6.1) of GmbH and 'NOISEMAP-2000' (Ver. 2.51a) of ATKINS group. Various other characteristics of the roads nearest to the location was studied simultaneously including traffic flow rate, traffic volume and composition, percent heavy vehicles, road type and site geometry, etc. These data were used in the flapping programme.

Results and Discussion

Noise levels were monitored on a continuous basis in three phases, Day (12 hours), Evening (4 hours) and Night (8 hours), in and around five sensitive (silence) zones. The zones were chosen as those locations which are classified as noise sensitive zones and moderate to high noise dose has detrimental effect on the activity of such locations, including school, colleges, courts, place of worship and health institutes like hospitals and nursing homes. The average noise level (Leq) along with other parameters like maximum-minimum noise levels; Day-Night (Ldn) and Day-Evening-Night (Lden); Percentile Noise level (L_{10} and L_{90}) are shown in Table 40.2. The study reveals that the noise levels in all the locations are in excess then those prescribed by the Central Pollution Control Board. The major noise levels in the sensitive areas are mostly due to traffic movement and all the locations are situated near major roads. Ignorance among the local people in addition to lack of awareness has caused no actions to be taken for the abatement of such noise exposure. Prolonged exposure to such noise on regular basis particularly in case of school children and patients may have significant physical and physiological health impacts. Lack of well maintained roads, unscientific traffic supervision, overpopulated side walks, encroachment of commercial activities along the roadsides and lack of greenery have intensified the problem.

The noise pollution is more during the summer as evident from the results as because during winter the movement of personal vehicles sharply falls after 5:00 p.m. due to the chill cold which sometimes plunges to 7°C in this region. The Noise Climate during summer ranges between 9.3 dB(A) and 19.9 dB(A), with the highest level at the Govt. Hospital site, situated along side the busy S.B.Gorai Road, is quite high. The Noise Climate varies between 6.1 dB(A) and 16.5 dB(A) during winter with highest level at the Asansol Court, situated on a very important road. The noise pollution (L_{np}) ranged between 72.5 dB(A) and 104 dB(A) in summer and between 63.4 dB(A) and 94dB(A) in winter. Both the ranges being quite high and more towards the alarming level. The Noise Climate and Noise Pollution level are represented graphically in Figure 40.2. The noise maps for the study sites using 18 hour L_{eq} system are given in Figures 40.3–40.7. The figures show the heavy impact of road traffic noise near the vicinity of the silence zones and its spreading. All the locations are subjected to moderate to high noise dose throughout the 24 hours. The lack of greenery and barriers in most cases prevents the abetment of the noise propagation towards these areas. This is evident from the high L_{dn} and L_{den} levels in and around these locations. The L_{dn} level had mean value of 77.6 dB(A) in summer and 76.1 dB(A) in winter, very high in comparison to standard set by EPA at 50 dB(A). The L_{den} had average value of 77.8 dB(A) and 75.8 dB(A) in winter. This is mainly attributed to the high growth rate in personal vehicles over the last five years although the road condition and capacity remained the same.

Table 40.2: Variation in Noise Parameters During Summer and Winter in the Five Monitoring Sites at Asansol

(a) Summer

Location	*Monitoring Date*	*Period*	*Leq*				*Mean L_{10}*	*Mean L_{90}*	L_{dn}	L_{den}
			Max	*Min*	*Mean*	*S.E.*				
RGN	10-04-2003	D	80.3	60.8	71.53	0.97	78.4	64.3		
	to	E	90	60.5	71.59	1.39	76.2	60.4	72.39	73.53
	12-04-2003	N	77	52	63.71	1.3	69.1	58.3		
ACO	18-04-2003	D	86.2	64.9	79.4	1.95	79.7	70.4		
	to	E	85.6	65.2	72.1	1.41	73.5	64.2	77.9	77.46
	21-04-2003	N	64.9	59.3	60.3	2.7	66.4	52.3		
EGC	14-04-2003	D	83.4	68.4	74.9	2.71	84.1	68.5		
	to	E	76.8	64.8	71	1.49	74.2	63.2	74.77	74.3
	17-04-2003	N	71.3	52.3	64.5	1.72	70.1	56.3		
GHOS	26-04-2003	D	91.3	69.2	87.3	1.11	90.1	81.4		
	to	E	85.6	70.1	77.3	0.71	86.9	70.2	85.88	85.13
	30-04-2003	N	80.1	58.6	69.4	1.46	73.1	53.2		
BMD	02-05-2003	D	89.4	67.3	78.4	2.3	87.2	74.2		
	to	E	87.2	70.1	78.2	0.93	79	68.2	76.97	78.64
	05-05-2003	N	65.4	56.2	60.4	0.81	64.2	52.1		

(b) Winter

Location	*Monitoring Date*	*Period*	*Leq*				*Mean L_{10}*	*Mean L_{90}*	L_{dn}	L_{den}
			Max	*Min*	*Mean*	*S.E.*				
RGN	19-12-2003	D	86.4	67.4	77.3	0.68	80.1	67.2		
	to	E	78.2	61.2	66.5	1.63	70.2	55.2	76.14	75.38
	21-12.2003	N	72.1	55.3	62	2.15	64.3	52.1		
ACO	22-12-2003	D	85.1	78.9	74.2	0.89	77.9	61.4		
	to	E	80.1	65.4	69.3	0.97	73.1	57.8	74.41	74.43
	25-12-2003	N	70.2	56.6	64.8	0.96	70.1	55.3		
EGC	27-12-2003	D	90.4	78.9	82.5	1.66	82.7	71.2		
	to	E	82.1	70.6	74.9	1.23	77.6	64.3	82.43	82.09
	30-12-2003	N	78.1	64.2	72.3	0.78	78.4	66.2		
GHOS	03-01-2004	D	86.9	67.4	79.8	0.56	84.6	71.2		
	to	E	81.4	70.7	74.3	0.88	79.2	64.1	78.83	78.73
	05-01-2004	N	70.3	54.9	65.8	2.13	67.1	52.8		
BMD	07-01-2004	D	79.7	68.8	69.3	1.46	73.4	59.4		
	to	E	73.2	60.3	63.7	1.8	68.4	58.7	68.73	68.62
	10-01-2004	N	61.3	51.2	57.3	0.87	60.2	54.1		

Note: All noise units in dB(A); D–Day (6 am to 6 pm), E–Evening (6 pm–10 pm) and N–Night (10 pm to 6 am); SE is standard Error; L_{eq} is continuous equivalent noise level, L_{den} if Day-evening-night equivalent noise level, L_{dn} is Day-Night equivalent noise level, L_{10} and L_{90} are percentile noise levels exceeded for the monitoring time.

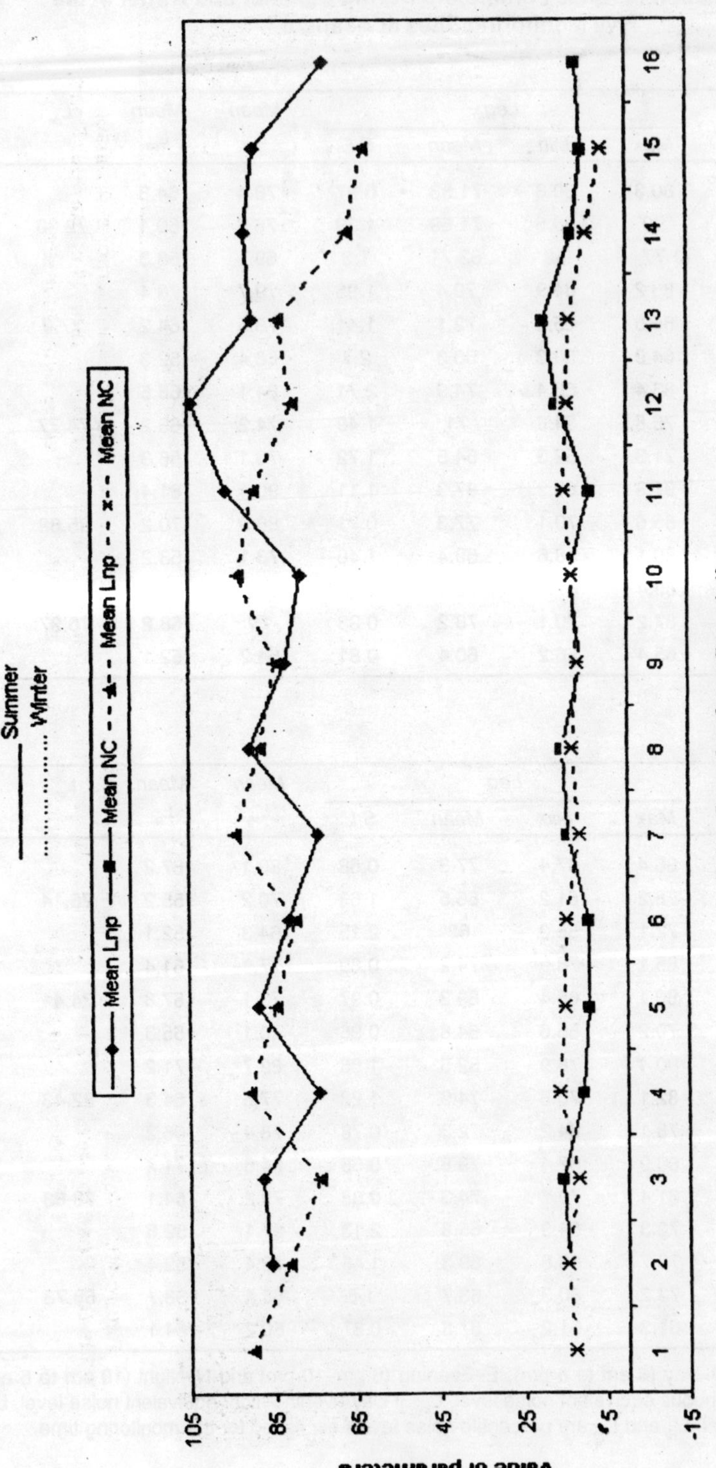

Figure 40.2: Seasonal (Summer and Winter) Variations in Noise Climate (NC) and Noise Pollution Level (L_{np}) in Five Monitoring Sites

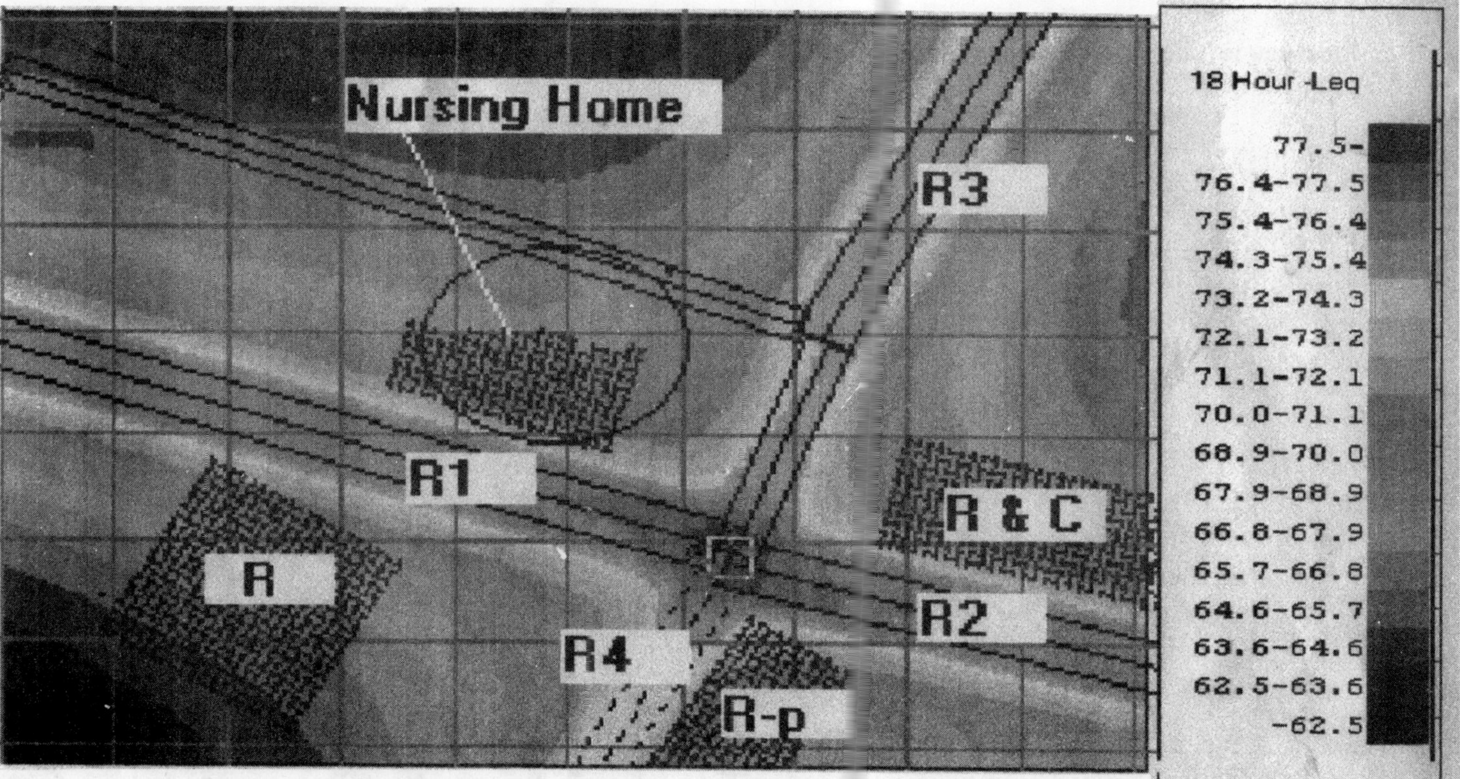

Figure 40.3: Noise Map Showing Ambient Noise Quality for Area Around Study Site (Nursing Home) at Asansol. R1 is G.T. Road to Barakar; R2 is G.T. Road to Durgapur; R3 is Connector of NH2 with G.T. Road; R4 is Burnpur Road, R denotes residential areas; C denotes commercial areas and R-p is Police Line Quarters.

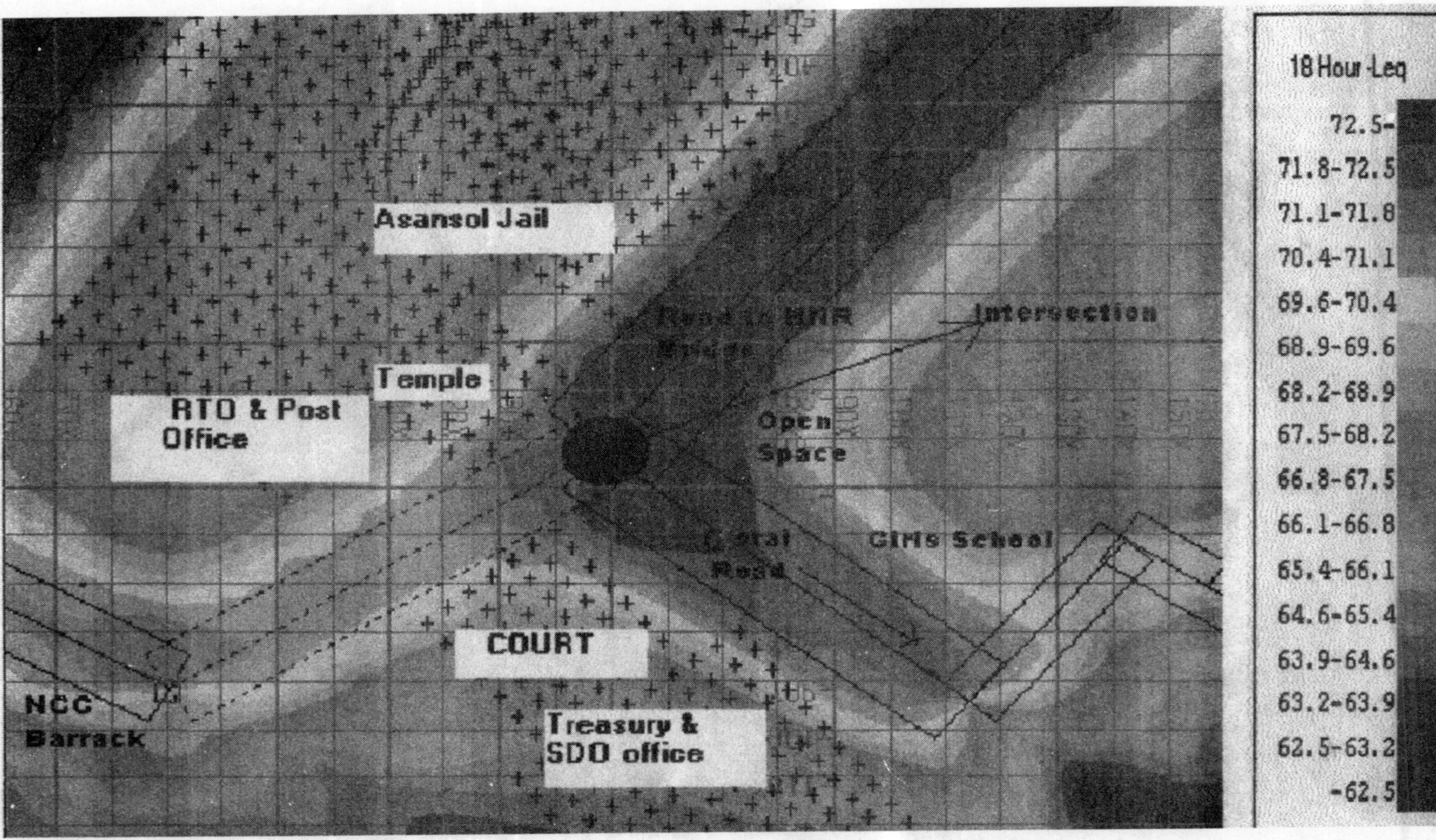

Figure 40.4: Noise Map Showing Ambient Noise Quality for Area Around Study Site (Asansol Court) at Asansol

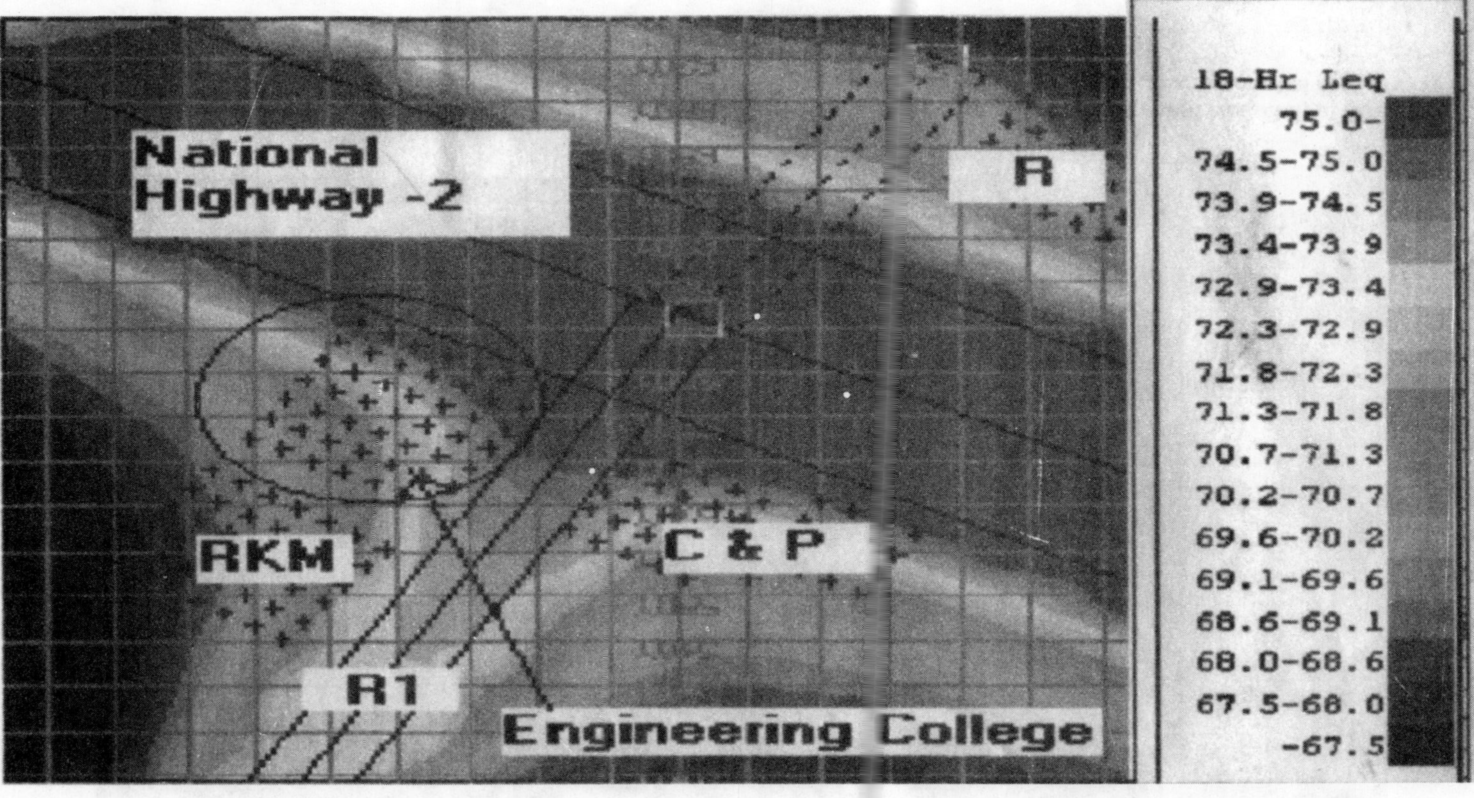

Figure 40.5: Noise Map Showing Ambient Noise Quality for Area Around Study Site (Asansol Engineering College) at Asansol. R1 is Connector NH2 with G.T. Road; R denotes residential areas; RKM denotes Ramakrishna Mission School Campus; C denotes commercial area and P is Indian Oil Petrol Station.

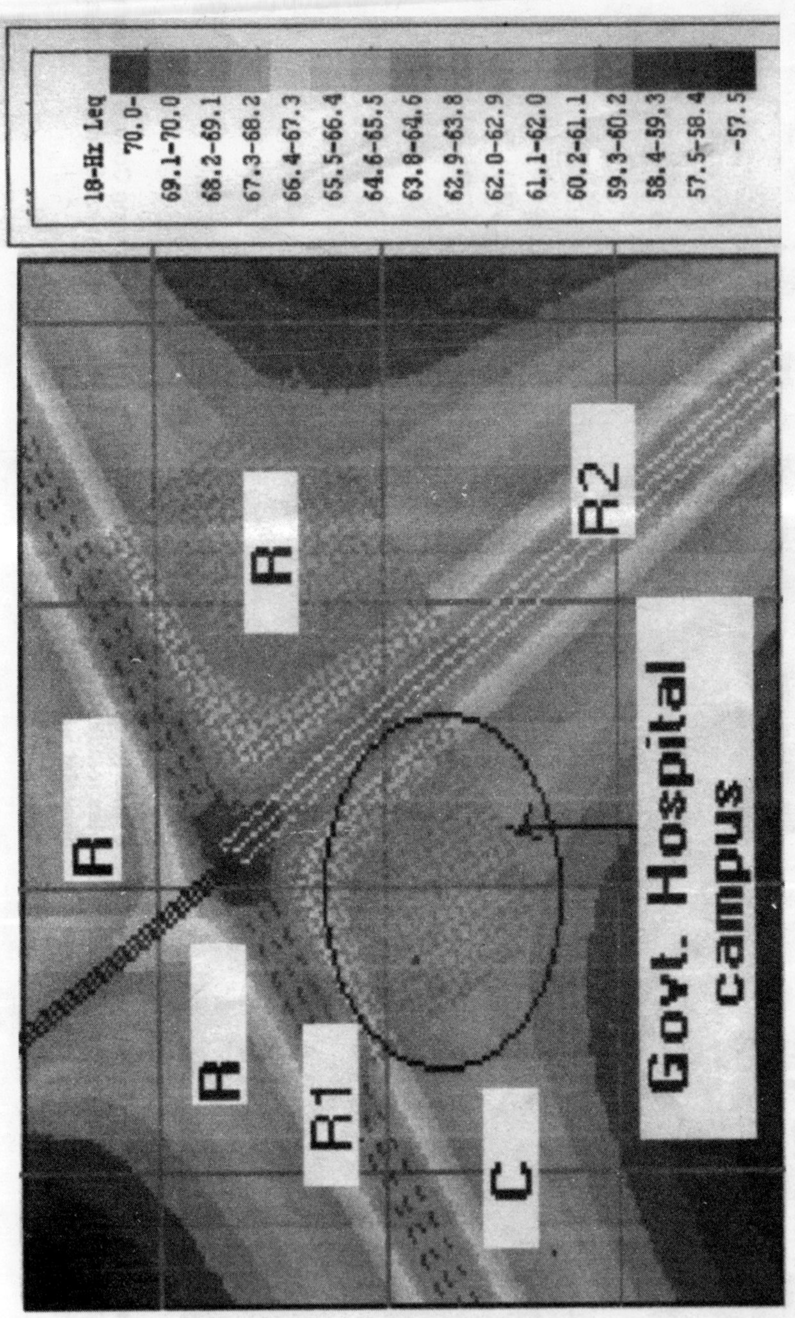

Figure 40.6: Noise Map Showing Ambient Noise Quality for Area Around Study Site (Asansol Government Hospital) at Asansol. R1 is Gorai Road Connecting with G.T. Road; R2 is Lane to Mohisilla Residential Colony; R denotes residential areas; C denotes commercial area.

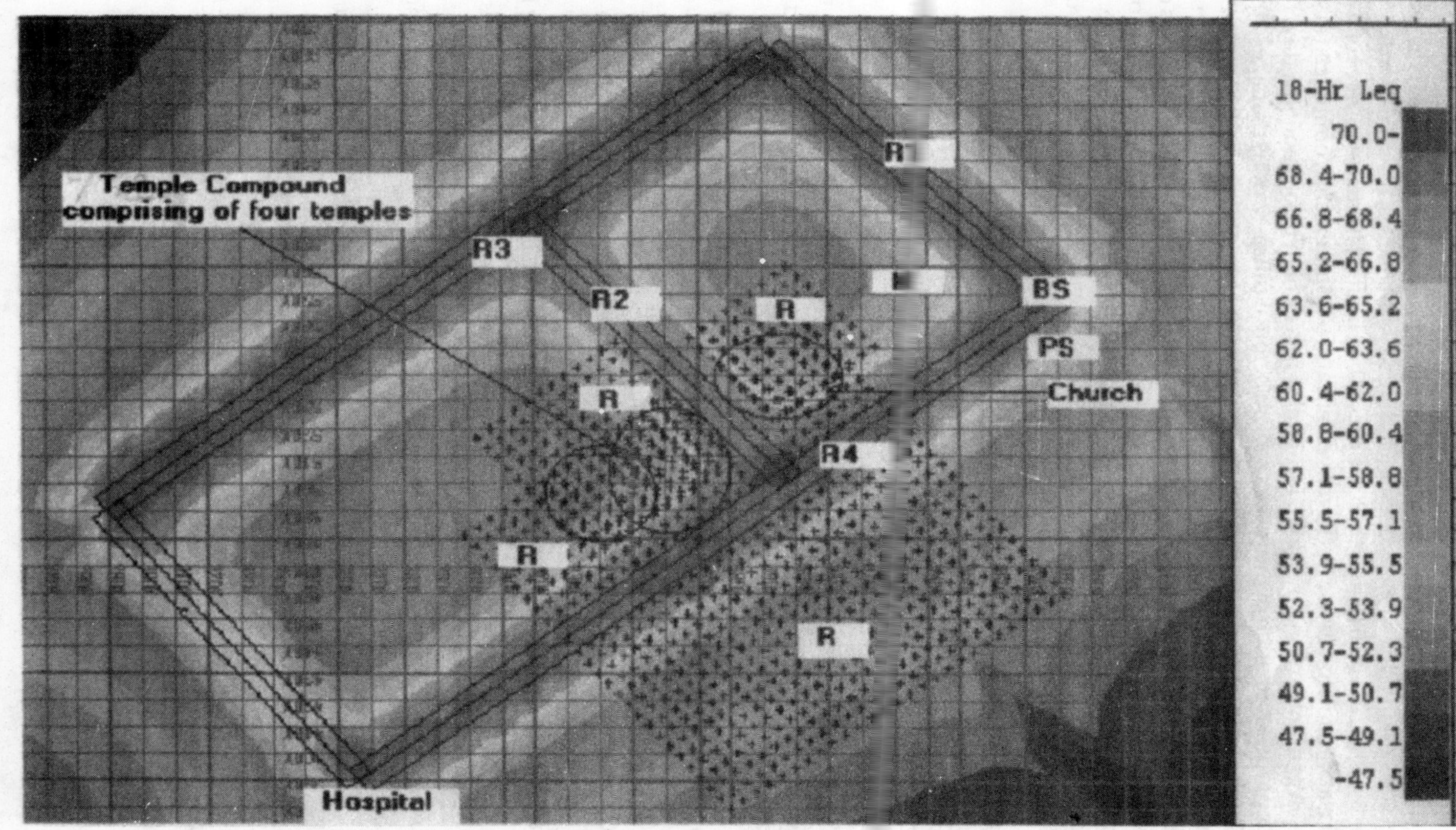

Figure 40.7: Noise Map Showing Ambient Noise Quality for Area Around Study Site (Bari-Maidan Temple Complex) at Asansol. R1 is Road connecting Abdul Bari Road to Burnpur road; R2 is Station Road; R3 is Burnpur Road; R4 is Abdul Bari Road; R denotes residential areas; M denotes Market Complex; BS is Burnpur Bus Stand and PS is Police Station.

Conclusion

On the basis of the study carried out in this industrial city of Asansol, it is evident that the ambient noise level surrounding the silence zones, including school, colleges, hospital, nursing homes, court and places of worship, is much higher than the prescribed limits. The major sources of the noise in these areas can be attributed to intense traffic movement and to some extent, commercial activities. The ambient noise level in and around the Govt. Hospital was very high throughout the day. There is immediate need for proper mitigation of the noise problem before it causes serious damage to the local population. Green belts are required all along the boundary of the hospitals as well as protection of sensitive receivers by use of noise insulated walls and window glasses are required in the schools and, colleges and health institutes. Diversion of heavy vehicles away from inside the city limits during the day and evening will decrease the noise impact and traffic jams. Reduction of the value of Lden will reduce the number of people that suffer from specific health effects as sleep disturbance, stress, hypertension or interference with communication. For optimum assessment and reduction of such specific effects the use of additional noise indicators may be necessary.

Acknowledgement

The authors express sincere thanks to the Principal, B.B. College, Asansol, and the Course-coordinator, B.B. College, Asansol, for their kind permission to publish the findings in the form of this research article.

References

Banerjee, D. Chakraborty, S.K. 2005. Mathematical modeling for assessment of urban road traffic noise. *Pollution Research*, Karad (In press).

Banerjee, D. 2004. An investigation of ambient noise level around an integrated iron and steel works. *International Journal of Ecology and Environmental Science*, New Delhi (In press).

Canter, L.W. 1996. *Environmental Impact Assessment*. Mc-Graw Hill, New York, p. 304–340.

Central Pollution Control Board (CPCB). 2000. *Ambient Air Quality in Respect of Noise*, Schedule-Part II, Sec-3(ii), CPCB, New Delhi.

Central Pollution Control Board (CPCB). 2001. *Noise Pollution Regulation in India*. CPCB, New Delhi.

Central Pollution Control Board (CPCB). 1992. *Pollution Control Acts, Rules and Notifications Issued Thereunder*, 1992, *PCLS/02/1992*, CPCB, New Delhi.

Leeuwen, H.V. 2003. *The Sound of Cars and Trucks or Road Traffic Noise*. Traffic Noise-Traffic Technology International, Netherlands.

Pandya, G.H. *et al.* 2002. *A Comprehensive Investigation of Noise Exposure in and Around an Integrated Iron and Steel Works*. American Industrial Hygiene Association, 63: 172–177.

Rosen, M.J. *Industrial Noise Pollution*. John Willey and Sons Inc, New York.

Sinha, S. and Sridharan, P.V. Present and future assessment of noise level in the Neyveli region. *Journal of Environmental Studies and Policy*, 2(1): 1–13.

Chapter 41

Physico-chemical Characteristics of Drinking Water in Selected Areas of Namakkal Town (Tamil Nadu), India: A Case Study

M. Rajasekara Pandian, G. Sharmila Banu*, G. Kumar** and K. H. Smila**

**Centre for Biotechnology, Muthayammal College of Arts and Sciences, Kakkaveri, Rasipuram, Namakkal – 637 408, Tamil Nadu, India*

***Department of Biochemistry, Selvamm Arts and Science College, Namakkal – 637 003, Tamil Nadu, India*

ABSTRACT

The physico-chemical and bacterial quality of drinking water of the town Namakkal was studied for 3 months from December 2001–February 2002. Water samples were collected at fortnight intervals from tap water from areas Kondichettipatti, S.P. Pudur, Ganesapuram, Periyapatti and Kosavampatti in Namakkal town. The values of Total solids (312.2–574.5 mg/l), pH (7.2–8.6), Chloride (24.5–58.4 mg/l), Dissolved oxygen (4.8–10.2 mg/l), BOD (0.28–2.7 mg/l), Nitrate (0.007–0.02 mg/l), Phosphate (0.004–0.015 mg/l) were recorded. The total coliform count varied 58–1125/100 ml and all the water samples showed a positive result.

Keywords: Drinking water, Physico-chemical constituents, Total coliforms.

Introduction

Water priceless gift of nature, is most essential for all living organism. Water from surface source provides sustenance to plant and animal and constitutes the habitat for aquatic organisms and meet importance of agricultural and industrial needs (Prasad and Gaur, 1992) The rapid industrialisation, growing urbanization and increasing use of chemicals in agriculture constitute some of the important factors responsible for various forms of pollution of water bodies (Tripathi and Pandy, 1990). Freeda *et al.* (2002) pointed out that the surface waters usually rich in turbidity suspended impurities of decaying organic matter, sand and finely divided clay, microorganisms and bacteria and small amount of mineral salts like Ca^{2+}, Mg^{2+}, Na^{+}, SO_4^{2-}, Cl^{-} etc. Natural waterways receive significant microbial load through domestic sewage, animal and human excreta and industrial wastes (Indrabai and George, 2002). Therefore, the present study was made to assess the quality of drinking water in selected areas of Namakkal town.

Materials and Methods

The physico-chemical and bacterial quality of drinking water of the town Namakkal was studied for 3 months from December 2001 to February 2002. The samples were collected from areas of Kondichettipatti, S.P. Pudur, Ganesapuram, Periyapatti and Kosavampatti in Namakkal town. The samples designated as T_1 to T_5, were collected from these sources at fortnight interval for about a period of 3 months (December 2002–February 2003) in 5 litre plastic cans and were analyzed tor air temperature, water temperature, pH, chloride, total solid, dissolved O_2, PO_4, NO_3, BOD, and Total coliform count by following standard methods (APHA, 1985; Trivedy and Goel, 1986; Chessbrough, 1991). The water quality index is obtained from the weighting curve charts (1–9) (Mitchell and Stapp, 1992).

Results and Discussion

Air temperature of tap water in different areas of Namakkal town varied from 22.5 to 25.5°C, and the water temperature ranged between 25 and 26°C and so no marked variation on temperature were observed during the period of December 2002–February 2003. Brown *et al.* (1967) pointed out that fluctuation in temperature was dependent on the types and concentration of polluted matter.

The pH values in sampling areas ranged from 7.2 to 8.5 and so alkaline in nature. According to WHO (1984), the desirable pH of drinking water is 7 to 8.5. The pH has no direct adverse effect on health, out at the same time alters the taste of water. Higher pH reduces the germicidal potentiality of Chlorine and induces the formation of toxic trihalomethanes (Trivedy and Gael, 1986). However in the present study the pH levels were with in the limits set for protecting of aquatic life 6.5 109.0 (USEPA, 1975), irrigation 5.5 to 9 and domestic use 7 to 9 (ICMR, 1975).

The concentration of total solid in the drinking water was found to range between 312.2 and 574.5 mg/l (Table 41.1). This may be due to the breakage of taps, improper purification of water, rust particles from the pipelines, addition of chloride. The maximum permissible quantity is 1500 mg/l (WHO, 1984) and recommended levels of TDS (250–2100 mg/l) for the protection of aquatic life (USEPA, 1975), irrigation (ISI, 1982) and domestic use (ICMR, 1975).

The Chloride content of tap water ranged from (24.5–58.4 mg/l). The chloride content was quite lower than permissible limit (250 mg/l) of ISI and WHO standards. The values are in agreement with the results reported by Chopra and Negi (1993).

Dissolved oxygen present in drinking water adds taste and it is highly fluctuating factor in water. According to European Economic Community, the permissible standard of drinking water for dissolved oxygen is 5mg/l. In this study the dissolved O_2 content varied between 4.8 to 10.2 mg/1 (Table 41.1) and the tap water in Ganesapuram showed dissolved O_2 content less than the permissible limit. The low levels of dissolved O_2 at Ganesapuram tap water (T_3) could be attributed to presence of high level of *E. coli.* Sakul and Chaudhuri (1999) also reported the low level of dissolved oxygen due to the presence of non-gas forming bacteria.

Table 41.1: Physico-chemical Characteristics of Drinking Water at Namakkal Town

Parameter	T_1	T_2	T_3	T_4	T_5
Air temperature (°C)	24±0.17	22.5±0.25	25.2±0.37	24.5±0.76	23±0.26
Water temperature (°C)	25.5±0.3	25±0.18	25.5±0.25	26±0.31	25.5±0.11
pH	7.2±0.19	7.5±0.13	8.3±0.31	8.6±0.16	8.4±0.31
Total solid (mg/l)	312.2±12.6	474.5±11.7	574.5±17.4	487.8±11.5	425.4±15.8
Chloride (mg/l)	24.5±3.8	30.8±4.3	58.4±7.1	39.56±3.7	42.6±2.1
Dissolved O_2 (mg/l)	10.2±0.3	9.2±0.4	5.4±0.3	6.8±0.4	7.4±0.1
BOD (mg/l)	0.28±0.05	0.3±0.024	1.4±0.039	2.7±0.072	2.2±0.01
Nitrate (mg/l)	0.007±0.001	0.009±0.002	0.015±0.001	0.02±0.002	0.014±0.001
Phosphate (mg/l)	0.004±0.002	0.015±0.004	0.005±0.003	0.009±0.004	0.005±0.001
MPN of coliform bacteria (Number/100 ml)	135.8±32.1	1115±22.3	1125±19.74	58.6±12.34	165.72±15.62
Water Quality Index	60.45±3.1	63.38±4.3	62.39±5.3	68.8±6.8	64.74±3.1

The present study recorded the BOD values at the range of 0.28 to 2.7 mg/l (Table 41.1). Drinking water usually has a BOD of less than 1 mg/l and water is considered to be fairly pure with BOD values reach 5 mg/l (Rao, 1997). In this study the BOD values were below 3 mg/l in all the sampling station. From this data it is axiomatic that all samples were not much polluted by organic matter and the water is suitable for drinking purposes.

The present study showed a very low amount of nitrate in tap water samples, which ranges between 0 and 0.02 mg/l (Table 41.1). According to WHO, the desirable concentration of nitrate in drinking water is 45 mg/l (Rao, 1997). However the content of nitrate is within oxygen permissible limit for the different samples tried

Phosphate was present in trace amount (0–0.015 mg/l) in the samples tried. Phosphate is guarded as of great importance in the determination of biological productivity in water. The low content of phosphate in our samples indicated that there was no such contamination.

Sandhy (1979) observed high bacteriological density in surface water. In the present study also very high bacteriological count in the drinking water samples were recorded. A high *E. coli* count in stations T_2 and T_3 and low in T_1, T_4 and T_5 were observed (Table 41.1). The high bacteriological load could be due to the frequent discharges of

The MPN count varied between 58/100 ml and 1125/100 ml and all the water samples showed positive results. This was mainly due to the improper sewage as well as untreated water system. According to WHO (1984) and ICMR (1975) the permissible amount of coliform is 1/100 ml.

The water quality index calculated from nine parameters studied showed the values varied between 60.45 and 68.8 and so they are of medium (poor) quality.

Disinfections of community taps, proper sewage-drainage system, periodical quality monitoring of corporation water, simple and economical water treatments *viz.*, filtration, boiling along with public awareness programmes on sanitation would prove beneficial to avoid water borne health hazards legal issues on pollution regulation and pollution in the town. Hence it is pertinent that innovative technique for water quality management and similar studies are carried out in other areas of Namakkal town.

References

APHA. 1985. *Standard Methods for the Examination of Water and Wastewater*, 16th edn. American Public Health Association, Washington D.C., p.245–246.

Brown, V.M., Jordan, D. H. M. and Tiller, B. A. 1967. The effect of temperature on the acute toxitown of phenol to rainbow trout in water. *Wat. Res.*, 1: 587–594.

Chessbrough, M. 1991. Medical laboratory manual for tropical countries. Microbiology. *Tropical Health Technology ELBS*. Vol II, 415: 216–220.

Chopra, A.K. and Negi, T.S. 1993. Study on potability of water from different sources of two villages of Hardwar. In: *Advances in Limnology*, Narendra Publishing House, New Delhi, p. 63–68.

Freeda Gnana Rani, D., Geetha, S. and Ebanazar, J. 2002. The drinking water quality characteristics of five rural places in around Thittagudi, Tamil Nadu. *Poll. Res.*, 21(2): 215–221.

ICMR. 1975. *Manual of Standards of Quality for Drinking Water Supplies*. Indian Council of Medical Research. Spl. Rep. Ser. No. 44.

Indirabai, W.P.S. and George, S. 2002. Assessment of drinking water quality in selected areas of Namakkal town, after flood. *Poll. Res.*, 21(2): 209–214.

ISI (Indian Standard Institution). 1982. *Draft Indian Standard Specification for Drinking Water DOC.* CDC, New Delhi.

Mitchell, K. and Stapp, B. 1992. *Field Manual for Water Quality Monitoring*. Thomson Shore Printers, Michigan, p. 27–28.

Prasad, D. and Gaur, H.S. 1992. *Environmental Pollution: Water*. Venus Publishing House, New Delhi, p. 1–2.

Rao, C.S. 1997. *Environmental Pollution Control Engineering*. New Age International Publishers Ltd., New Delhi, p. 294–330.

Sakul, S. and Chaudhuri, S. 1999. Formulation of Kargil medium for detection of Coliform bacteria on water and other products. *Ecotoxicol. Environ. Monit.*, 9 (3): 217–219.

Sandhy, S.S. 1979. Magnitude of pollution indicator organisms in rural potable water. *App. Env. Microbiol.*, 37: 44.

Tripathi, A.K. and Pandey, S.N. 1990. *Water Pollution*. Ashish Publishing House, New Delhi, p. 18–19.

Trivedy, R.K. and Goel, P.K. 1986. *Chemical and Biological Methods for Water and Pollution Studies.* Environmental Publications, Karad, p. 35–38.

USEPA (United Stated Environmental Protection Agency). 1975. *Quality Criteria for Water*, (Ed. Train, R.E.). Caste House Publication Ltd., Great Britain.

WHO. 1984. *Guidelines for Water Quality*, Vol 1: Recommendations. World Health Organizations, Geneva.

Chapter 42

Assessment of Copper Concentrations in Two Freshwater Reservoirs of Nanded, Maharashtra State

*G. Gyananath**#*, S.V. Shewdikar*, T.A. Kadam*, S.K.G.K Charyulu** and R.S. Rao****

**School of Life Sciences, **School of Earth Sciences, Swami Ramanand Teerth Marathwada University, Vishnupuri, Nanded – 431 606, Maharashtra, India*

****Department of Chemical Engineering, CBIT, Hyderabad*

ABSTRACT

Heavy metals of vital importance are widely distributed in the environment. Several studies have been carried out in recent years to understand the distribution of heavy metals in different water bodies. Among the metals, Copper (Cu) is of environmental interest and an assessment of Cu in two freshwater reservoirs of Nanded, using Flame Atomic Absorption Spectroscopy. The results indicated that Cu was detected in both Amduri and Shikarghat reservoirs of raw and treated water samples. The changes in copper concentrations in natural waters and its possible effects on water quality are discussed. This study suggests that the source of copper in two reservoirs may be dissimilar.

Keywords: Copper, Trace amounts, Environment, Seasons, Freshwater Reservoirs.

Introduction

Copper is one of the significant metals to mankind and is being used since prehistoric times Copper reaches the aquatic environment through wet or dry deposition, mining activities, surface

\# Email: gyananath@yahoo.com.

runoff, industrial, domestic and agricultural waste disposal. The various studies shows that Cu is ubiquitously distributed in the environment and its occurrence in significant amounts in the water samples of reservoirs may suggest the availability of free Cu ions in the water. Since higher levels may show adverse biological and toxic effects, the water samples need constant monitoring of Cu concentration to protect the aquatic environment.

The variations in copper concentrations in water are different in different aquatic environments and are discussed by many scientists earlier with the possible health disorders due to copper. Copper is toxic to aquatic biota and the toxicity of the metal varies with factors such as pH, hardness, presence of other metals and Copper species present in water (Taylor and Demayo, 1981). Chakoumakos, (1977) attributed alkalinity and pH are to be the major factors controlling Copper speciation. McKnight *et al.* (1983) suggested that the aqueous chemistry of copper is dominated by its tendency to form complexes principally with legends. Patel *et al.* (1985) reported dissolved Cu in Bombay harbor waters which receives large amounts of industrial and municipal wastes. Ajmal and Uddin (1986) reported elevated levels of Cu in the treated hand pump water samples compared to the running water from the City of Aligarh (India) attributing it to non-compliance with the standards. deLeon *et al.* (1986) and Whitehead *et al.* (1988) reported Cu in river environments, Mississippi river and Mediterranean River. Arens (1999) elaborated various factors to be considered concerning the corrosion of Cu^{++} tubes in drinking water.

The occurrence of Cu has caused a series of severe systematic diseases in Germany (Cu induced Liver Cirrhosis), besides Cirrhosis another type of disease has occurred which like wise appeared to be induced by Cu in tap water (Eife *et al.*, 1999). Buchanan *et al.* (1999) made an assessment of copper levels in drinking water to determine the association of copper with Gastrointestinal (GI) illness. Pizarro *et al.* (1999) reported that there was no association between aggregate Cu in drinking water within the range of 0 to 50 µg/l, but a level of 30 µg/l of ionized Cu was associated with nausea, vomiting or abdominal pain.

The present study is an attempt to understand the distribution of a heavy metal, copper based on seasonal variations in two reservoirs since seasonal variations of Cu in aquatic environment may prove useful to understand long-term implications of Cu in the reservoirs. Since the reservoirs are primarily used for irrigation and drinking water, study of Copper contamination of source water may provide useful database for understanding of environmental health effects in humans.

Methodology

Water samples from different stations of two fresh reservoirs, Shikarghat and Amduri reservoirs, of Nanded district, Maharashtra were collected from June 1998 to May 1999 in the screw cap polyethylene bottles for heavy metal analysis. The bottles were cleaned with the tap water rinsed and soaked for 24 hrs in 1 per cent Nitric acid (HNO_3) and again rinsed with high purity (AIMIL; Water Purification System). The bottles were then dried and cooled to room temperature, capped and labeled. Nitric acid (1 ml acid 1100 ml water sample) was added to the bottle and each bottle was filled to brim with the water sample to avoid any air space. The water samples, before and after treatment, obtained and analysed from both Shikarghat and Amduri reservoirs for each month for the entire year. The samples collected were transported to the laboratory and the analysis was carried out using Perkin-Elmer model 'Analyst 100 and 300' Atomic Absorption Spectrophotometer (AAS). Air-acetylene flame was used exclusively and instrument was operated according to the optimal experimental conditions.

Results and Discussion

Variations in Copper Levels of Amduri and Shikarghat Reservoirs

Monthly changes in the Copper concentrations of raw water samples of both Amduri (6 stations) and Shikarghat (one station) reservoirs along with the variations of one treated sample from each reservoir are shown in Figure 42.1. Figures 42.1 (A) and (B) shows the monthly copper variations of both treated and raw waters of Amduri and Shikarghat reservoirs. The average copper levels at all

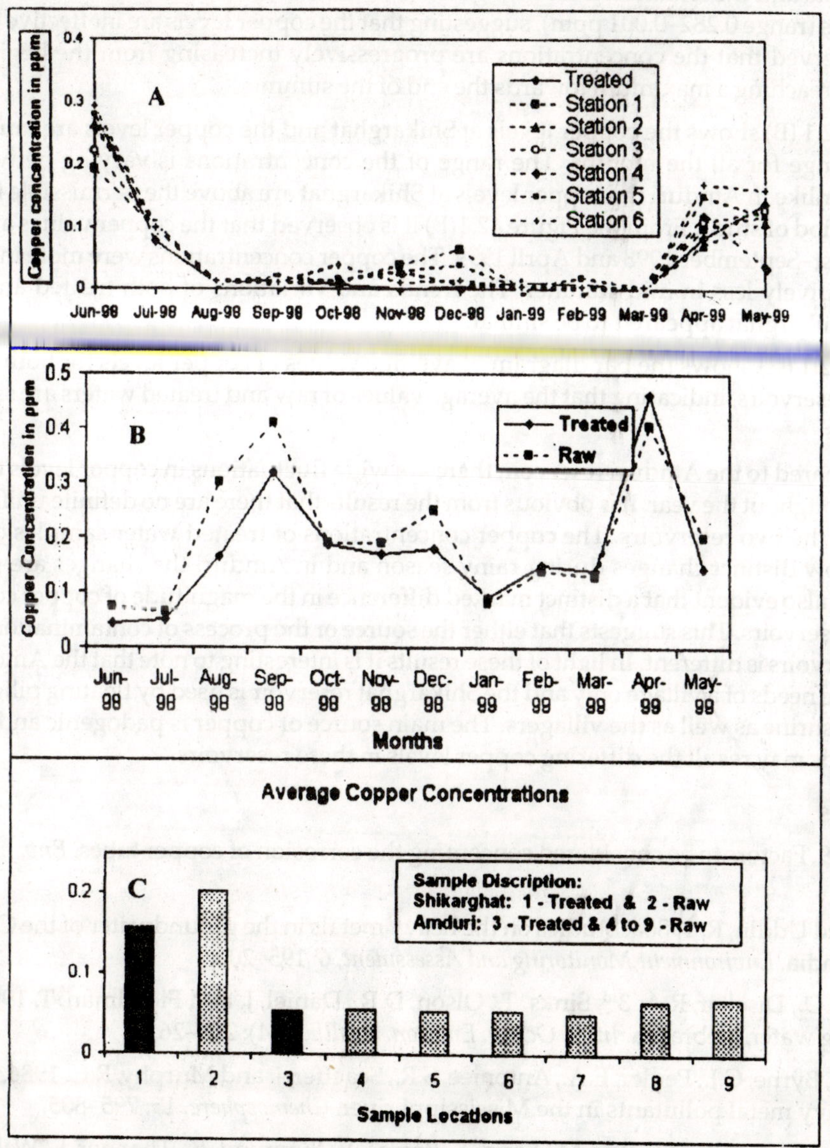

Figure 42.1: Copper Concentrations (ppm) of Treated and Raw Water Samples (A) Monthly Variations of Amduri Reservoir; (B) Monthly Variations of Shikarghat Reservoir and (C) Average Values in Both Amduri and Shikarghat Reservoirs

stations were separately shown as bar graph in Figure 42.1 (C). The copper values at all the stations were compared with the Bureau of Indian Standards (BIS).

Copper was above the detectable limits at all the stations of Amduri and the range of the concentrations is between 0.295 to 0.001 ppm (Figure 42.1A). The copper levels at Amduri are above the permissible limits during the months of June–July 1998 and April–May 1999 *i.e.*, almost in summer. For the remaining months though the copper values are below the permissible limits, all are above detection limits and the situation is alarming. The variations are similar at all stations including the treated waters (range 0.282–0.001 ppm), suggesting that the copper levels are ineffective to treatment. It is also observed that the concentrations are progressively increasing from the beginning of the summer and reaching a maximum towards the end of the summer.

Figure 42.1 (B) shows the copper levels at Shikarghat and the copper levels are much above the detectable range for all the months. The range of the concentrations is varying between 0.462 to 0.042 ppm. Unlike in Amduri the copper levels at Shikarghat are above the permissible limits during the entire period of study. From the Figure 42.1(B) it is observed that the copper values are maximum during August–September 1998 and April 1999. The copper concentrations were more in rainy season and comparatively less in mid summer. The trends and variations of both treated and raw water samples of Shikarghat appeared to be similar.

Figure 42.1 (C) shows the bar diagram of average values of copper levels for both Amduri and Shikarghat reservoirs, indicating that the average values of raw and treated waters are similar in both the reservoirs.

As compared to the Amduri reservoir, there are wide fluctuations in copper levels in Shikarghat reservoir throughout the year. It is obvious from the results that there are no definite uniform seasonal variations at the two reservoirs. The copper concentrations of treated water samples of Shikarghat reservoir show distinct changes during rainy season and in Amduri the changes are prominent in summer. It is also evident that a distinct marked difference in the magnitude of copper concentrations at both the reservoirs. This suggests that either the source or the process of contamination of copper in both the reservoirs is different. In light of these results it is interesting to note that the Amduri reservoir is catering the needs of a village only and the Shikarghat reservoir is used by floating piligrims of Holy Gurudwara shrine as well as the villagers. The main source of copper is padogenic and the different catering needs may result the differing copper levels in these reservoirs.

References

Arens, P. 1999. Factors to be considered concerning the corrosion of copper tubes. *Env. J. Med. Res.*, 28: 4(6).

Ajmal, M. and Uddin, R. 1986a. Studies on the heavy metals in the groundwater of the City of Aligarh (U.P.), India. *Environment Monitoring and Assessment*, 6: 195–205.

Buchanan, S.D., Disekar, R.A. 3rd, Sinks, T., Olson, D.R., Daniel, J. and Floodman, T. 1999. Copper in drinking water, Nebraska. *Int. J. Occup. Environ. Health.*, 5(4): 256–261.

DeLeon, J.R., Byrne, C.J., Peuler, E.A., Antoniee, S.R., Seaeffer, J. and Murphy, R.C. 1986. Trace organic and heavy metal pollutants in the Mississippi river. *Chemosphere*, 15: 795–805.

Eife, R., Weiss, M., Muller-Hocker, M., Lang, T., Barros, V., Sigmund, B., Thanner, F., Welling, P., Lang, H., Wolf, W., Rodeck, B., Kittel, J., Schramol, P. and Reiter, K. 1999. Chronic poisoning of Copper in tap water: II: Copper intoxication with predominantly systemic symptoms. *Env. J. Med. Res.*, 4(6): 224–228.

McKinght, D.M., Feder, G.L., Thurman, E.M., Wershaw, R.L. and Westall, J.C. 1983. Complexation of Copper by aquatic humic substances from different environments. *Science and Total Environment*, 28: 65–76.

Patel, B., Bangera, V.S., Patel, S. and Balani, M.C. 1985. Heavy metals in the content of drinking water, Bombay Harbour area. *Marine Pollution Bull.*, 16: 22–28.

Pizarro, F., Olivares, M., Speisky, H., Lonnerdal, B. and Uauy, R. 1999. *Copper of Infant Nutrition Safety of World Health Organization Provincial Guideline Value for Copper.*

Samanidou, V. and Fytianos, K. 1987. Partitioning of heavy metals into selective chemical fractions in sediments and rivers in Northern Greece. *Science of the Total Environment*, 67: 279–285.

Taylor, M.C. and Demayo, A. 1981. Copper. In: *Guidelines for Surface Water Quality*, Vol. I: Inorganic Chemical substances, inland waters. Directorate, Environment, Canada.

Whitehead, N.E. and Huynh-Nhoc Arson, S.R. 1988. Trace metals in two Northern Mediterranean waters. *Air and Soil Pollution*, 42: 7–18.

Chapter 43

Limnological Studies of Ponds of Chikmagalur, Karnataka

S.G. Malammanavar and N. Ramesh

Department of Environmental Science, Kuvempu University, Jnana Sahyadri, Shankaraghatta – 577 451

ABSTRACT

Limnological studies on two ponds of Chikmagalur have been carried out for a period of six months (September 97 –February 98). The Nygaards phytoplankton quotient values proved that these water bodies are in eutrophic range. Both Kudlur pond and Pundenahalli pond were organically polluted. A list of algae resistant to organic pollution has been provided.

Introduction

Chikmagalur district extends between 12°, 54′, 42″ and 13°, 53′, 53″ North latitude and 75°, 04′, 46″ and 76°, 21′, 50″ east longitude. It receives an average rainfall of 1,989 mm. The maximum and minimum temperature fluctuate between 30.7° and 19°C. There are several ponds whose ecology has not been studied.

Hence the present limnological study was carried out on two ponds of Chikmagalur situated at a distance of 37 km. away from Kuvempu University campus on Bangalore–Honnavar state highway. Both the ponds, Kudlur pond and Pundenahalli pond are used by the people for washing clothes and bathing of domestic animals. Kudlur pond also receives over flow water from Pundenahalli pond.

Materials and Methods

Monthly collections of surface water samples were made for the study of physical, chemical parameters and phytoplankton. Methods prescribed by standard methods for the examination of water and waste water (APHA, 1992) were employed for the estimation of various factors. Temperature and pH were noted on the spot and remaining factors were analyzed in the laboratory without lapse of much time. Average values of these factors for the two ponds studied are given in Table 43.1.

Table 43.1: Average Values of Physical and Chemical Parameters of the Two Ponds of Chikmagalur

Parameters	*Kudlur Pond*	*Pundenahalli Pond*
Water temperature °C	24.50	21.50
pH	8.37	7.25
Oxidizable organic matter ppm	9.40	8.50
Nitrate ppm	4.16	1.20
Free carbondioxide ppm	2.64	7.25
Dissolved oxygen ppm	7.32	6.88
Hardness ppm	163.00	320.00
Total alkalinity ppm	926.80	699.15
Phosphate ppm	0.04	0.07

A litre of water sample was collected separately every time for the phytoplankton study, the sedimentation of which was made in acid lugol's solution. The sediment was concentrated to 30 ml by decantation and centrifugation. Phytoplankton frequency was estimated following the method given by Hegde (1988) examining 10 drops from each collection and for each drop 10 fields were observed.

Results

Kudlur pond has dense bloom formation as compared to Pundenahalli pond and it is free from macrophytes. Whereas, Pundenahalli pond had rich growth of Typha and a huge amount of leaves enter the pond. Both the ponds have the continuous disturbance by people due to washing of clothes and bathing of domestic animals and make the pond water organically rich. Kudlur pond water is deep green in colour while it is moderate to deep green in Pundenahalli pond.

The physical and chemical parameter of the two ponds were studied during winter season. Comparatively Kudlur pond had high values of DO, total alkalinity, NO_3 and oxidizable organic matter. At the same time it had less concentration of CO_2, hardness and PO_4 and the water temperature and pH remained high (Table 43.1). A total of 21 taxa of algae were encountered in Kudlur pond belonging to Chlorophyceae (5), Cyanophyceae (6), Bacillariophyceae (7) and Euglenophyceae (3) and the bloom of Cyanophycean algae *Raphidiopsis indica* and secondarily by *Macrocystis aueroginosa.* Pundenahalli pond had the growth of 17 taxa of algae belonging to Chlorophyceae (4), Cyanophyceae (2), Bacillariophyceae (7), Euglenophyceae (4) and the bloom of Euglenoid alga *Euglena elastica.*

Discussion

The phytoplankton community on which whole aquatic population depends, is largely influenced by interaction of a number of physical, chemical and biological factors. From the present observation it is clear that comparatively more species diversity existed in Kudlur pond than Pundenahalli pond. Amongst these *Raphidiopsis indica* and *Microcystis aeuroginosa* of Cyanophyceae had more frequency of occurrence. Kudlur pond received water from Pundenahalli pond, which had more concentration of CO_2, hardness and PO_4 which are possibly made use of by the plankters in the former for their growth. The pond had more of NO_3 and oxidizable organic matter which are responsible for the growth of algae an observation similar to that of Munawar (1974) and Hegde (1988). Similarly algal blooms present in the pond results the precipitation of carbonate of calcium and magnesium from bicarbonate causing the higher alkalinity. This confmns the observations of Gonzalves and Joshi (1946). According to Munawar(1974) PO_4^{--} is released into media when it had low concentration of oxygen and high concentration of CO_2. Similarly, Pundenahalli pond had high concentration of PO_4^{--} with low concentration of oxygen and high concentration of CO_2. Possibly this may favour the dominate growth of *Euglena elastica*, an observation similar to that of Hegde and Bharati (1986) and Puttaiah and Somashekar (1987).

As per the Nygaards phytoplankton quotient (Sheath and Munawar, 1974) both ponds are in the eutrophic range (Table 43.2). It can be concluded that both ponds with more oxidizable organic matter and nitrate are organically polluted.

Table 43.2: Nygaards Phytoplankton Quotients for the Two Ponds of Chikmagalur

Phytoplankton Quotients	*Kudlur Pond*	*Pundenahalli Pond*
Myxophycean	3.0	3.0
Chlorophycean	2.5	1.0
Diatom	1.0	1.0
Euglenine	0.8	0.4
Compound	7.0	4.0

Table 43.3 gives the average frequency of occurrence of dominant algae in the two ponds. These are catagorised as most pollution tolerant species and only those whose frequency is above 50 individuals per 100 fields have been considered. These are arranged in the order of decreasing emphasis.

Table 43.3: Average Frequency Occurrence of Pollution Resistant Algae of Chikmagalur Ponds

Sl.No.	*Algae*	*Average Frequency Occurrence*	*Group*
1.	*Raphidiopsis indica*	964	Cyanophyceae
2.	*Microcystis aeuroginosa*	639	Cyanophyceae
3.	*Euglena elastica*	417	Euglenophyceae
4.	*Synedra acus*	201	Bacillariophyceae
5.	*Ankistrodesmus falcatus*	186	Chlorophyceae
6.	*Chlorella ellipsoidea*	96	Chlorophyceae
7.	*Scenedesmus armatus*	62	Chlorophyceae

References

American Public Health Association. 1992. *Standard Methods for the Examination of Water and Wastewater*, 18th edn. APHA, AWWA and WPCF, Washington, USA.

Gonzalves, E.A. and D.B. Joshi. 1946. Fresh water algae near Bombay I. The seasonal succession of the algae in a tank at Bhandra. *J. BHNS*, 46: 154–176.

Hegde, G.R. 1988. Limnological studies of temple ponds of Dharwad, Karnataka state. *Proc. Nat. Symp. Past. Present and Future of Bhopal Lakes*, p. 247–252.

Hegde, G.R. and Bharati, S.G. 1986. Ecological studies in ponds and lakes of Dharwad: Occurrence of Euglenoid bloom. *Phykos*, 25: 62–67.

Munawar, M. 1970. Limnological studies on fresh water ponds of Hyderabad, India. I. The Biotope. *Hydrobiologia*, 35: 127–162.

Munawar, M. 1974. Limnological studies on fresh water ponds of Hyderabad India. IV. The Biocenose. Periodicity and species composition of unicellular and colonial phytoplankton in polluted and unpolluted environments. *Hydrobiologia*, 45: 1–32.

Puttaiah, E. T. and R.K. Somashekar.1987. Distribution of Euglenoids in lakes of Mysore city. *Phykos*, 26: 39–46.

Sheath, R. and Munawar, M. 1974. Phytoplankton composition of a small subarctic lake in the Northwest Territories, Canada. *Phycologia*, 13: 149–161.

Chapter 44

Heavy Metal Concentrations in the Edible Crab *Scylla serrata* in the Malancha Region of Indian Sundarbans

Kakoli Banerjee, Abhijit Mitra**, Rajib Chakraborty**, Anumita Das**, Debarati Mukherjee****

**WWF-India Secretariat, Tiger and Wildlife Programme, Canning Field Office, South 24 Parganas – 743 329, W.B.*

***Department of Marine Science, University of Calcutta, 35, B.C. Road, Kolkata – 700 019, W.B.*

**** Department of Theoretical Physics, Indian Association for the Cultivation of Science, Jadavpur, Kolkata – 700 032, W.B.*

ABSTRACT

Concentrations of Zn and Cu were estimated in the muscle of the edible crab *Scylla serrata* inhabiting the Malancha region of Indian Sundarbans in different seasons. Highest concentrations of these metals were detected during the pre-monsoon. In the monsoon months of low aquatic salinity, pH and high dilution factor, the metal level in the body tissue of the detritivorous crabs decreased, which may be as per the biologically available heavy metals in the surface sediments.

***Keywords**: Scylla serrata, Edible crab, Metals, Salinity, pH, Detritivorous.*

Introduction

Production of heavy metals and their discharge into the aquatic environment have increased greatly in recent years. This results in the deposition of heavy metals on the underlying sediment bed. Heavy metals find their way into the marine and estuarine environment from various sources:

1. Pesticides and insecticides from agricultural fields carried by streams, rivers and estuaries.
2. Industrial effluents of diverse types discharged directly or along with domestic sewage.
3. Radioactive wastes discharged from nuclear power stations.
4. Waste heat discharged from thermal power stations.
5. Solid wastes and military wastes.
6. Debris from ocean bed exploration for oil and minerals.

A fairly good amount of these discharged metals accumulate in the body tissues of benthic organisms and hence they are often considered as potential indicators of pollution. The availability and magnitude of heavy metals in the body tissue is dependent on several factors such as nature of chemical species (Waldichuk *et al.*, 1974), salinity (Bryan and Uysal, 1978), temperature (Sastry and Moiler, 1981), pH (Ramamoorthy, 1988) etc.

It is again recommended that the damage of estuarine ecosystem is mainly a function of biologically available metallic fraction rather than the total amount of metal present in water or in sediment (Chester and Stoner, 1975; Chester and Voutsinou, 1981).

Based on these informations, an attempt has been made in the present programme to determine the seasonal pattern of metal accumulation by the edible crab *Scylla serrata* collected from Malancha region of North 24 Parganas, Sundarbans, West Bengal, India. The study was conducted for a stretch of two years during 2002 and 2003.

Materials and Methods

The estuarine system of West Bengal (India) is one of the most biologically productive, taxonomically diverse and aesthetically celebrated ecotone of the country and is supporting the world's largest mangrove block, the Sundarbans. This system is a unique nursery ground for a wide variety of fin fish and shell fish juveniles. Based on the culture period of crabs in the study area, crabs were collected monthly during March, 2002 to December, 2002 and March, 2003 to December, 2003 from the Malancha region of Indian Sundarbans. Precaution was taken to maintain a standard size during crab collection so as to avoid any error due to size difference.

Soft parts were carefully removed after quick rinse with double distilled water, and the body parts were oven dried at 110°C. The metals (Zn and Cu) were estimated in the dried samples after digestion with conc. H_2SO_4 and conc. $HClO_4$ in the ratio of 3 : 1. The resulting solution was digested and made upto a constant volume with 0.05 N HNO_3. The analytical blanks were prepared and treated with the same reagents. The digested samples were analysed for Zn and Cu through a Perkin Elmer Atomic Absorption Spectrophotometer (AAS Model 3030) equipped with a HGA-500 graphite furnace atomizer and a deuterium background corrector.

Results and Discussion

Concentration of metals in the body tissues varied markedly, showing a definite pattern with season. In general, high concentrations were observed during the pre-monsoon months (March to June), the period characterized by highest salinity and pH of the overlying aquatic phase. This condition of the water favours the process of precipitation of heavy metals on the sediment as observed by earlier workers (Mitra *et al.*, 1992; Trivedi *et al.*, 1995; Choudhury *et al.*, 1994; Gupta *et al.*, 1994). As the crabs are detritivores, therefore the body burden of the selected heavy metals also increased in the pre-monsoon season (Tables 44.1 and 44.2; and Figures 44.1–44.4). *Scylla serrata* being an important edible

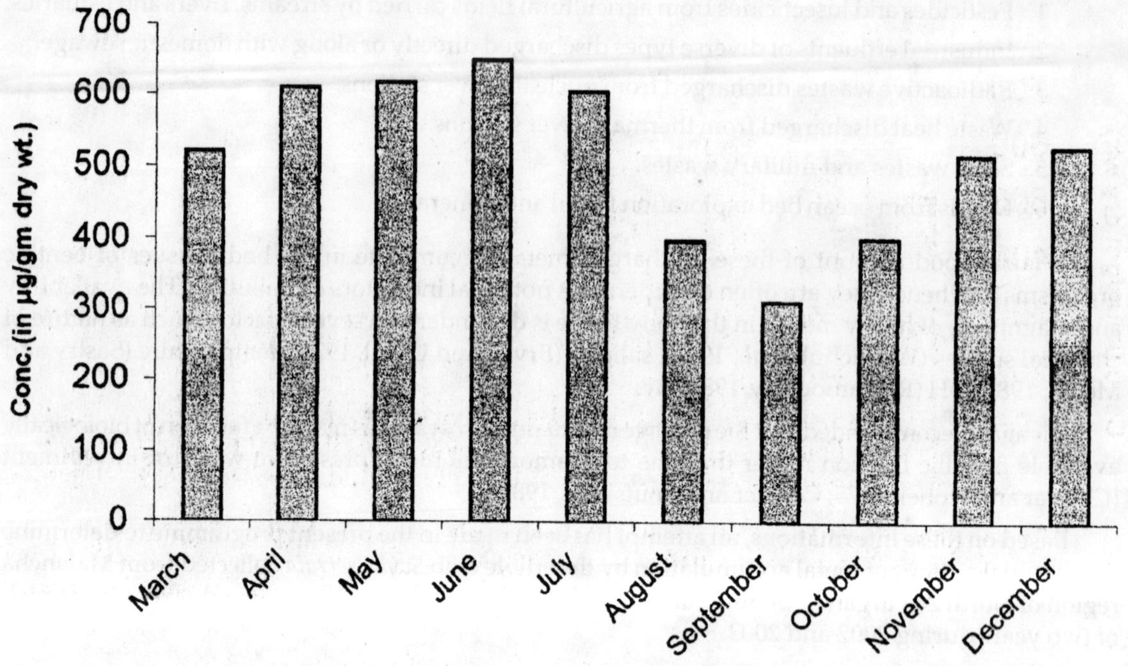

Figure 44.1: Monthly Variation of Zn (in µg/gm dry wt.) During 2002

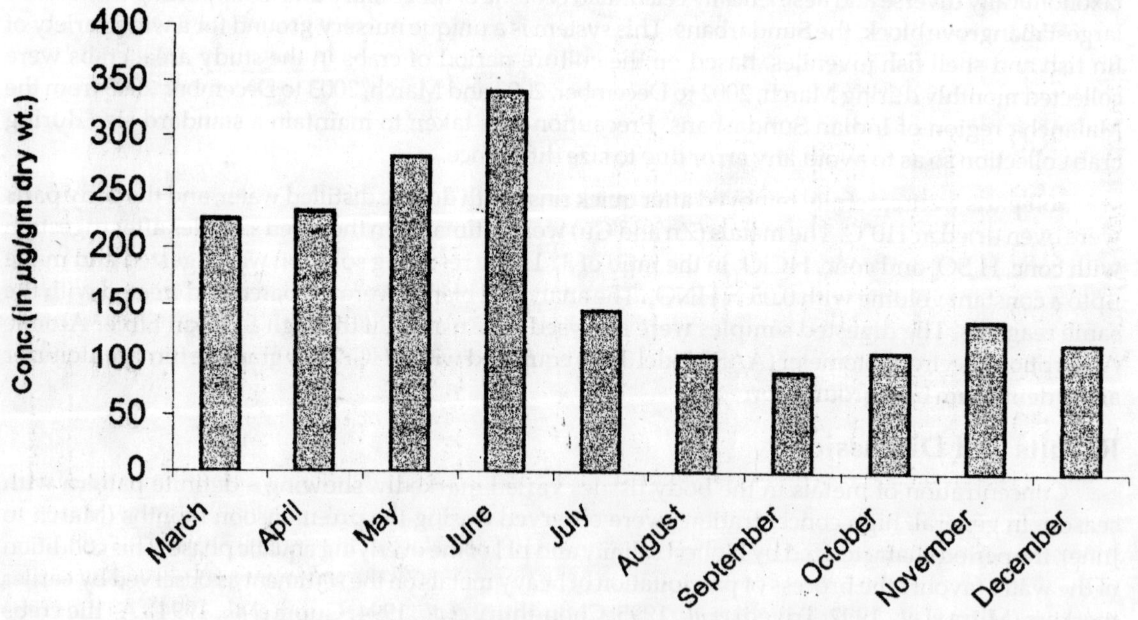

Figure 44.2: Monthly Variation of Cu (in µg/gm dry wt.) During 2002

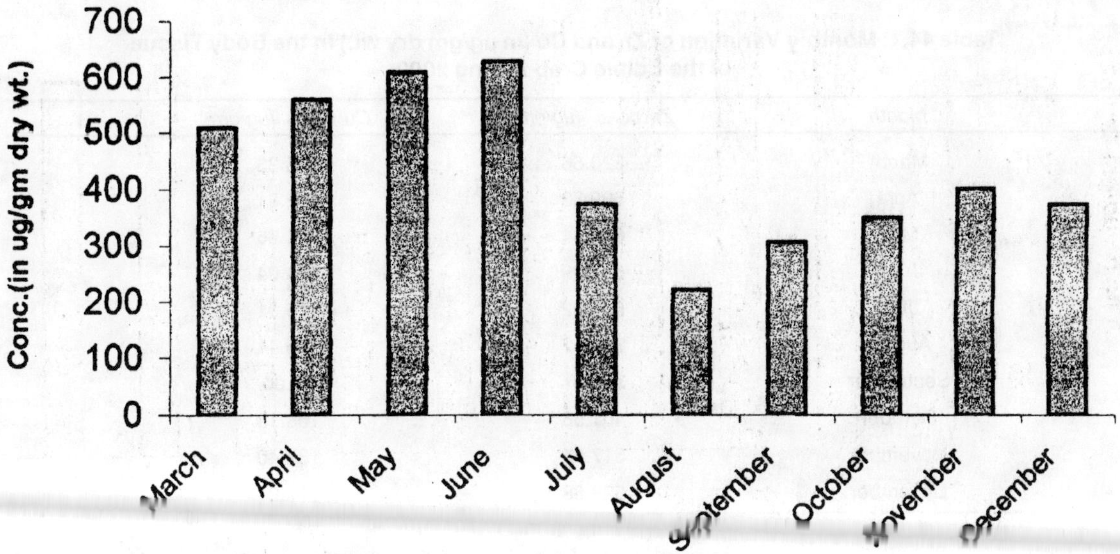

Figure 44.3: Monthly Variation of Zn (in µg/gm dry wt.) During 2003

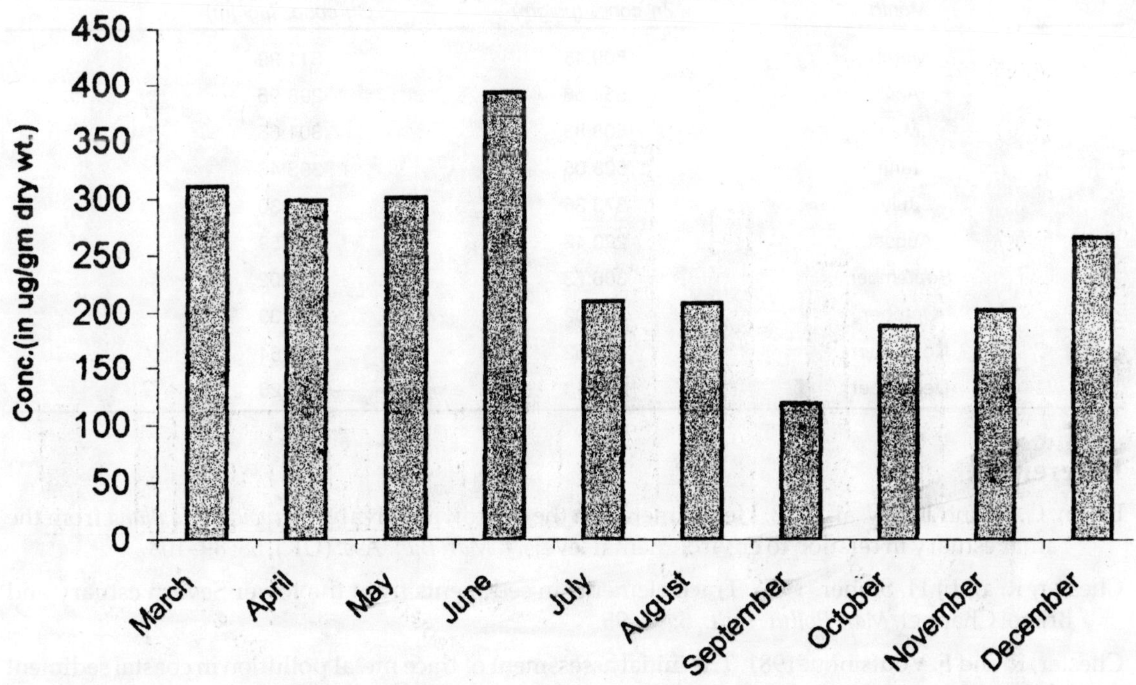

Figure 44.4: Monthly Variation of Cu (in µg/gm dry wt.) During 2003

species fetches high price in the international market; hence such study is of utmost importance from the quality point of view.

Table 44.1: Monthly Variation of Zn and Cu (in µg/gm dry wt.) in the Body Tissue of the Edible Crab During 2002

Month	*Zn conc. (µg/gm)*	*Cu conc. (µg/gm)*
March	520.56	225.35
April	609.83	232.81
May	616.84	281.66
June	648.29	341.04
July	605.02	146.31
August	398.42	105.44
September	311.37	90.00
October	401.56	108.13
November	517.24	137.40
December	528.88	117.04

Table 44.2: Monthly Variation of Zn and Cu (in µg/gm dry wt.) in the Body Tissue of the Edible Crab During 2003

Month	*Zn conc. (µg/gm)*	*Cu conc. (µg/gm)*
March	509.45	311.29
April	559.58	298.96
May	608.83	301.65
June	628.05	393.46
July	373.36	210.30
August	220.42	208.59
September	306.73	119.02
October	351.02	188.39
November	402.52	202.54
December	373.43	266.28

References

Bryan, G.W. and H. Uysal. 1978. Heavy metals in the burrowing bivalve *Serobicularia plana* from the Tamar estuary in relation to environmental levels. *J. Mar. Biol. Ass.* (UK), 58: 89–108.

Chester, R. and J.H. Stoner. 1976. Trace elements in sediments from the lower Severn estuary and Bristol Channel. *Mar. Pollut. Bull.*, 5: 92–95.

Chester, R. and F. Voutsinou. 1981. The initial assessment of trace metal pollution in coastal sediment of Hooghly estuary. *Environ. Ecol.*, 7: 427–430.

Choudhury, Abhijit, Mitra, Abhijit, Trivedi, Subrata, Gupta, Ananda and Amalesh Choudhury. 1994. Phosphate and nitrate status in the east coast of the Indian subcontinent. *Proceedings of Seminar on Our Environment: Its Challenges to Development Projects*. American Society of Civil Engineers–India Section. p. AC-1–AC-9.

Gupta, Ananda, Mitra, Abhijit, Mitra, Trivedi, Subrata, Choudhury, Abhijit and Amalesh Choudhury. 1994. Simple Statistical approach to underst and fate of pollutants in the Hooghly estuary in different seasons. *Proceedings of Seminar on Our Environment: Its Chullenges to Development Projects*. American Society of Civil Engineers–India Section. p. AT-1–AT-6.

Mitra, Abhijit, Choudhury, Amalesh and Yusuf Ali Zamaddar. 1992. Effects of heavy metals on Benthic Molluscan Communities in Hooghly estuary. *Proc. Zool. Soc., Kolkata*, 45: 481–496.

Ramamoorthy, S. 1988. Effect of pH on speciation and toxicity of aluminium to rainbow trout (*Salmo qairdneri*). *Can. J. Fish. Aquat. Sci.*, 45: 634–642.

Sastry, A.N. and Don C. Moller. 1981. Application of Biochemical and Physiological Responses to Water Quality Monitoring. In: *Biological Monitoring of Marine Pollutants*, (Eds.) J. Vemberg, A. Calabrese, F.P. Thurberg and W.B. Verberg. Academic Press. New York, p. 265–294.

Trivedi, Subrata, Mitra, Abhijit, Bag, Manigrib, Ghosh, Indranil and Amalesh Choudhury. 1995. Heavy metal concentration in Mud Skipper [illegible] of Nayachar Island, India. *Indian J. Env. Hlth., NEERI*. 37(2): 120–122.

Waldichuk, M. 1974. Some biological concerns in heavy metals pollution. In: *Pollution and Physiology of Marine Organisms*. Academic Press, New York, p. 1–58.

Chapter 45

Population Structure of *Calotes versicolor* (Daudin) in an Industrial Area in Vadodara District of Gujarat State, India

Rushita Adhikari, B. Suresh* and Bonny Pilo

Department of Zoology, Facuity of Science, The Maharaja Sayajirao University of Baroda, Vadodara – 390 002, Gujarat, India

Introduction

The distribution of the various reptiles in India is well documented (ZSI, 1992). However, it is very essential to record the populations of Indian reptiles in the wild by surveying a substantial number of areas periodically. Such data forms the baseline for monitoring the population dynamics, and thereby helps in formulating ideal conservation strategies.

In the present study an attempt was made to understand the demography of one of the most common agamid lizard of India, the *Calotes versicolor.* The garden lizard, as it is commonly called, is a medium-sized, arboreal lizard with oval head and laterally compressed body. The body is brown or sand grey above which is often studded with a pattern of spots and bars on the back and sides. The adult male has a muscular cheek and during breeding season they develop scarlet colouration in head, shoulder and parts of the forelegs. Calotes is a diurnal lizard often seen in gardens, hedges, scrubland and forest. They appear to adapt well to humans and are thus survive well in spite of acute anthropogenic pressure. The study area recorded a good number of this species possibly due to the adequate food availability in the form of insects notably ants. Apart from insects they also relish spiders, small frogs and rodents. These lizards breed during April to end September (Daniel, 2002).

* Corresponding author.

A survey on available literature prompted one to develop a baseline data on the population size in a given area for *Calotes versicolor,* since no attempt was made so far in this direction. The present study therefore was carried out to understand the status of the population of *Calotes* at an industrial area and to present a base index for further monitoring studies.

Study Area and Methods

The field survey was carried out in an industrial area in Vadodara (22°17'N, 73°15'E), covering an expanse of 3,15,000 m^2. The habitat at the study site is composed of soil with cracks and crevices. Trees like *Acacia nilotica, Lagerstromia* sp., *Azadirachta indica, Delonix regia* and *Peltophorum pterocarpum* dominated the vegetation community along with the presence of shrubs like *Bougainvillea spectibilis, Ixora coccinea, Cassia* sp., *Capparis* sp. and *Calotropis procera.* There was a greater incidence of these lizards on rough-barked native trees than on smooth-barked exotics.

On the basis of vegetation characteristics the study area was divided into three sites (Figure 45.1). Six fixed-width transects were laid at each site (Krebs, 1999). The transects laid covered an area of 400 m^2, 300 m^2 and 100 m^2 at sites A, B and C respectively. Surveys were carried out in summer season (May–June) on five different days. Surveys were undertaken during the day and during the evening. The time chosen for these surveys was 0800–1000 hours and 1600–1800 hours. The prime reason for selection of this season and time was that the garden lizard is known to breed during this period, and during morning hours it exhibits maximum activity (Dunham, 1988), thereby providing for their easy visibility. Specimens were mainly located opportunistically during visual surveys of all the three sites. Search techniques included visual scanning of terrain and refuge examination *viz.*, lifting logs and boulders, scraping through leaf litter, tapping on the bark etc.

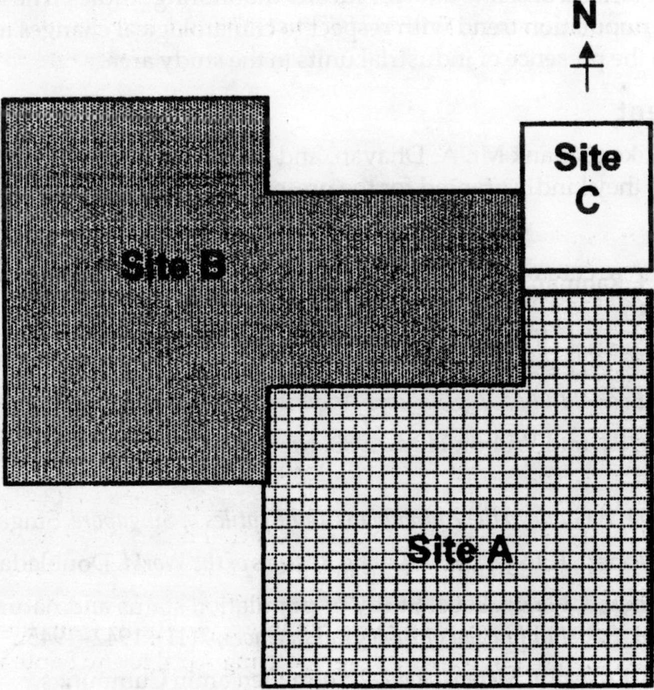

Figure 45.1: Study Area

Results and Discussion

The maximum number of adult *Calotes versicolor* recorded at Site A, Site B and Site C, on all the five days was consistent. Site A recorded 32, 29, 31, 30 and 33 individuals on five different days respectively. Site B showed the presence of 18, 22, 16, 25 and 23 individuals respectively, whereas Site C recorded 9, 11, 10, 8 and 4 lizards respectively on five different days. No significant difference in the number of males and females could be found. The average numbers of individuals found at the three sites within the transects were 31, 20.8 and 8.4 respectively.

The population of lizards at the three different sites was calculated in terms of individuals present per acre area. The results were as follows: 209/acre at Site A, 280/acre at Site B and 340/acre at Site C.

The number of these lizards sighted was greatest during morning hours, the average air temperature being 31.33°C±1°C. As temperature increased during afternoon hours, very few of them could be sighted. But, during evening hours also, a substantial number of these organisms were seen, but they did not account to more than those sighted during morning hours.

The specimens sighted were adults in breeding stage. Hardly any juvenile forms were encountered. The male's head and shoulders turned bright orange to crimson and his throat black, marking the breeding season. The lizards were engaged in feeding, basking and other breeding related activities during morning hours. During noon time, they were restricted to burrows and other shady and cooler areas. This made their sighting very difficult. In evening, they displayed only a moderate amount of activity.

The results of the survey do not provide information about the changing trend in the population of these lizards, but it forms a baseline data for further monitoring studies. The results can be utilized to study the varying population trend with respect to climatological changes as well as the adverse effects caused due to the presence of industrial units in the study area.

Acknowledgement

Authors would like to thank Mr. A. Dhavan, and Mr. S.N. Shah M/s ABB Baroda for financial support and facilities they kindly granted for the present study.

References

Auffenberg, W. and H. Rahman. 1993. Studies on Pakistan reptiles, Part 3: Calotes versicolor. *Asiatic Herp. Res.*, 5: 14–30.

Cogger Harold, G. *et al.*, 1993. *Encyclopedia* of *Animals.*

Daniel, J.C. 2002. *The Book of Indian Reptiles and Amphibians.* Bombay Natural History Society.

Dunham, A.E., Morin, P.J. and Wilbur, H.M. 1988. Methods for the study of reptile populations. In: *Biology of Reptilia.*

Frances, L.K. Lim. 1992. *A Guide to the Amphibians and Reptiles of Singapore.* Singapore Science Centre.

Karl, P. Schmidt and Robert, F. Inger. 1957. *Living Reptiles of the World.* Doubleday and Company, Inc.

Khan, M.Z. and Mahmood, Nazia. 2004. Study of population status and natural history of agamid lizards of Karachi. *Pakistan Journal of Biological Sciences,* 7(11): 1942–1945.

Krebs, Charles J. 1999. *Ecological Methodology,* 2nd edn. Benjamin Cummings.

Lazaroff, Cat. 2000. Reptiles. In: *Trouble Around the World*. Environment News Service (ENS).

Robert, G. Sprackland, Jr. 1977. *Neptune City, All About Lizards*. T.F.H. Publications, Inc., Ltd.

Shanbhag, B.A. *et al.*, 2003. Demography of fan-throated lizard, *Sitana ponticeriana* (Cuvier) in a cotton field in Dharwad District of Karnataka State, India. *Current Science*, p. 1363–1366.

ZSI. 1992. *Handbook of Indian Lizards*. Calcutta.

Index

B

C

F

G

H

I